RNA Methodologies

SECOND EDITON

RNA Methodologies

A LABORATORY GUIDE FOR

ISOLATION AND CHARACTERIZATION

SECOND EDITION

Robert E. Farrell, Jr.
Exon-Intron, Inc.
Columbia, Maryland

Academic Press
Harcourt Brace & Company, Publishers

San Diego   London   Boston   New York   Sydney   Tokyo   Toronto

Front cover photographs: (Inset) Effect of dirty rollers in a Polaroid camera. (For more details, see Chapter 10, Figure 3.) (Background) Image © 1996 PhotoDisc, Inc.

This book is printed on acid-free paper. ⊚

Academic Press
a division of Harcourt Brace & Company
525 B Street, Suite 1900, San Diego, California 92101-4495, USA
http://www.apnet.com

Academic Press Limited
24-28 Oval Road, London NW1 7DX, UK
http://www.hbuk.co.uk/ap/

Library of Congress Catalog Card Number: 98-84574

International Standard Book Number: 0-12-249695-7

PRINTED IN THE UNITED STATES OF AMERICA
98 99 00 01 02 03 EB 9 8 7 6 5 4 3 2 1

For Catherine Ann,
Sean Patrick, Emma Catherine, and Liam Michael

Contents

Messenger RNA

Resilient Ribonucleases

RNA Isolation Strategies

Isolation of Polyadenylated RNA

CHAPTER   10

Photodocumentation and Image Analysis

CHAPTER   11

The Northern Analysis

Nucleic Acid Probe Technology

Practical Nucleic Acid Hybridization

Principles of Detection

RT PCR

Messenger RNA Differential Display

Quantitative Polymerase Chain
Reaction Techniques

CHAPTER 18

Quantification of Specific Messenger RNAs by Nuclease Protection

CHAPTER 19

Analysis of Nuclear RNA

Electrophoresis: Principles, Parameters, and Safety

Polyacrylamide Gel Electrophoresis

Internal Controls

Trade Citations   505

Selected Suppliers of Equipment, Reagents, and Services   507

Preface

Why study RNA? This question was posed as the first edition of *RNA Methodologies* went to press in 1993. The isolation of biologically competent and chemically stable ribonucleic acid continues to be a central though somewhat unappreciated procedure in molecular biology and has even greater relevance today. Regulation of tissue-specific gene expression remains an area of intense interest, and its study is possible, in part, using RNA as a parameter of gene expression. Scrutiny of mRNA biogenesis affords a unique look at the cellular biochemistry from the perspective of the exquisite orchestration of RNA complexity. Coupled with traditional methods for assay, the polymerase chain reaction now affords a means of unparalleled sensitivity and resolution for the characterization of transcriptional activity in the cell. mRNA biogenesis, maturity, and function are influenced transcriptionally and posttranscriptionally and together constitute myriad regulation control points within the cell. Thus, RNA stability and processing clearly play central regulatory roles in the cell.

There remains an academic imperative to unify the numerous facets of RNA characterization in a coherent start-to-finish format. The purification of high-quality RNA from biological sources is a fundamental starting point for investigations designed to give clearer definition to at least one aspect of gene expression regulation. While many of the more traditional techniques, including Northern analysis, have been overshadowed by several semiquantitative PCR-based approaches, the thoughtful and expedient isolation of RNA is equally important as the subsequent analytical and preparative manipulations of the RNA in purified form. Whether isolated from cells in culture or directly from whole tissue, RNA must be meticulously handled to support these techniques.

This manual is a collection of tried, tested, and optimized laboratory protocols for the isolation and characterization of eukaryotic RNA. The more noteworthy additions to this volume include a variety of RNA-based PCR techniques, such as differential display PCR and transcript quantification by competitive PCR. Methods such as these are especially important today in view of the unprecedented interest in the absolute abundance of specific transcripts in the cell. Clearly, the focus here is the study of RNA as a parameter of gene expression with emphasis on how these methodologies can be used in the study of transcriptional and posttranscriptional regulation of eukaryotic genes.

This text is written for the principal investigator, bench scientist, physician, veterinarian, lab technician, graduate student, and anyone else capable of performing basic research techniques. This resource is intended to provide a rationale to assist in the decision-making process for individuals at all levels of sophistication, from the novice to the well-seasoned, and at the same time to present realistic alternatives for achieving the same experimental goals. Many of the incorporated notations and hints are based upon personal experience and pave the way for the expedient recovery of RNA from biological sources, with particular regard to those biological sources enriched in ribonuclease. Day-in and day-out, unsound tactics for RNA characterization result in wasted resources due to an obvious failure to understand the "what" and the "why" from the onset of the study. While this text is by no means a comprehensive review of the fundamentals of gene expression, the recurrent theme herein is the correct way to handle RNA. These pages also demonstrate clearly how a selected technique fits into the grand scheme of nucleic acid research. While this author would hope that the text be studied from cover to cover, one may pick and choose salient protocols without loss of continuity. For those readers who are new to the study of cellular biochemistry from the perspective of RNA analysis, it may be beneficial to first read chapter 20 (An RNA Paradigm).

Collectively, the chapters work to embellish the RNA story, each presenting clear take-home lessons. The liberal incorporation of flow charts, tables, and graphs likewise facilitates learning and assists in the planning phases of a project. Investigators are limited only by their own ingenuity.

* * *

The author acknowledges, with sincere thanks and appreciation, the intellectual encouragement of the many colleagues and friends who, in some way, supported the preparation of the manuscript. The support and patience of the author's family are also publicly acknowledged and are very much appreciated.

Initium sapientiae timor Domini

RNA and the Cellular Biochemistry Revisited

Why Study RNA?

All cell and tissue functions are ultimately governed by gene expression; therefore, the reasons for electing to study the modulation of RNA as one parameter of cellular biochemistry may be as diverse as the intracellular RNA population itself. The goals in any experimental design involving RNA generally revolve around one or more fundamental themes, including but not limited to the following:

1. Measurement of the steady-state[1] abundance of cellular RNA transcripts. This is the most commonly studied parameter of gene expression. Both quantitative and qualitative profiles of a population of RNAs can then be generated by Northern analysis, nuclease protection assay, or polymerase chain reaction (PCR).

2. Measurement of the *rate* of transcription of gene sequences or the pathway of processing of RNAs. This may be deduced, at least in part, by the nuclear runoff assay in which labeled ribonucleotide precursors are incorporated into nascent transcripts, in direct proportion to the abundance of each RNA being transcribed. In addition, when used in conjunction with Northern analysis, the level of regulation of sequences can often be assigned as transcriptional or due to a posttranscriptional event(s).

3. Mapping of RNA molecules, including the 5′ end, 3′ end, and size and location of introns. This can be derived from partial digestion of the RNA molecules of interest by nuclease protection assay and/or by PCR.

4. Cell-free *in vitro* translation of purified messenger RNA (mRNA). The resulting polypeptide may be further characterized by immunoprecipitation and/or Western analysis. This is often quite helpful in the identification of newly isolated genes.

5. Synthesis of complementary DNA (cDNA). Purified mRNA can serve as the template for the enzymatic conversion of these relatively unstable, single-stranded molecules into much more stable, double-stranded cDNA molecules. Among the more common reasons for performing cDNA synthesis are the amplification of the sequence by PCR, often for some "quantitative purpose," and the direct ligation of the cDNA into a vector, or both. After cloning, the resulting cDNA library (clone bank) may be propagated for long-term storage and analysis. In so doing, one constructs a permanent record of the cellular biochemistry at the time of

---

[1]The final accumulation of RNA in the cell, or in a subcellular compartment such as the nucleus or the cytoplasm, is referred to as its steady-state level.

cell lysis. Historically, the synthesis of cDNA is one of the most important yet technically challenging methodologies in the molecular biology laboratory.

RNA is chemically and biologically more labile than DNA, particularly at elevated temperatures (> 65°C) and in the presence of alkali. These natural difficulties are further compounded by the endonucleolytic and exonucleolytic activities of a variety of resilient ribonucleases (RNase), the apparent ubiquity of which is undisputed. The isolation of RNA from biological sources enriched in RNase is especially challenging. These factors mandate rapid recovery of RNA with reagents that are free of RNase contamination. Likewise, extraction tactics must be formulated to include sufficient management of cellular RNase activity upon disruption of the organelles and vacuoles that normally sequester these enzymes. Failure to observe scrupulously procedures for eliminating potential sources of RNase contamination will almost always yield a useless sample of degraded RNA. As far as *which* RNA extraction procedure to use, it is incumbent upon the investigator to always think two steps ahead and ask "What is to be done with this RNA after it has been purified?" Contaminants, even trace quantities, carried over from some isolation procedures may compromise the efficiency of subsequent reactions. Moreover, due to their naturally short half-life, experiments involving RNA are most judiciously planned around the availability of cell cultures or tissue samples and the actual date of RNA isolation.

What Is RNA?

RNA is a long, unbranched polymer of ribonucleoside monophosphate moieties joined together by phosphodiester linkages, and both eukaryotic and prokaryotic RNAs are essentially single-stranded molecules. The basic, unassembled building blocks of RNA (and DNA) are called nucleotides. These building blocks consist of three key components: a pentose (5-carbon sugar or ribose), a phosphate group, and a nitrogenous base (Figure 1). A nitrogenous base joined to a pentose sugar is known as a nucleoside. When a phosphate group is added the composite, a phosphate ester of the nucleoside, is referred to as a nucleotide.

The key chemical differences between DNA and RNA are (1) the presence of a hydroxyl group (–OH) joined to the number 2 carbon (2′ position) of the ribose sugar (the absence of this –OH group in DNA is the underlying basis of the name of the sugar deoxyribose); and (2) the absence of the base thymine in RNA, substituted in RNA by the closely re-

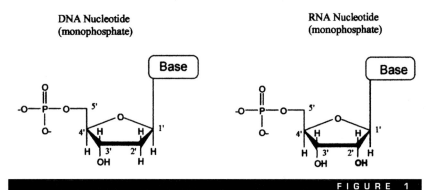

Common Nitrogenous Bases

Adenine   Guanine

Cytosine   Thymine   Uracil

Essential Nucleotide Structure

DNA Nucleotide
(monophosphate)

RNA Nucleotide
(monophosphate)

FIGURE   1

Comparative nucleotide structure.

lated pyrimidine uracil. More specifically, RNA is assembled from ribonucleotide precursors and DNA is assembled from deoxyribonucleotide precursors. The principal features and nomenclature of these building blocks are summarized in Tables 1 and 2.

Nitrogenous bases and the pentose sugar components that make up nucleotides are both cyclic. To avoid confusion when referring to the constituent carbon atoms of the sugar and/or base that constitute a particular nucleotide, a special numbering system has been devised. By convention, the numbering system for the atoms that make up the bases is 1, 2, 3, etc., and that for the constituent carbon atoms of the sugar (ribose or deoxyribose) is $1'$, $2'$, $3'$, etc.

Comparison of Ribonucleotides and Deoxyribonucleotides

	RNA nucleotides	DNA nucleotides
Key nucleotide components	5-Carbon ribose Phosphate group Nitrogenous base	5-Carbon deoxyribose Phosphate group Nitrogenous base
Common nitrogenous bases	Adenine Cytosine Guanine Uracil	Adenine Cytosine Guanine Thymine

The ribonucleotide triphosphates may be collectively (and sometimes singularly) referred to as NTP. The placement of a lowercase "d" before a nucleotide triphosphate, as in dNTP, denotes the presence of the deoxy form of the ribose sugar in the backbone of the molecule, that is, a deoxyribose polynucleotide or DNA. It is the triphosphate forms of these nucleotides that serve as precursors in nucleic acid synthesis. The phosphate nearest the nucleoside moiety, that is, closest to the sugar, is the α phosphate, followed by the β phosphate, followed by the γ phosphate, which is furthest from the nucleoside (Figure 2). During nucleic acid polymerization, the β and γ phosphates are cleaved from the nucleotide, and the resulting nucleoside monophosphate is then incorporated into the nascent polynucleotide chain.

Guanosine 5′-triphosphate (GTP). The three constituent phosphate groups are designated α, β, and γ based on the proximity of each group to the nucleoside (base + sugar) component of the molecule.

Nomenclature of Common RNA and DNA Nucleosides and Nucleotides

	RNA	DNA
Nucleosides (sugar + base)	Adenosine	2'-Deoxyadenosine
	Cytidine	2'-Deoxycytidine
	Guanosine	2'-Deoxyguanosine
	Uridine	2'-Deoxythymidine
Nucleotides (nucleoside + phosphate) triphosphate precursors	Adenosine 5'-triphosphate (ATP)	2'-Deoxyadenosine 5'-triphosphate (dATP)
	Cytidine 5'-triphosphate (CTP)	2'-Deoxycytidine 5'-triphosphate (dCTP)
	Guanosine 5'-triphosphate (GTP)	2'-Deoxyguanosine 5'-triphosphate (dGTP)
	Uridine 5'-triphosphate (UTP)	2'-Deoxythymidine 5'-triphosphate (dTTP)

Assembly of Polynucleotides

The synthesis of RNA proceeds via the activity of enzymes known as RNA polymerases (Chapter 2). Any enzyme capable of DNA or RNA polymerase activity has the ability to assemble nucleic acid molecules from precursor nucleotides. In a polynucleotide chain adjacent nucleotides are linked by phosphodiester bonds. The formation of these bonds involves the hydrophilic attack by the 3′ hydroxyl group of the last nucleotide added to nascent polynucleotide on the 5′ triphosphate of the incoming nucleotide.

Assuming that the intracellular biochemistry can support initiation and elongation of RNA molecules, there are two fundamental requirements that must be fulfilled and maintained *in vivo* and *in vitro* to support continued polymerization:

1. There must be a template strand to direct the polymerase-mediated insertion of the correct (complementary) nucleotide into the nascent chain.
2. For initiation and elongation, there must be a free 3′ hydroxyl group to which the next nucleotide in the chain can be joined via a phosphodiester linkage.

Most of the enzymes used in molecular cloning that exhibit polymerase activity have nearly identical template and 3′ hydroxyl primer requirements.

Polynucleotide elongation involves the incorporation of a nucleoside monophosphate, and the formation of this linkage is accompanied by the release of pyrophosphate. The synthesis of nucleic acid, both RNA and DNA, always proceeds from the 5′ direction toward the 3′ direction, resulting in a polynucleotide consisting of 5′–3′ linkages between adjacent nucleotides.

Upon completion, nucleic acid molecules are assembled in such a way that:

1. The ends of the molecule are structurally different from one another. The first nucleotide of the molecule has a free (tri)phosphate, the so-called 5′ phosphate, while the last nucleotide manifests a free 3′ hydroxyl group.
2. The backbone of the molecule consists of an alternating series of sugar and phosphate groups. Frequently referred to as the phosphodiester backbone of the molecule, it imparts a net negative charge to the molecule by virtue of its constituent phosphate groups.
3. The bases protrude away from the backbone of the molecule. This stereochemistry makes the bases very accessible for hydrogen bonding

(base pairing) to a complementary polynucleotide sequence, which, of course, is the very heart of molecular hybridization.

The nitrogenous bases are categorized as either purines (adenine and guanine) or pyrimidines (cytosine, thymine, and uracil), both of which are flat aromatic molecules. The specificity of base pairing (purine::pyrimidine) is maintained by the preference of adenosine and cytosine for the amino over the imino form and the preference of guanine and thymine (and uracil) for the keto rather than the enol form. Hydrogen bonding between complementary bases occurs when an electropositive hydrogen atom is attracted to an electronegative atom such as oxygen or nitrogen; these bonds are highly directional. Because of the manner in which bases protrude from their respective phosphodiester backbones, base pairing or hybridization of complementary strands in an **antiparallel** (opposite) configuration is favored. Thus, the 5′ end of one strand is directly opposite the 3′ end of the complementary strand to which it is base-paired:

```
5′ ----------------------- 3′
   | | | | | | | | | | | | |
3′ ----------------------- 5′
```

The obvious structural differences at the 5′ and 3′ ends of a molecule support a convention by which one may unambiguously refer to the position of any topological feature of a nucleic acid molecule in relation to any other topological feature:

Upstream means that the salient structure or feature is toward the 5′ end of the molecule relative to some other point of reference.

Downstream means that the salient structure or feature is toward the 3′ end of the molecule relative to some other point of reference.

For the sake of simplicity, upstream and downstream may be used to mean "in the opposite direction of expression" and "in the direction of expression," respectively. This nomenclature may be especially useful when describing the salient features of a double-stranded nucleic acid molecule.

The actual base sequence of a polyribonucleotide is known as the primary structure of the RNA, though there is a tremendous proclivity for regions within a single molecule to exhibit intramolecular base pairing, perhaps better known as secondary structure. The variety of possible interactions within the phosphodiester backbone are often described using such colorful nomenclature as RNA hairpin, stem, interior loop, bulge loop, multibranched loop, and pseudoknot. Higher order three-dimensional folding (the so-called tertiary structure) that RNA molecules exhibit is best described as the collection of 2° structural elements required by an RNA molecule to perform its biological function; much has been suggested about the role of folding by careful study of transfer RNAs, an example of folding par excellence. To a great extent, much of the 2° and 3° structure of RNA may be attributed to noncanonical[2] intramolecular base pairing. For most laboratory applications, higher order folding must often be disrupted before an experiment can be performed using a purified RNA sample, as described below.

Stringency: Conditions That Influence Nucleic Acid Structure

The individual strands of a double-stranded nucleic acid molecule are dynamic indeed. Several variables, all of which can be manipulated in the laboratory, affect the ability of two single-stranded nucleic acid molecules to form stable hydrogen bonds between their complementary bases. In the laboratory, it is essential that the investigator be able to influence the double-strandedness or single-strandedness of a polynucleotide pair, and the term **stringency** is commonly used to describe a defined set of conditions. Stringency is a measure of the likelihood that a double-stranded nucleic acid molecule will dissociate into its constituent single strands. Stringency is also a measure of the ability of single-stranded nucleic acid molecules to discriminate between other molecules that share a high degree of complementarity and those that share a lesser degree of complementarity. Thus, stringency is all about the ability of double-stranded nucleic acid molecules to remain base-paired. In practice, high-stringency conditions favor stable hybridization only between nucleic acid molecules with a high degree of homology. As the stringency in a system is lowered, a proportional increase in nonspecific hybridization is favored.

In vitro, the variables that are commonly manipulated to either promote or prevent hybridization from occurring include pH, ionic strength, temperature, and formamide (Table 3). These extrinsic variables act as a function of time. For intrinsic qualities, the length of the molecules and their degree of complementarity influence the stability of the duplex at any given stringency. Practically speaking, manipulating these variables

---

[2]Canonical base pairs: G:C and A:U; noncanonical base pairs: A:C, G:U, U:C, U:U, G:G, and A:A:U trimers.

TABLE 3

Factors That Influence Nucleic Acid Structure

Low stringency	High stringency
High salt	Low salt
Slightly acidic pH	Alkaline pH
Low temperature	High temperature

Note. The formation of double-stranded molecules is promoted in a low-stringency environment, while dissociation into constituent single strands is favored as the stringency is increased. The addition of formamide permits maintenance of high-stringency conditions at temperatures lower than would otherwise be required.

is really fine-tuning the ability of nucleic acid molecules to discriminate between other molecules that have a high degree of complementarity and those that exhibit a lesser degree of complementarity. Moreover, organic solvents such as formamide (and to an extent, formaldehyde) are frequently used to lower the melting temperature (T_m) of base-paired molecules, that is, that temperature at which 50% of the double-stranded molecules will dissociate into constituent single-stranded molecules. Organic solvents interfere with hydrogen bonding and thus lower the temperature necessary for conducting high-stringency hybridizations, affording the investigator more flexibility with respect to experimental parameters.

Among the easier methods for modulating stringency is to adjust the salt concentration and/or the temperature of the system. Conditions of low stringency favor or promote double-strandedness while conditions of high stringency promote or favor single-strandedness in a mixture of nucleic acid molecules. Other manipulations, such as including organic solvents or changing the pH, also influence stringency, and a discussion of these more complex parameters is found in Chapter 13.

Effect of Salt on Stringency

Physiological ionic strength varies from 10 to 20 mM. In an experimental setting the natural electrostatic repulsion that would prevent two complementary, negatively charged strands from base pairing can be neutralized with low, moderate, or high concentrations of monovalent cation, including Na^+ and Li^+. These act as positively charged counterions that minimize the tendency for natural electrostatic repulsion between two negatively charged phosphodiester backbones. In low concentrations or in the absence of salt, electrostatic repulsion may easily exceed the strength of the hydrogen bonds that exist between complementary base pairs, resulting in dissociation of double-stranded molecules. In the cell, intracellular cationic proteins act, at least in part, to neutralize the negative charges of nucleic acids.

Effect of pH on Stringency

pH influences the degree of ionization of a nucleic acid molecule. At near neutral physiological pH, nucleic acids are ionized to an extent; hybridization can occur between polynucleotides even though the individual molecules demonstrate a net negative charge. *Slightly* below physiological pH, a more acidic pH will neutralize the phosphates and cause nucleic acids, especially double-stranded DNA (dsDNA), to become condensed. Above physiological pH, a more alkaline pH will increase the ionization of the phosphate groups to the extent that electrostatic repulsion can easily exceed the thermodynamic stability of a double-stranded molecule, resulting in strand dissociation. Although adding NaOH to denature DNA is a fairly common practice, when working with RNA it is of paramount importance to remember the exquisite susceptibility of these molecules to degradation by alkaline hydrolysis.

Effect of Temperature on Stringency

In general, temperature has an effect opposite that of salt concentration. The thermodynamic stability of a duplex is a function not only of the number of nucleotides that participate in duplex formation but also of the guanine and cytosine (G+C) content (remember that three hydrogen bonds naturally occur between the bases guanine and cytosine and only two hydrogen bonds exist between adenine and thymine or adenine and uracil), the concentration of salt in the hybridization buffer, as well as the presence of organic solvents such as formamide. By applying more heat to the system, one will eventually exceed the T_m of the duplex, causing dissociation of the strands. The exact experimental conditions that define stringent hybridization and posthybridization washes are addressed in Chapter 13.

Effect of Formamide on Stringency

The T_m of a double-stranded molecule is largely a function of the number of base pairs involved and the G+C content, especially when the molecule is less than 100 base pairs. In cases such as these, the thermodynamic stability of the duplex is exquisitely sensitive to temperature changes of as little as 1°C. Any molecular biologist can attest to the fact that oligonucleotides used as hybridization probes or primers for PCR often work (or don't work!) precisely because the hybridization/annealing temperature is off by only a few degrees. With longer probes, however, very high temperatures would be required to maintain a comparable level of stringency. Often, the calculated hybridization temperature (discussed in Chapter 13) is simply too high to permit base pairing under any circumstances. In instances such as these, the investigator may choose to incorporate formamide into a hybridization recipe. The purpose of this technique is to lower the T_m while maintaining high stringency, the net result of which is high stringency at a temperature that is low enough to permit base pairing to occur. This ensures high-fidelity hybridization without the concerns associated with temperature extremes. In general, one may expect to see the T_m lowered by 0.75°C for each 1% formamide added.

Types of Double-Stranded Molecules

Beyond the truism that recombinant DNA methodology is a technology of enzymology, one of the principal tools of the molecular biologist is hybridization technology, that is, the ability to promote the formation of double-stranded molecules from two complementary single-stranded nucleic acid molecules. *In vitro,* this technology allows the identification and recovery of unique sequences from a mixed population of DNA or RNA molecules. Given that a pair of complementary nitrogenous bases will form hydrogen bonds irrespective of the nature of the backbone (polyribonucleotide (RNA) or polydeoxyribonucleotide (DNA)), the molecular biologist can promote hybridization (duplex formation) between any two complementary polynucleotide strands:

DNA:DNA   DNA:RNA   RNA:RNA

Thermodynamically speaking, double-stranded DNA molecules are the least stable of the three. Whereas DNA:RNA hybrids are more stable than double-stranded DNA (DNA:DNA duplexes), the most stable hybrid forms between two complementary strands of RNA. While RNA molecules are believed to exist primarily in single-stranded form *in vivo,* the fact that extensive intramolecular base pairing occurs is extremely well documented. In an experimental context, the formation of double-stranded RNA molecules permits more stringent hybridization and wash conditions; for this reason RNA probes are often desirable as a means of increasing the signal to noise ratio in an experiment.

References

Lewin, B. (Ed.). (1997). "Genes VI." Oxford University Press, New York.
Watson, J. D., *et al.* (Eds.). (1987). "Molecular Biology of the Gene." Cummings, Menlo Park, CA.
Wolfe, S. L. (1993). "Molecular and Cellular Biology." Wadsworth, Belmont, CA.

Transcription and the Organization of Eukaryotic Gene Sequences

Transcription and the Central Dogma

According to the central dogma[1] of molecular biology, the expression of hereditary information flows from genomic sequences (DNA), through a messenger RNA (mRNA) intermediate, to ultimate phenotypic manifestation in the form of a functional polypeptide. Whereas this design mirrors what occurs in both prokaryotic and eukaryotic cells, certain "violations" have been observed in nature (1), accompanying the discovery of the retroviral enzyme reverse transcriptase (RNA-dependent DNA polymerase), by which viral genomic RNA is replicated through the production of a double-stranded DNA intermediary, the goal being integration of viral sequences into the genome of the host (lysogeny) or the production of virion progeny, and (2) the discovery of RNA editing (Benne *et al.*, 1986), also known as RNA sequence alteration, in which the transcribed sequence is subject to change. For review, see Chan and Nishikura, 1997

Transcription is that process by which a single-stranded RNA molecule is synthesized from a defined sequence (locus) of double-stranded DNA. All phases of transcription are subject to variation and are potential control points in the regulation of gene expression. A transcriptional unit, therefore, is a DNA sequence that manifests appropriate signals for the initiation and termination of transcription and is capable of supporting polymerization of a primary RNA transcript. The process of transcription is so named because the transfer of information from DNA to RNA is in the same language, namely the language of nucleic acids. In contrast, the process known as **translation** is so named because nucleic acid instructions in the form of RNA are used to direct the assembly of a primary polypeptide from amino acid precursors: the nucleic acid instructions are executed in the language of proteins.

Regulatory Elements

Transcription is mediated by enzymes known as RNA polymerases. These enzymes recognize very specific and highly conserved **promoter,** or initiation, sequences within the enormous complexity of genomic DNA. The exact sequence and precise geometry of these regulatory elements can either promote or prevent the onset of transcription. Knowledge of these promoter/consensus sequences is due largely to experiments involving standard recombinant DNA methodologies and DNA

---

[1]Coined by Francis Crick in 1957 in a paper presented to a symposium of The Society for Experimental Biology.

sequencing. In prokaryotic systems the promoter sequences are commonly known as the Pribnow box, centered about 10 bases upstream from the transcription start site (–10 sequence) and the so-called –35 sequence (about 35 base pairs upstream from the start site of transcription). The eukaryotic promoter counterparts are known as the Hogness, or TATA box, so named because of the prevalence of adenine- and thymine-containing nucleotides, and the CAAT box, found in several but not all promoters, and so named because of the conservation of the nucleotide motif that constitutes the sequence. These eukaryotic sequences, centered around –25 bases (Hogness box) and –75 bases (CAAT box) upstream from the transcription start site, may be involved in the control of initial binding of the RNA polymerase and the choice of starting point, respectively (Lewin, 1997).

Eukaryotic promoters do not necessarily function alone. In many cases, the efficiency of transcription in eukaryotic cells can be influenced profoundly by the presence of an **enhancer** sequence. The precise location and orientation of enhancer sequences, with respect to the gene promoter, varies from one gene to the next. Some genomic sequences, such as the immunoglobulin genes, even carry enhancers within the structural portion (the body) of the gene. Removal of enhancer sequences can compromise tremendously the transcriptional efficiency of a locus normally under the influence of an enhancer sequence. Moreover, the transcription of genes that are not naturally associated with an enhancer element can be increased tremendously if an enhancer is ligated to the DNA, usually in no particular orientation, and hundreds, if not thousands of base pairs away from the transcription start site. Whereas the transcriptional influence of upstream and downstream enhancer sequences on cellular promoters is well documented, the mechanisms by which enhancers exert their effect remain obscure.

DNA Template

During transcription, both strands of the DNA sequence being transcribed have different names and different roles. The strand that actually serves as the template upon which RNA is polymerized is properly referred to as the **template strand.** The other strand, which does not act in a template capacity, is called the **coding strand.** In publishing a gene sequence, the convention is to show the sequence of the coding strand, written 5′ to 3′, from left to right. The implication is that the other strand, the template strand, is base paired to and lying 3′ to 5′, or antiparallel, with respect to the coding strand. The template strand is so named because the precise sequence of nucleotides inserted into the nascent RNA transcript is determined by and complementary to the template strand nucleotide sequence. It is important to realize that the coding strand and template strand may switch roles depending upon the placement of transcriptional promoter sequences. The quintessential example of this phenomenon *in vitro* is the cloning of a double-stranded DNA between two different RNA promoters, often the bacteriophage polymerase promoters S6, T3, or T7. Constructions such as these are frequently employed to accommodate *in vitro* transcription of large amounts of sense and/or antisense RNA for use as nucleic acid probes (see Chapter 12).

Gene Organization

In order to understand the significance of the products of transcription, it is first essential to understand the organization of the genes themselves. The typical prokaryotic genome exhibits little extraneous material. Frequently, genes that encode polypeptides associated with a common metabolic pathway are clustered together, as suggested by the operon model. The RNA molecule that results from the transcription of such an operon is usually polycistronic, meaning that more than one polypeptide is encoded in a single RNA transcript. Within a polycistronic mRNA, the coding information for *each* polypeptide is contiguous, meaning that there are no interruptions in coding sequences by extraneous or noncoding information. In fact, the kinetics of prokaryotic gene expression are so rapid that bacterial mRNA is usually being transcribed, undergoing translation, and being degraded simultaneously. Moreover, all of these activities occur in the same cellular compartment. These intrinsic factors have in the past frustrated valiant attempts to clone prokaryotic mRNA. While a great many improvements that favor the isolation of useful RNA from both gram-negative and gram-positive bacteria have been made, the isolation of prokaryotic RNA remains something of a challenge.

In contrast to prokaryotic systems, virtually all eukaryotic mRNAs are monocistronic. Although a single polypeptide species results from the translation of any monocistronic eukaryotic mRNA molecule, most polysomes consist of 10–50 members, a system that accommodates repeated translation of the same mRNA molecule as long as the transcript remains biologically and chemically competent. Of course, occasional deviations from this norm have been observed (for review, see Bag, 1991).

Close examination of eukaryotic genes reveals that for a vast majority of genes there are considerably more nucleotides constituting a particular locus than are necessary to direct the synthesis of the corresponding polypeptide; that is, the DNA sequence and the amino acid sequence are not colinear over the span of the locus. This size differential can also be observed at the level of the mature mRNA in the cytoplasm, which also is usually shorter than the DNA sequence from whence it is derived. Upon further scrutiny, this discrepancy can be resolved at the level of the organization of the gene itself, the sequences within which fall into one of two categories:

1. **Exons** are regions of DNA that are represented in the corresponding mature cytoplasmic RNA; exons may or may not have a peptide coding function.

2. **Introns** are regions of DNA that are transcribed but are not represented in the corresponding mRNA; introns are systematically removed by splicing together the exon sequences that lie adjacent to them. Introns usually do not direct polypeptide synthesis, though some fungal mitochondrial RNAs may be an exception (for reviews, see Tereba, 1985; Cech and Bass, 1986).

The number and length of exons and introns associated with a gene are highly variable depending upon locus and this variability even pertains to loci that are highly conserved across evolutionary time. Some genes, such as α-interferon, lack intron sequences completely. There is no significant homology found among intron sequences, although introns generally have termination codons in all reading frames. This is not entirely unexpected because the noncoding function of introns favors the accumulation of multiple mutations that might otherwise be lethal were they to occur within an exon sequence.

Examination of the splicing junctions of introns reveals a strict conservation of a dinucleotide consensus sequence (Breathnach and Chambon, 1981; Mount, 1982). Proceeding from the 5' end of the RNA, introns are found to begin with a GU dinucleotide and end with an AG dinucleotide;[2] this phenomenon is believed to occur nearly 100% of the time:

$$\ldots \text{Exon} \ldots /\text{GU} \ldots \text{Intron} \ldots \text{AG}/ \ldots \text{Exon} \ldots$$

The nucleotides immediately adjacent to the dinucleotide intron boundaries (lying within exon and intron domains) are also conserved to an extent (65–75%). A point mutation at any splice site generally results in the inactivation of the splice site. This is often accompanied by the production of an aberrant mRNA through the use of an alternative splice site,

---

[2]The so-called GT–AG rule, describing exon–intron splice sites, refers to the DNA sequence. The transcript itself manifests GU at the 5' end of an intron.

usually located within the intron (Triesman *et al.*, 1982), as in certain β-thalassemic individuals. Whereas this interrupted gene organization is probably common to all higher eukaryotes, the consensus phenomenon does not apply to mitochondrial or chloroplast loci, nor to yeast tRNA genes (Lewin, 1997).

The actual mechanics of intron removal/splicing are mediated in part by a highly conserved family of small nuclear RNAs; these molecules exist as the RNA–protein complexes commonly known as small nuclear ribonucleoproteins (snRNPs), such as U1, U2, and U4–6. Methodologies for efficiently characterizing these essential splicing cofactors have been described (Hamm and Mattaj, 1990). A similar, abundant group of small cytoplasmic RNAs (scRNA) are found in the eukaryotic cytoplasm. As with snRNAs, scRNA molecules have been shown to exist as RNA–protein complexes. The role of the scRNA molecules in regulating protein synthesis and mRNA degradation remains speculative (reviewed by Brunel *et al.*, 1985). It should also be noted that the excision of certain introns is the result of RNA self-cleavage. Such catalytic RNAs are generically referred to as ribozymes (for reviews, see Lewin, 1997; Michel and Ferat, 1995)

Although transcription produces an RNA molecule whose sequence correlates precisely with the DNA from which it is derived, the primary product of transcription is only an RNA precursor; collectively, all of the precursor RNAs in the nucleus are known as heterogeneous nuclear RNA (hnRNA). hnRNA and specific nuclear proteins form heterogeneous nuclear ribonucleoprotein complexes (hnRNPs), an extremely abundant component of the nucleus (for review, see Dreyfuss, 1986). For the RNA to function as a template for the synthesis of its encoded protein, the introns must first be removed by a process known as RNA splicing. This results in the concatenation of exon sequences and is but one of several modifications of the precursor RNA on the road to RNA maturation. It appears that maturation of precursor RNA occurs in the nucleus and that only thoroughly processed transcripts are exported in the form of mature mRNA. Intervening intron sequences are reserved and then degraded in the nucleus.

RNA Polymerases and the Products of Transcription

Genes are transcribed by enzymes known as RNA polymerases, and the major RNA types produced are known as ribosomal RNA (rRNA), transfer RNA (tRNA), and mRNA. Eukaryotic genes are transcribed by one of three nuclear RNA polymerases that are among the largest and most complex proteins in the cell and consist of more subunits than their prokaryotic counterpart. The eukaryotic enzymes are known as RNA polymerases I, II, and III, each of which is responsible for transcribing a different class of genes (Table 1); bacteria, in contrast, exhibit only one type of RNA polymerase, which transcribes all three classes of RNA. RNA polymerases are only active in the presence of DNA, require the nucleotides ATP, CTP, GTP, and UTP as precursors, and generally function in a Mg^{2+}-dependent manner. The transcriptional products of each of the three eukaryotic RNA polymerases may be distinguished by their differential sensitivity to the bicyclic octapeptide fungal toxin α-amanitin[2] (Marzluff and Huang, 1984; Roeder, 1976). See Chapter 19 for applications.

As with all nucleic acid molecules, RNA transcripts are synthesized only in the 5′ to 3′ direction. Transcription involves three distinct phases, initiation, elongation, and termination, all of which have been described in great detail elsewhere (Lewin, 1997; Olave et al., 1997). Briefly, initiation involves the attachment of RNA polymerase to a DNA template and the acquisition of what will be the first ribonucleotide in the RNA molecule, all of which are mediated by proteins known as initiation factors. Elongation involves the sequential addition of ribonucleotides to the nascent chain, a process also involving accessory protein elongation factors. Termination is the completion of RNA synthesis, whether appropriately or prematurely, and the disengagement of both RNA and enzyme from the DNA template. Transcription termination, as with initiation and elongation, is influenced by the presence of small proteins, in this case known as termination factors. The nucleotide sequence of the resulting RNA molecule is essentially identical to the coding strand of the DNA from which it is derived, the only difference being the substitution of the base uracil for thymine. The natural fate of the products of transcription is the subject of Chapter 3.

[2]Derived from the poisonous mushroom *Amanita phalloides.*

TABLE 1

Eukaryotic RNA Polymerase Enzymes and Their Respective Products and Sensitivities to α-Amanitin

Eukaryotic enzyme	Products	Sensitivity to α-amanitin
RNA polymerase I	rRNA (28S, 18S, 5.8S)	−
RNA polymerase II	hnRNA →→ mRNA	+ +
RNA polymerase III	tRNA, 5S rRNA, snRNA	+

With respect to eukaryotic cells, transcription of the genes encoding ribosomal RNA is catalyzed by RNA polymerase I. The primary product of RNA polymerase I transcription, a large unspliced 45S precursor RNA, yields the smaller 28S, 18S, and 5.8S rRNAs after processing. RNA polymerase III is responsible for transcribing the genes that encode tRNA molecules, the 5S rRNA, certain repetitive elements (Fornace and Mitchell, 1987), and a class of short RNA molecules (100–500) bases) that might best be thought of as "structural RNAs." These include small nuclear RNA (snRNA) and scRNA. Small nuclear RNAs are known to be involved in the removal of intron sequences and the joining together of exon sequences, and at least one type of snRNA is produced by RNA polymerase III transcription. Those short RNAs that are localized in the cytoplasm are, of course, known as scRNAs. Of particular interest to the molecular biologist are the transcriptional products of RNA polymerase II, namely mRNA and its unspliced nuclear precursor, hnRNA.

The typical mammalian cell contains approximately $1–5 \times 10^{-5}$ μg RNA, a majority of which (70–80%) represents the transcriptional products of RNA polymerase I (50–70%) and RNA polymerase III (10–12%). Between 80 and 85% of cellular RNA is found in ribosomes, predominantly 28S and 18S rRNA, which then complexed with myriad ribosome-specific proteins, constitute the 60S and 40S ribosomal subunits, respectively. The prokaryotic counterparts are the 23S and 16S rRNAs, and 50S and 30S ribosomal subunits, respectively. As demonstrated in Chapter 9, these two highly abundant RNA species serve as excellent natural molecular weight size markers when either total cellular or total cytoplasmic RNA is electrophoresed.

Historically, rRNA was first believed to have a template role in the synthesis of proteins. The first indication that a new, separate class of RNA, mRNA, acted as the template molecule emerged from studies involving T4 phage-infected *Escherichia coli* cells (Volkin and Astrachan, 1956; Brenner *et al.*, 1961; Gros *et al.*, 1961; Hall and Spiegelman, 1961). Later, multiple 5′ ends were observed for the late SV40 mRNAs (Ghosh *et al.*, 1978) and late polyoma mRNAs (Flavell *et al.*, 1979), the first indications of the heterogeneous nature of the RNA polymerase II transcription initiation.

Unlike the fairly undiversified transcription products of RNA polymerases I and III, the transcription products of RNA polymerase II are as diverse as the cellular biochemistry itself. This is not at all unexpected because it is, essentially, the mRNA constituency that drives the phenotype of the cell. The relative contribution of RNA polymerase II transcription products is generally between 20 and 40% depending on cell type and cell state, though mature mRNA represents only 1–4% of the total. It is certainly worth noting that about 75% of all hnRNA is degraded in the nucleus; acknowledgment of this fact sets the stage for a posttranscriptional regulatory paradigm. While 90% of genomic DNA is thought to be transcriptionally silent, the typical eukaryotic cell is transcribing tens of thousands of different genes at any given point in the cell cycle (Williams, 1981). Given the complexity of cellular biochemistry, this observed heterogeneity within the mRNA population is necessary to satisfy even basal-level requirements for viability.

References

Bag, J. (1991). mRNA and mRNP. *In* "Translation in Eukaryotes," pp. 71–79. CRC Press, Ft. Lauderdale, FL.

Benne, R., van den Berg, J., Brakenhoff, J. P., Sloof, P., Van Boom, J. H., and Tromp, M. C. (1986). Major transcript of the *coxII* gene from trypanosome mitochondria contains four neucleotides that are not encoded in the DNA. *Cell* **46**, 816.

Breathnach, R., and Chambon, P. (1981). Organization and expression of eukaryotic split genes coding for proteins. *Annu. Rev. Biochem.* **50**, 349.

Brenner, S., Jacob, F., and Meselson, M. (1961). An unstable intermediate carrying information from genes to ribosomes for protein synthesis. *Nature* **190**, 576.

Brunel, C., Sri-Widada, J., and Jeanteur, P. (1985). snRNPs and scRNPs in eukaryotic cells. *Prog. Mol. Sub. Cell. Biol.* **9**, 1.

Cech, T. R., and Bass, B. L. (1986). Biological catalysis by RNA. *Annu. Rev. Biochem.* **55**, 599.

Chan, L., and Nishikura, K. (1997). Mammalian RNA editing. *In* "mRNA Metabolism and Post-Transcriptional Gene Regulation" (J. B. Harford and D. R. Morris, Eds.). Wiley-Liss, New York, NY.

Darnell, J. E., Philipson, L., Wall, R., and Adesnik, M. (1971). Polyadenylic acid sequences: Role in conversion of nuclear RNA into messenger RNA. *Science* **174**, 507.

Dreyfuss, G. (1986). Structure and function of nuclear and cytoplasmic ribonucleoprotein particles. *Annu. Rev. Cell Biol.* **2**, 459.

Flavell, A. J., Cowie, A., Legon, S., and Kamen, R. (1979). Multiple 5' terminal cap structures in late polyoma virus RNA. *Cell* **16**, 357.

Fornace, A. J., Jr., and Mitchell, J. B. (1986). Induction of B2 RNA polymerase III transcription by heat shock: Enrichment for heat shock induced sequences in rodent cells by hybridization subtraction. *Nucleic Acids Res.* **14**, 5793.

Ghosh, P. K., Reddy, V. B., Swinscoe, J., Lebowitz, P., and Weissman, S. M. (1978). Heterogeneity and 5' terminal structures of the late RNAs of simian virus 40. *J. Mol. Biol.* **126**, 813.

Gros, F., Hiatt, H., Gilbert, W., Kurland, C. G., Risebrough, R. W., and Watson, J. D. (1961). Unstable ribonucleic acid revealed by pulse labelling of *Escherichia coli. Nature* **190**, 581.

Hall, B. D., and Spiegelman, S. (1961). Sequence complementarity of T2 DNA and T2-specific RNA. *Proc. Natl. Acad. Sci. USA* **47**, 137.

Hamm, J., and Mattaj, I. W. (1990). Structural analysis of U small nuclear ribonucleoproteins by *in vitro* assembly. *Methods Enzymol.* **181,** 273.

Lewin, B. (Ed.). (1997). "Genes VI." Oxford University Press, New York.

Marzluff, W. F., and Huang, R. C. C. (1984). Transcription of RNA in isolated nuclei. *In* "Transcription and Translation: A Practical Approach" (B. D. Hames and S. J. Higgins, Eds.). IRL Press, Washington, DC.

Michel, F., and Ferat, J. L. (1995). Structure and activities of group II introns. *Annu. Rev. Biochem.* **64,** 435.

Mount, S. M. (1982). A catalogue of splice junction sequences. *Nucleic Acids Res.* **10,** 459.

Olave, I., Drapkin, R., and Reinberg, D. (1997). Transcription and translation control: An overview. *In* "mRNA Metabolism and Post-Transcriptional Gene Regulation" (J. B. Harford and D. R. Morris, Eds.). Wiley-Liss, New York, NY.

Roeder, R. G. (1976). *In* "RNA Polymerase" (R. Losick and M. Chamberlain, Eds.). Cold Spring Harbor Laboratory Press, Cold Spring Harbor, NY.

Tereba, A. (1985). Chromosomal localization of protooncogenes. *Int. Rev. Cytol.* **95,** 1.

Triesman, R., Proudfoot, N. J., Shander, M., and Maniatis, T. (1982). A single-base change at a splice site in a β^0-thalassemic gene causes abnormal RNA splicing. *Cell* **29,** 903.

Volkin, E., and Astrachan, L. (1956). Phosphorus incorporation of *E. coli* ribonucleic acid after infection with bacteriophage T2. *Virology* **2,** 146.

Williams, J. G. (1981). The preparation and screening of a cDNA clone bank. *In* "Genetic Engineering" (R. Williamson, Ed.), Vol. 1, pp. 1–59. Academic Press, New York.

Messenger RNA

A great many genes are transcribed constitutively by RNA polymerase II, though it is clear that large quantities of heterogeneous nuclear RNA (hnRNA) are turned over in the nucleus. In eukaryotic cells, messenger RNAs (mRNA) are derived from precursor hnRNA through a series of modifying reactions, which include formation of the 5' cap, methylation, splicing, 3' end processing, and frequently, polyadenylation. Only 1–4% of the total RNA in the cytoplasm of a typical eukaryotic cell is mature mRNA: these mRNAs may then be further subdivided, depending on the cytoplasmic prevalence of a particular mRNA species. There are three official such categories, high abundance, medium abundance, and low abundance mRNAs, and, in the mind of this author, the unofficial *very low* abundance category.

Highly abundant transcripts are present in hundreds of copies per cell. These are most often observed when a cell is producing an enormous quantity of a particular protein or is high specialized or **differentiated** to perform some function. Medium abundance transcripts are best thought of as being present in dozens of copies per cell, and often many genes with "housekeeping" functions manifest their mRNAs at this level of prevalence in the cell. Low abundance mRNAs are prevalent in 14 or fewer copies per cell and often are difficult to assay by many of the older, classical techniques, such as Northern analysis (Chapter 11), without some form of enrichment in order to increase the statistical probability that such rare messages will be detectable. Very low abundance mRNAs are those present in fewer than 1 copy per cell, necessitating a mixed population of cells (i.e., a tissue) rather than a cell line to induce and observe transcription of that particular gene. Most importantly, it is essential to realize that the prevalence or abundance of a transcript in a cell is subject to change of monumental proportions; such changes may occur in response to natural changes in the cellular milieu or due to experimental manipulation.

Topology of a Typical mRNA Molecule

A typical fibroblast cell contains approximately 1 picogram (pg) of mRNA, which is equivalent to about 10^6 molecules transcribed from 10,000 to 25,000 different genes. Although this mRNA heterogeneity reflects the equally diverse population of proteins that these mRNAs encode, a typical eukaryotic mRNA molecule shares several topological features with nearly all other mRNA molecules (Figure 1).

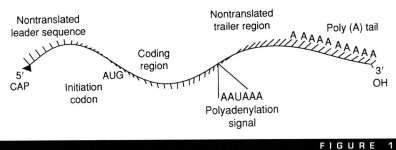

Topology of a typical mammalian mRNA molecule.

5′ Cap

Mature eukaryotic mRNA molecules are essentially monocistronic polyribonucleotides and the 5′ terminus of the message appears to correspond to the transcription start site. Phosphodiester bonds join adjacent ribonucleotides 5′–3′, with the exception of the first two nucleotides, which together constitute what is known as the 5′ cap (Reddy *et al.*, 1974; for reviews, see Shatkin, 1976; Banerjee, 1980); the third nucleotide may be involved as well. The immediate product of RNA polymerase II transcription, a precursor hnRNA molecule, displays the following structure at the 5′ end of the molecule

$$5'\ pppA/GpNpNpNp \ldots 3'$$

meaning that the first *transcribed* nucleotide is usually adenosine *or* guanosine.

The cap itself consists of a rare 5′–5′ triphosphate linkage between the first transcribed nucleotide with a 7-methylguanosine (m^7G) nucleotide. The m^7G is added to the 5′ end of the molecule very soon after the initiation of transcription, and nascent polynucleotides with fewer than 100 bases are found to be capped (Babich *et al.*, 1980). Capping of eukaryotic mRNAs also precedes internal mRNA adenylate methylation to form N^6-methyladenosine, a modification that also occurs early in mRNA biogenesis; the resulting N^6-methyladenosine residues are conserved during RNA processing (Shatkin, 1976; Chen-Kiang *et al.*, 1979).

Addition of the 5′ terminal G is mediated by the nuclear enzyme guanylyl transferase resulting in the following structure

$$5'\ GpppA/GpNpNp \ldots 3'$$

where the terminal guanosine nucleotide is linked 5′–5′ to what was the first transcribed nucleotide. This structure is the 5′ cap, which is then subjected to one or more methylations, catalyzed by the enzymes guanine-7-methyltransferase and 2′-*O*-methyltransferase. In the literature, this structure is also represented as

$$m^7G(5')ppp(5')X$$

where methylation of the terminal guanosine occurs in the base, while subsequent methylations may involve the base and/or sugar of the penultimate and subpenultimate nucleotides. The extent of methylation is a function of mRNA species, and higher organisms usually have more extensively methylated caps. Cap structures with multiple methyl groups have also been observed in the small nuclear RNA species (Furuichi and Shatkin, 1989). The donor of the methyl groups in the cap is *S*-adenosylmethionine.

5′ Capping is only one example of a posttranscriptional modification associated with cellular gene expression and is observed in virtually all eukaryotic mRNAs. The presence of the 5′ cap is responsible for the translational efficiency of mRNAs at the level of initiation. Formation of the translation apparatus is mediated in part by cap-binding proteins (for reviews, see Shatkin, 1985; Dreyfuss, 1986): the assembly of the ribosomal subunits requires the presence of initiation factors, at least one of which is a cap-binding protein.

Capping apparently also confers transcript stability by protecting against phosphatase attack and 5′–3′ exonucleolytic degradation (Furuichi *et al.*, 1977). In contrast, prokaryotic mRNAs, which naturally lack a 5′ cap structure, are degraded exonucleolytically from the 5′ end during translation. Nor is the 5′ cap a characteristic of mitochondrial or chloroplast mRNAs. Most animal viruses that replicate in the eukaryotic cell manifest 5′ capped mRNAs, two noteworthy exceptions being polio virus and picornavirus.

Leader Sequence

The first nucleotides immediately 3′ to the eukaryotic cap structure constitute the nontranslated leader sequence or, more formally, the 5′ untranslated region (UTR). In eukaryotic systems, the typical leader sequence average is 30–50 bases long, though leader sequences as long as hundreds of bases or as short as 3 bases have been observed. This variable-length region is not translated, as its name implies, and is believed to function in the proper alignment of the mature mRNA with the protein translation machinery, mediated in part by interaction with mRNA-binding proteins (Leibold and Munro, 1988).

Coding Region

The structural portion of the mRNA, that is, the coding region, begins with the first AUG codon, which is also known as the initiation codon. Far less frequently, translation may be initiated from non-AUG start codons (Hann *et al.*, 1988; Florkiewcz and Sommer, 1989). In addition to this highly conserved initiation codon, it has also been demonstrated that a purine, usually A, must be present at the position three nucleotides upstream from the initiation codon in order for the initiation codon to be recognized by the ribosome (Kozak and Shatkin). This is presumed to be an indicator pointing to the initiation codon. Every three nucleotides after the initiation codon specifies the placement of another amino acid into the nascent polypeptide, though it has been suggested that eukaryotic peptide elongation may stall along specific stretches of the mRNA template (Woolin and Walter, 1988). Elongation continues in this manner until a nonsense codon (UAA, UAG, or UGA) is encountered, which is the signal to halt translation. Thus, the 5' end of the mRNA coding region corresponds to the amino terminus and the 3' end corresponds to the carboxy terminus of the primary polypeptide that results from translation of the mRNA. On average, typical eukaryotic mRNA coding regions range from 200 to 1500 nucleotides long, though extremely short and extremely long coding regions have been identified in some mRNAs.

Trailer Sequence

Beyond the nonsense codon lies another nontranslated area, the trailer sequence or 3' UTR. This sequence ranges from 50 to 150 nucleotides long, although trailer sequences as long as 1000 bases have been reported. Further, different mRNA molecules encoding the same polypeptide may differ only with respect to the length of their 3' trailer region (e.g., the dihydrofolate reductase mRNAs). Although no function has yet been ascribed to the trailer region, it is believed to influence the stability of the molecule in the cytoplasm. It should be noted, however, that most trailer regions manifest a highly conserved AAUAAA sequence that specifies the correct placement of the poly(A) tail.

Poly(A) Tail

Most eukaryotic mRNA species are further characterized by a 50- to 250-nucleotide tract of polyadenylic acid at the 3' end of the molecule (Lim and Cannelakis, 1970; Edmonds *et al.*, 1971; Brawerman, 1976). This structure is known as the poly(A) tail, the precise length of which is a function of mRNA species and translational status in the cell (Kuge and

Richter, 1995; reviewed by Baker, 1997). The poly(A) structure is added enzymatically to precursor hnRNA very soon after transcription, by one of a family of nuclear enzymes known as polyadenylate (poly(A)) polymerases. Historically, polyadenylate polymerases were known long before discovery of the 3′ poly(A) tail. All poly(A) polymerases show tight specificity for ATP and require a polyribonucleotide with a free 3′-hydroxyl group to which AMP residues can be attached (for review, see Edmonds and Winters, 1976; Edmonds, 1982).

The mRNA fraction that exhibits this feature is collectively described as poly(A)$^{+}$ mRNA. While greater than 97% of all eukaryotic mRNAs are believed to exhibit this poly(A) feature, fully one-third of the total mass of cellular mRNA lacks a 3′ poly(A) tail, of which the predominant components are the histone mRNAs (Lewin, 1997). The absence of this structure, however, does not appear to influence the nucleocytoplasmic transport or translatability of these molecules; naturally nonpolyadenylated mRNAs are clearly exported from the nucleus without a poly(A) tail and can be translated with efficiencies equal to that of the polyadenylated message. The modulation of polyadenylation of many mRNAs has been observed and a rapidly growing body of evidence supports a role for the poly(A) tail both in the stability of a particular species in the cytoplasm as well as its translational efficiency. It is likely that the specialized 5′ and 3′ features of eukaryotic mRNA may protect these molecules, at least in part, from exonuclease digestion. On the other hand, sequence information near the 5′ and/or 3′ end may facilitate degradation of these molecules in response to various cytoplasmic cues (Jackson and Standart, 1990; Spirin, 1994; Standart and Jackson, 1994; Curtis *et al.,* 1995).

Transcription apparently does not terminate at the site of polyadenylation, but at some distance downstream (that is, in the 3′ direction). This phenomenon was first observed in the adenovirus model system (Nevins and Darnell, 1978; Fraser *et al.,* 1979). The actual site of polyadenylation is generated by stringently regulated splicing within the trailer region of the primary transcript, directly 3′ to which the poly(A) tail is added. Splicing and polyadenylation efficiency are heavily dependent on the highly conserved AAUAAA motif, known more commonly as the poly(A) signal, and another somewhat conserved, GU-rich sequence, located downstream from the polyadenylation start site. The precise spatial relationship of these sequences and sequence identity direct the cleavage of hnRNA downstream (3′) to the AAUAAA sequence, and subsequent poly(A) polymerization to the 3′-hydroxyl group, which is generated by that cleavage event. These posttranscriptional modifications have been shown to be mediated, at least in part, by small nuclear ribonucleoproteins (snRNPs) (Birnstiel *et al.*, 1985; Berget and Robberson, 1986; Black and Steitz, 1986; Hashimoto and Steitz, 1986).

The importance of the polyadenylation signal is easily demonstrated (Fitzgerald and Shenk, 1981; Manley *et al.*, 1985): mutations and relocations have been shown to interfere profoundly with the formation of poly(A)$^+$ mRNA. The polyadenylation signal, while highly conserved and generally within 10–30 nucleotides upstream from the start of the poly(A) tail (Proudfoot and Brownlee, 1976), is not a universal signal. For example, the polyadenylation signal AUUAAA has been observed in chicken lysozyme mRNA (Jung *et al.*, 1980) and in a one of several adenovirus mRNAs (Ahmed *et al.*, 1982); the sequence AAUAUA has likewise been found in a minor form of pancreatic α-amylase mRNA (Tosi *et al.*, 1981). Further, a motif for short-lived mRNAs, AUUUA, has been observed in certain oncogenes and cytokines (Shaw and Kamen, 1986). This sequence, now known as ARE (AU-rich element), is commonly observed in mRNAs and is believed to facilitate the degradation of mRNAs so-endowed. This pentanucleotide sequence, localized in the nontranslated trailer region, may be repeated in tandem several times. ARE-mediated destabilization occurs by deadenylation, i.e., cleavage of the poly(A) tail, followed by endonuclease-mediated degradation. The role of the poly(A) tail has been investigated using cordycepin, a drug known to block addition of poly(A) as well as the accumulation of mRNA in the cytoplasm. While the highly conserved nature of the polyadenylation signal suggests a central role in the cell, the physiological role of the poly(A) tail and associated polyadenylate-binding protein[1] remains somewhat obscure. It is clear, however, that poly(A) length correlates well with translational efficiency in eukaryotes.

Organellar mRNAs

Although the mechanics of transcription and translation are highly conserved, certain structural differences distinguish mRNA species confined to organelles. Mitochondrial mRNA, though lacking the 5′ cap structure, is polyadenylated at the 3′ terminus. Further, mitochondrial mRNAs frequently utilize AUA and AUU initiation codons found close to the 5′ terminus. In contrast, mRNAs found in the mitochondria of yeast are more closely related, structurally speaking, to typical cytoplasmic mRNAs than to other mitochondrial species; the poly(A) sequence, however, is usually quite short.

---

[1]Methodologies for characterization of polyadenylate-binding proteins are described by Sachs and Kornberg (1990) and Görlach *et al.* (1994).

Stability in the Cytoplasm

Once in the cytoplasm, both polysomal and nonpolysomal mRNAs exist as messenger ribonucleoprotein (mRNP) complexes.[2] Although the majority of polysomes have been shown to be attached to the cytoskeleton (Lenk et al., 1977; Howe and Hershey, 1984), no absolute requirement for this association and concomitant translation has been shown. The poly(A) tail itself and its length appear to play a role in mRNA stability: shortening of the poly(A) tail can cause destabilization (Decker and Parker, 1994; Beelman and Parker, 1995). For example, upon enzymatic removal of the poly(A) tract, globin mRNA rapidly loses translatability in frog oocytes, due to rapid degradation (Huez et al., 1974; Marbaix et al., 1975). Further, through the use of cordycepin to inhibit the addition of the poly(A) tail, the ability of normally poly(A)$^+$ mRNAs to accumulate in the cytoplasm is compromised (Zeevi et al., 1982). Treatment with this drug has no apparent effect on transcription itself or nucleocytoplasmic transport. More recent studies have demonstrated the role of 3′ AREs; deletion of these sequences greatly reduces the rate of readenylation, thereby prolonging mRNA in the cytoplasm (Wilson and Treisman, 1988; Shyu et al., 1991; Decker and Parker, 1993; Chen and Shyu, 1994).

Levels of Regulation

Historically, experiments performed using both adenovirus and SV40 (reviewed by Nevins, 1983) have shed light on the biochemical events associated with the production of biologically functional mRNA; these characteristics are undoubtedly common to all eukaryotic mRNAs. The number of potential points of gene regulation in living cells is virtually infinite. In the *broadest* sense, these myriad potential regulatory points may be classified under one of four main headings (Figure 2):

1. Regulation at the transcriptional level;
2. Regulation at the posttranscriptional level;
3. Regulation at the translational level;
4. Regulation at the posttranslational level.

At the subcellular level, RNA molecules are produced by transcription and processed in the nucleus. This is the primary regulation level in the cell. From there, RNA molecules, which have been appropriately modi-

---

[2]mRNP complexes can be characterized *in vitro* by crosslinking the constituent RNA and protein moieties, followed by oligo(dT)-cellulose affinity chromatography. For details, see Greenberg (1980, 1981); Wagenmaker et al. (1980); reviewed by Bag (1991); Görlach et al. (1994).

fied, and not the intron sequences, which have been removed by excision, may be exported from the nucleus into the cytoplasm. Details as to the nature of this extremely discriminating nucleocytoplasmic transport mechanism, involving movement to and then through nuclear pore complexes, have been reviewed (Piñol-Roma and Dreyfuss, 1993; Maquat, 1997). Mature messenger RNA may or may not become associated with the protein translational apparatus. Indeed the chemical stability of cytoplasmic RNA is itself a regulation point of gene expression. It is important to recall here that although a majority of the mRNA is found in polysome complexes, the formation of a nontranslatable mRNA:protein complex in the cytoplasm could potentially be one possible translation

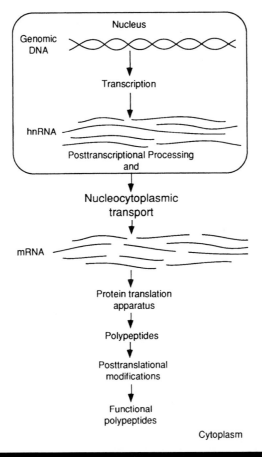

FIGURE 2

Relationship between genetic information encoded in genomic DNA and subsequent expression in the form of the primary hnRNA transcript and—following nucleocytoplasmic transport—the mature mRNA molecule.

regulation control point. The term "informosome" has been used to describe such nontranslatable cytoplasmic mRNA:protein complexes (Preobrazhensky and Spirin, 1978; Bag, 1991). If an mRNA molecule does undergo translation, then the primary product of translation, a polypeptide, is usually subjected to extensive posttranslational modifications. Perhaps the most common of these is cleavage of the peptide signal sequence. Other common posttranslational modifications include, but are not limited to, peptide cleavage, methylation, carboxylation, glycosylation, phosphorylation, acetylation, and hydroxylation.

Any variable that influences the efficiency of transcription or prevents transcription from occurring altogether is said to act at the transcriptional level (transcriptional regulation). Any event that influences the splicing of hnRNA (precursor to mRNA), hnRNA stability in the nucleus, nucleocytoplasmic transport, or stability in the cytoplasm is said to act posttranscriptionally (posttranscriptional regulation). For example, while as much as one-half of the mRNA in mammalian cells in tissue culture may have a half-life of up to 6 h, many mRNA species have half-lives of 10 min or less, the brevity of which is frequently taken to be a natural modulator of gene expression. It is also clear that environmental stimuli can influence profoundly the stability of a particular mRNA species (Buckingham *et al.*, 1974; Baumbach *et al.*, 1984; Knight *et al.*, 1985). Stating that gene regulation occurs at a translational or posttranslational level usually implies that a mature mRNA in the cytoplasm is somehow experiencing translational interference at the protein synthesis level or beyond.

Clearly, the biochemical status of a population of cells can be qualitatively and quantitatively defined as a function of mRNA complexity at the time of cellular lysis. For example, an investigator may wish to determine whether the observed modulation of gene expression in a model system is regulated transcriptionally or by a posttranscriptional event(s). In such a case, the method of cellular disruption and subsequent RNA purification must permit analysis of the nuclear abundance of salient gene transcripts independently of the cytoplasmic abundance of the same. In addition, transcription rate studies require the isolation of intact nuclei that are able to support elongation of, and label incorporation into, initiated transcripts (see nuclear runoff assay, see Chapter 19). Therefore, the conditions under which the RNA is purified from biological sources must support the chemical stability of the RNA and ensure RNase inactivation throughout the procedure. Scrupulous, if not compulsive, attention to these details is often the deciding factor in the success or failure of an RNA isolation procedure and subsequent evaluation.

References

Ahmed, C. M. I., Chanda, R. S., Snow, N. D., and Zain, B. S. (1982). The nucleotide sequence of mRNA for the M_r 19000 glycoprotein from early gene block III of adenovirus 2. *Gene* **20,** 339.

Babich, A., Nevins, J. R., and Darnell, J. E. (1980). Early capping of transcripts from the adenovirus major late transcription unit. *Nature* **287,** 246.

Bag, J. (1991). mRNA and mRNP. *In* "Translation in Eukaryotes," pp. 71–79. CRC Press: Ft. Lauderdale, FL.

Baker, E. J. (1997). mRNA polyadenylation: Functional implications. *In* "mRNA Metabolism and Post-Transcriptional Gene Regulation" (J. B. Harford and D. R. Morris, Eds.). Wiley-Liss, New York, NY.

Banerjee, A. K. (1980). 5′-Terminal cap structure in eucaryotic messenger ribonucleic acids. *Microbiol. Rev.* **44,** 175.

Baumbach, L. L., Marashi, F., Plumb, M., Stein, G., and Stein, J. (1984). Inhibition of DNA replication coordinately reduces cellular levels of H1 histone mRNAs requirement for protein synthesis. *Biochemistry* **23,** 1618.

Beelman, C. A., and Parker, R. (1995). Degradation of mRNA in eukaryotes. *Cell* **81,** 179.

Berget, S. M., and Robberson, B. L. (1986). U1, U2, and U4/U6 small nuclear ribonucleoproteins are required for in vitro splicing but not for polyadenylation. *Cell* **46,** 691.

Birnstiel, M. L., Busslinger, M., and Strub, K. (1985). Transcription termination and 3′ end processing: The end is in site! *Cell* **41,** 349.

Black, D. L., and Steitz, J. A. (1986). Pre-mRNA splicing in vitro requires intact U4/U6 small nuclear ribonucleoprotein. *Cell* **46,** 697.

Brawerman, G. (1976). Characteristics and significance of the polyadenylate sequence in mammalian messenger RNA. *Prog. Nucleic Acid Res. Mol. Biol.* **17,** 117.

Buckingham, M. E., Caput, D., Cohen, A., Whalen, R. G., and Gros, F. (1974). The synthesis and stability of cytoplasmic messenger RNA during myoblast differentiation in culture. *Proc. Natl. Acad. Sci. USA* **71,** 1466.

Chen, C. Y. A., and Shyu, A. B. (1994). Selective degradation of early-response-gene mRNAs: Functional analysis of sequence features of the AU-rich elements. *Mol. Cell. Biol.* **14,** 8471.

Chen-Kiang, S., Nevins, J. R., and Darnell, J. E. (1979). N-6-Methyl-adenosine in adenovirus type 2 nuclear RNA is conserved in the formation of messenger RNA. *J. Mol. Biol.* **135,** 733.

Curtis, D., Lehmann, R., and Zamore, P. D. (1995). Translational regulation in development. *Cell* **81,** 171.

Decker, C. J., and Parker, R. (1993). A turnover pathway for both stable and unstable mRNAs in yeast: Evidence for a requirement for deadenylation. *Genes Dev.* **7,** 1632.

Decker, C. J., and Parker R. (1994). Mechanisms of mRNA degradation in eukaryotes. *Trends Biochem. Sci.* **19,** 336.

Dreyfuss, G. (1986). Structure and function of nuclear and cytoplasmic ribonucleoprotein particles. *Annu. Rev. Cell Biol.* **2,** 459.

Edmonds, M. (1982). *In* "The Enzymes (P. Boyer, Ed.), Vol. 15, p. 217. Academic Press, New York.

Edmonds, M., Vaughn, M. H., Jr., and Nakazato, H. (1971). Polyadenylic acid sequences in the heterogeneous nuclear RNA and rapidly labeled poly-ribosomal RNA of HeLa cells: Possible evidence for a precursor relationship. *Proc. Natl. Acad. Sci. USA* **68,** 1336.

Edmonds, M., and Winters, M. A. (1976). Polyadenylate polymerases. *Prog. Nucleic Acid Res. Mol. Biol.* **17,** 149.

Fitzgerald, M., and Shenk, T. (1981). The sequence 5'-AAUAAA-3' forms part of the recognition site for polyadenylation of late sv40 mRNAs. *Cell* **24,** 251.

Florkiewcz, R. Z., and Sommer, A. (1989). Human basic fibroblast growth factor gene encodes four polypeptides: Three initiate translation from non-AUG codons. *Proc. Natl. Acad. Sci. USA* **86,** 3978.

Fraser, N. W., Nevins, J. R., Ziff, E., and Darnell, J. E. (1979). The major late adenovirus type-2 transcription unit: Termination is downstream from the last poly(A) site. *J. Mol. Biol.* **129,** 643.

Furuichi, Y., LaFiandra, A., and Shatkin, A. J. (1977). 5'-Terminal structure and mRNA stability. *Nature* **266,** 235.

Furuichi, Y., and Shatkin, A. J. (1989). Characterization of cap structures. *Methods Enzymol.* **180,** 164.

Görlach, M., Burd, C. G., and Dreyfus, G. (1994). The mRNA poly(A)-binding protein: Localization, abundance, and RNA-binding specificity. *Exp. Cell Res.* **211,** 400.

Greenberg, J. R. (1980). Proteins cross-linked to messenger RNA by irradiating polyribosomes with ultraviolet light. *Nucleic Acids Res.* **8,** 5685.

Greenberg, J. R. (1981). The polyribosomal mRNA-protein complex is a dynamic structure. *Proc. Natl. Acad. Sci. USA* **78,** 2923.

Hann, S. R., King, M. W., Bentley, D. L., Anderson, W. C., and Eisenman, R. N. (1988). A non-AUG translational initiation in C-myc exon 1 generates an N-terminally distinct protein whose synthesis is disrupted in Burkitt's lymphomas. *Cell* **52,** 185.

Hashimoto, C., and Steitz, J. A. (1986). A small nuclear ribonucleoprotein associated with AAUAAA polyadenylation signal in vitro. *Cell* **45,** 581.

Howe, J. G., and Hershey, J. W. B. (1984). Translation initiation factor and ribosome association with the cytoskeletal framework from HeLa cells. *Cell* **37,** 85.

Huez, G., Marbaix, G., Hubert, E., Leclercq, M., Nudel, U., Soreq, H., Salomon, R., Lebleu, R., Revel, M., and Littauer, U. Z. (1974). Role of the polyadenylate segment in the translation of globin messenger RNA in *Xenopus* oocytes. *Proc. Natl. Acad. Sci. USA* **71,** 3143.

Jackson, R. J., and Standart, N. (1990). Do the poly(A) tail and 3' untranslated region control mRNA translation? *Cell* **62,** 15.

Jung, A., Sippel, A. E., Grez, M., and Schutz, G. (1980). Exons encode functional and structural units of chicken lysozyme. *Proc. Natl. Acad. Sci. USA* **77,** 5759.

Knight, E., *et al.* (1985). Interferon regulates *c-myc* expression in Daudi cells at the post-transcriptional level. *Proc. Natl. Acad. Sci. USA* **82,** 1151.

Kozak, M. (1986). Point mutations define a sequence flanking the AUG initiator codon that modulates translation in eukaryotes. *Cell* **44,** 283.

Kuge, H., and Richter, J. D. (1995). Cyplasmic 3' poly(A) addition induces 5' cap ribose methylation: inplications for translational control of maternal mRNA. *EMBO J.* **14,** 6301.

Leibold, E. A., and Munro, H. N. (1988). Cytoplasmic proteins binds *in vitro* to a highly conserved sequence in the 5' untranslated region of ferritin heavy and light-subunit mRNAs. *Proc. Natl. Acad. Sci. USA* **85,** 2171.

Lenk, R., Ransom, L., Kaufman, Y., and Penman, S. (1977). A cytoskeletal structure with associated polyribosomes obtained from HeLa cells. *Cell* **10,** 67.

Lewin, B. (Ed.). (1997). "Genes VI." Oxford University Press, New York.

Lim, L., and Cannelakis, E. S. (1970). Adenine-rich polymer associated with rabbit reticulocyte messenger RNA. *Nature* **227,** 710.

Maquat, L. E. (1997). RNA export from the nucleus. *In* "mRNA Metabolism and Post-Transcriptional Gene Regulation" (J. B. Harford and D. R. Morris, Eds.). Wiley-Liss, New York, NY.

Manley, J. L., Yu, H., and Ryner, L. (1985). RNA sequence containing hexanucleotide AAUAAA directs efficient mRNA polyadenylation *in vitro. Mol. Cell. Biol.* **5,** 373.

Marbaix, G., Huez, G., Burny, A., Cleuter, Y., Hubert, E., Leclercq, M., Chantrenne, H., Soreq, H., Nudel, U., and Littauer, Z. (1975). Absence of polyadenylate segment in globin messenger RNA accelerates its degradation in *Xenopus* oocytes. *Proc. Natl. Acad. Sci. USA* **72,** 3065.

Nevins, J. R. (1983). The pathway of eukaryotic mRNA formation. *Annu. Rev. Biochem.* **52,** 441.

Nevins, J. R., and Darnell, J. E. (1978). Steps in the processing of Ad2 mRNA: Poly(A)$^+$ nuclear sequences are conserved and poly(A) addition precedes splicing. *Cell* **15,** 1477.

Piñol-Roma, S., and Dreyfus, G. (1993). hnRNP proteins: Localization and transport between the nucleus and the cytoplasm. *Trends Cell Biol.* **7,** 151.

Preobrazhensky, A. A., and Spirin, A. S. (1978). Informosomes and their protein components: The present state of knowledge. *Prog. Nucleic Acid Res. Mol. Biol.* **21,** 1.

Proudfoot, N. J., and Brownlee, G. G. (1976). 3′ Non-coding region sequences in eucaryotic messenger RNA. *Nature* **263,** 211.

Reddy, R., Ro-Choi, T. S., Henning, D., and Busch, H. (1974). Primary sequence of U-1 nuclear ribonucleic acid of Novikoff hepatoma ascites cells. *J. Biol. Chem.* **249,** 6486.

Sachs, A. B., and Kornberg, R. D. (1990). Purification and characterization of polyadenylate-binding protein. *Methods Enzymol.* **181,** 332.

Shatkin, A. J. (1976). Capping of eucaryotic mRNAs. *Cell* **9,** 645.

Shatkin, A. J. (1985). mRNA cap binding proteins: Essential factors for initiating translation. *Cell* **40,** 223.

Shaw, G., and Kamen, R. (1986). A conserved AU sequence from 3′ untranslated region of GM-CSF mRNA mediates selective messenger RNA degradation. *Cell* **46,** 659.

Shyu, A. B., Belasco, J. G., and Greenberg, M. E. (1991). Two distinct destabilizing elements in the *c-fos* message trigger deadenylation as a first step in rapid mRNA decay. *Genes Dev.* **5,** 221.

Tosi, M., Young, R. A., Hagenbuchle, O., and Schibler, U. (1981). Multiple polyadenylation sites in a mouse α-amylase gene. *Nucleic Acids Res.* **9,** 2313.

Wagenmaker, A. J. M., Reinders, R. J., and Venrooij, W. J. (1980). Cross-linking of mRNA proteins by irradiation of intact cells with ultraviolet light. *Eur. J. Biochem.* **112,** 323.

Wickens, M., and Stephenson, P. (1984). Role of the conserved AAUAAA sequence: Four AAUAAA point mutants prevent messenger RNA 3′ end formation. *Science* **226,** 1045.

Woolin, S. L., and Walter, P. (1988). Ribosome pausing and stacking during translation of a eukaryotic mRNA. *EMBO J.* **7,** 3559.

Zeevi, M., Nevins, J. R., and Darnell, J. E., Jr. (1982). Newly formed mRNA lacking polyadenylic acid enters the cytoplasm and the polyribosomes but has a shorter half-life in the absence of polyadenylic acid. *Mol. Cell Biol.* **2,** 517.

Resilient
Ribonucleases

Rationale

The difficulties associated with the isolation of full-length, intrinsically labile RNA are further compounded by the ubiquity of ribonuclease (RNase) activity. RNases are a family of enzymes present in virtually all living cells. These enzymes can degrade RNA molecules through both endonucleolytic and exonucleolytic activity. Particularly resilient are RNases of the pancreatic variety. These small, remarkably stable enzymes maintain their tertiary configurations by virtue of four disulfide bridges (Blackburn and Moore, 1982) that allow these enzymes to renature quickly, following treatment with most denaturants (Sela *et al.*, 1957), even after boiling. RNases have minimal cofactor requirements and are active over a wide pH range. Clearly, it is incumbent upon the investigator to ensure that both equipment and reagents are purged of nucleases from the onset of the experiment. For most RNA-minded molecular biologists, saying that a reagent or apparatus is sterile is more than likely a statement that it is RNase-sterile, that is, RNase-free.

Many investigators often worry less about RNase degradation of their samples than in the past. This is due, in part, to the ready availability of reagents that favor expedient isolation of RNA from biological sources of all types. As such, the investment of time required by many of the traditional techniques for the control of nuclease activity is often minimized and RNase is often not a concern of central importance. This is a very poor attitude. While no one would question the inhibitory nature of many reagents used to purify RNA, of even greater concern is the stability of the RNA once the isolation reagent has been removed. For example, purified RNA is never more susceptible to nuclease attack than when stored in aqueous buffer, such as TE (10 mM Tris, 1 mM EDTA, pH 7.5) or water. Unfortunately, losing a valuable sample is often a pivotal lesson on the need to maintain the stability of the RNA before, during, and after its isolation from the cell.

Elimination of Ribonuclease Activity

The method selected for controlling RNase activity must, first and foremost, demonstrate compatibility with the cell lysis procedure. Nuclease inhibitors are most often added to relatively gentle lysis buffers when subcellular organelles (especially nuclei) must be purified intact. For example, one objective of a study might be the examination of nuclear RNA and cytoplasmic RNA independently of one another, perhaps to assess the level of gene regulation. The method of nuclease inhibition must support the integrity of the RNA throughout the subsequent fractionation or purification steps, some of which may be quite time consuming. In addition, the reagents used to inhibit RNase activity must be easily removed from purified RNA preparations so as not to interfere with subsequent manipulations.

These realities mandate that one view the control of RNase activity from two perspectives:

1. Extrinsic RNase activity must be controlled. Extrinsic or external sources of potential RNase contamination must be identified and neutralized from the onset of the experiment and all reagents maintained RNase-free at all times. Extrinsic or external sources include, but are not limited to, bottles and containers in which chemicals are packaged, non-diethyl pyrocarbonate-treated water, gel boxes and combs, and microbial and fungal contamination of buffers. The single greatest source, however, is the oil from the fingertips, which is rich in RNase activity.

2. Intrinsic RNase activity must be controlled. Beyond the potential for accidental contamination of an RNA preparation with RNases from the lab environment, one must be acutely aware of the fact that intracellular RNases, normally sequestered within the cell, are liberated upon cellular lysis. When this occurs RNases are free to initiate degradation of the RNA that the investigator is attempting to isolate, unless they are inhibited without delay. As endogenous RNase activity varies tremendously from one biological source to the next, the degree to which action must be taken to inhibit nuclease activity is a direct function of cell or tissue type. Knowledge of the extent of intrinsic nuclease activity is derived from two principal sources: the salient literature and personal experience.

In all cases, and especially when characterizing a new cell type or model system for the first time, actions taken to control nuclease activity should be aggressive if not compulsive. Failure to do so is likely to yield a useless sample of degraded RNA. Laboratories isolating RNA with the highest degree of fidelity generally maintain a "private stock" of RNA reagents and supplies that are not used for any purpose other than RNA recovery and subsequent manipulation. Such materials should definitely not be in general circulation in the laboratory.

Types of Ribonuclease Inhibitors

Compounds employed for the control of RNase activity fall into two broad categories, nonspecific and specific inhibitors of RNase activity.

Some of these compounds are added directly to lysis and reaction buffers to control RNase activity, whereas others are used in the standard preparation of reagents *before use* in order to purge them of nuclease activity. The effectiveness of an RNase inhibitor in a particular system can be ascertained by a test incubation of RNase A, test RNA, and the inhibitor of interest. The postincubation products may then be visualized on a simple agarose gel (Chapter 7, Figure 1). If the RNA is degraded, a change in protocol is clearly indicated.

Specific Inhibitors

Vanadyl Ribonucleoside Complexes (VDR, VRC)

Vanadyl ribonucleoside complexes are frequently added to lysis buffers, which alone are ineffective in controlling RNase activity, before and at the time of cellular lysis. VDR consists of complexes formed between the oxovanadium ion and any or all of the four ribonucleosides (Leinhard *et al.*, 1971; Berger and Birkenmier, 1979) in which vanadium takes the place of phosphate. These complexes then function as transition-state analogs. In the absence of VDR, RNase-mediated cleavage of the phosphodiester backbone of RNA results in the transient formation of a dicyclic transition-state intermediate, which is subsequently opened up by reaction with a water molecule. In its capacity as an RNA analog, the vanadium complex forms a highly stable dicyclic species to which the enzyme remains irreversibly bound. Thus, VDR inhibits nuclease activity by locking RNase and "pseudo-substrate" in the transition state. VDR binds tightly to a broad spectrum of cellular RNases and is compatible with a variety of cell fractionation methods, including sucrose gradient centrifugation (Benecke *et al.*, 1978; Nevins and Darnell, 1978). Succinctly, VDR inhibits RNase by binding it irreversibly. VDR should be used selectively, however, as even trace carryover quantities are sufficient to inhibit *in vitro* translation of purified mRNA and can also interfere with reverse transcriptase activity (see RNA PCR, Chapter 15) during the first-strand synthesis of cDNA (Berger and Birkenmier, 1979; Berger *et al.*, 1980) unless it is completely removed. Thus, if purified RNA is to be subjected to either of these applications, the use of VDR is not recommended.

VDR is routinely prepared as a 200 mM stock solution (see Protocol section for synthesis) and used at a final concentration of 5–20 mM, depending on the anticipated degree of RNase activity. It has been shown to suppress RNase activity in cells with significant levels of endogenous RNase (Berger and Birkenmier, 1979). VDR is particularly effective against pancreatic RNase A and RNase T$_1$, but not against RNase H. It is also inhibitory against the family of enzymes commonly known as the poly(A) polymerases, which may or may not be significant in the context of the types of experiments being undertaken. After it has neutralized RNase activity following cellular lysis, the VDR can be removed systematically from the aqueous cytosol (discussion to follow) by repeated extraction with organic extraction buffer containing 0.1% 8-hydroxyquinoline (w/v), which is itself a partial inhibitor of RNase. Alternatively, VDR can be removed by the addition of 10 equivalents of EDTA. The investigator is cautioned, however, that 1 equivalent of EDTA is more than sufficient to dissociate the VDR complex. Thus, when VDR is present EDTA-containing buffers are to be avoided until another method of controlling RNase activity has been employed, for example, the addition of phenol:chloroform to the lysate. While the habitual use of VDR is one of the more cost-effective means of controlling nuclease activity, in the world of PCR it is often not an appropriate choice.

RNasin/RNAguard

As an alternative to VDR, the commercially available proteins RNasin (Promega) and RNAguard (Pharmacia Biotech) may be used to eliminate nuclease activity and circumvent some of the problems commonly associated with the use of VDR. Originally purified from human placenta and rat liver, the cloning of these and homologous genes has ensured an ample supply of these peptides. RNasin is a 51,000 M_r protein, the activity of which is highly dithiothreitol (DTT) dependent. RNasin inactivates RNase by competitive, noncovalent binding to the enzyme with a binding constant $K_i = 4.4 \times 10^{14}$ (Blackburn, 1979; Blackburn *et al.*, 1977). RNAguard is a noncompetitive inhibitor of RNase A-type ribonuclease. Purified from hog liver, this porcine peptide offers no risk of human DNA contamination, which may be worthwhile for some clinical and diagnostic research involving *Homo sapiens.*

RNasin and RNAguard are routinely used at a concentration of 250–1000 U/ml. Both RNasin and RNAguard inhibit RNase A, RNase B, and RNase C; they do not inhibit RNase T1, S1 nuclease, or RNase from *Aspergillus* (Promega and Pharmacia Biotech product inserts). A major advantage of these products as a means of controlling nuclease activity is their compatibility with a variety of *in vitro* reactions, which is much greater than that of VDR. These reactions include protection of template messenger RNA (mRNA) during the synthesis of cDNA (de Martynoff, 1980); *in vitro* transcription, in which the yield of RNA and its integrity are maintained (Scheider *et al.*, 1988); and *in vitro* translation, in which the yield of large polypeptides is improved as is the activity of purified polysomes (Scheele and Blackburn, 1979). Care must be taken to avoid powerful denaturing conditions, such as high concentrations of urea or heating to 65°C, as these have been shown to cause the uncoupling of RNase–RNasin complexes and accompanying restoration of RNase degradative activity (Promega product insert).

Nonspecific Inhibitors

The use of a variety of different types of nonspecific inhibitors has been reported. These compounds purportedly function by eliminating all intrinsic enzymatic activity and have been incorporated into RNA isolation procedures with varying degrees of success. Although none of these has been found to be completely satisfactory, one or more of these compounds may provide a means of controlling RNase activity that would otherwise not be possible. These nonspecific inhibitors include heparin (Palmiter *et al.*, 1970; Cox, 1979) at a working concentration of 0.2–2 mg/ml, iodoacetate, polyvinyl (dextran) sulfate, the clays macaloid and bentonite (reviewed by Blumberg, 1987), and, more recently, cationic surfactant (Dahle and Macfarlane, 1993). In addition to its pharmaceutical applications, the use of bentonite in winemaking (Blade and Boulton, 1988) is nearly universal, dating back to the mid 1930s (Saywell, 1934). Although no longer widely used today, the preparation of these clays for the elimination of RNase activity has been described elsewhere (Fraenkel-Conrat *et al.*, 1961; Poulson, 1973; Blumberg, 1987).

Perhaps closer to home is the ready availability of dilute solutions (3%) of hydrogen peroxide (H_2O_2) and mixtures of NaOH (0.5 N) and sodium dodecyl sulfate (SDS; 0.1–0.5%). As described below, either of these may provide very satisfactory removal of RNase from the surfaces of equipment and supplies commonly used in the manipulation of RNA.

Preparation of Equipment and Reagents

Rule number one when working with RNA is to wear gloves during the preparation of reagents and equipment and especially during the actual RNA extraction procedure. Finger greases, especially those of overly zealous lab technicians, are notoriously rich in RNase and are generally accepted as the single greatest potential source of RNase contamination. Further, one should not hesitate to change gloves several times during the course of an RNA-related experiment, since door knobs, micropipettors, refrigerator door handles, and telephone receivers are also potential sources of nuclease contamination. It is the strict policy of this investigator to remove his gloves when stepping away from the bench and don a fresh pair when returning to the bench and the RNA protocol.

With respect to laboratory plasticware, individually wrapped serological pipettes are always preferred for RNA work as no special pretreatment is required. Conical 15- and 50-ml tubes, both polypropylene and polystyrene, are considered sterile if already capped and racked by the manufacturer. In general, plasticware that is certified as tissue culture sterile is not likely to be a source of nuclease activity unless subsequently contaminated (be certain to wear gloves when handling these tubes for RNA work). Bulk-packed polypropylene products are potential sources of nuclease contamination, mainly due to handling and distribution from a single bag. This pertains to microcentrifuge (microfuge) tubes and polypropylene micropipette tips, because they can become contaminated and, in turn, contaminate stock solutions. Thus, any plastic product that comes into contact with an RNA sample at any time, either directly or indirectly, and which can withstand autoclaving, should be so treated. Bulk packages of microfuge tubes are best distributed with gloved hand into several 500-ml glass beakers; these can be covered with aluminum foil and autoclaved. These microfuge tubes are then handled only with gloves and set aside exclusively for RNA work.

When glassware must be used, as may be required for the manipulation of organic extraction buffers, which typically contain mixtures of the organic solvents phenol and chloroform, individually wrapped borosilicate glass pipettes are strongly preferred. Glassware that must be reused most certainly should be reserved exclusively for RNA work and not be in general circulation in the laboratory. Contrary to popular belief, the temperature and pressure generated during the autoclaving cycle are usually not sufficient to eliminate all RNase activity. Fortunately, RNases can be destroyed quite effectively by baking in a dry-heat oven. Glassware should be cleaned scrupulously, rinsed with RNase-free water (discussed below), and then baked for 3–4 h at approximately 200°C. This is a very effective method for purging glassware of RNase activity. It is important to note here that not all laboratory implements can withstand the heat generated in a dry-heat oven. If there is a question about a particular type of plastic or other material, be certain to check with the manufacturer (technical assistance toll-free numbers can usually provide the necessary information) before baking.

Diethyl Pyrocarbonate

CAUTION   Diethyl pyrocarbonate is carcinogenic and should be handled with extreme care. Be sure to observe all safety precautions indicated by the manufacturer.

The treatment of solutions, laboratory plastics, and other apparatuses by autoclaving does not ensure the complete elimination of RNase activity; buffers treated solely by autoclaving frequently maintain residual RNase activity. Lack of attention to this detail may or may not pose a severe impediment to the generation of data that are both representative and reproducible. For this reason, many stock solutions and buffers are treated, directly or indirectly, with the potent chemical RNase inhibitor diethyl pyrocarbonate (DEPC).

DEPC is an efficient, nonspecific inhibitor of RNase.[1] It is used to purge reagents of nuclease activity prior to use in an experiment involving the purification of RNA. Historically it has also been used in the preparation of nondisposable glassware prior to use for RNA work, though much easier and far less toxic methods are currently available (described below). For the preparation of solutions and buffers, DEPC is typically added to a final concentration of 0.05–0.1% (v/v); buffers are then shaken for several hours on an orbital platform (e.g., Lab-Line Instruments) or stirred vigorously with a magnetic stirrer for 20–30 min. Following incubation, DEPC must be destroyed completely. This is accomplished by autoclaving or by maintaining the solutions at 60°C overnight, a process that should be carried out in a chemical fume hood.[2] Alternatively, DEPC can be removed from DEPC-treated (sterile) water by boiling it for 1 h in a fume hood. Not only is DEPC carcinogenic, which alone would justify its complete removal and careful handling, but it also has a strong affinity for adenosine in RNA. Even trace amounts of residual DEPC will result in chemical modification of adenine residues (Henderson *et al.*, 1973), thereby changing the physical properties of RNA and compromising its utility for *in vitro* translation (Ehrenberg *et al.*, 1974) and other applications, including standard blot analysis and

---

[1]Because of the efficient inhibition of certain classes of proteins by histidine and tyrosine modification, DEPC has been used to facilitate the destruction of proteins in previous plasmid isolation procedures. This is a hazardous technique that should be avoided.

[2]Important: Do not autoclave solutions for RNA work that are not normally autoclaved, such as those containing SDS, NP-40, or NaOH. These components may be added after autoclaving in order to complete a particular buffer formulation.

PCR (personal observations). This is also the reason that DEPC is not added directly to cell suspensions or lysates containing RNAs to be purified. At autoclaving temperature and pressure, DEPC degrades into carbon dioxide and ethanol, both of which are quite volatile under these conditions.

Complete removal of DEPC can be further promoted by rapidly stirring the hot solutions with a nuclease-free magnetic stir bar. Frequently, if autoclaving time was not adequate, one notices the distinctive odor of residual DEPC. In this laboratory, solutions are routinely kept in the warm autoclave chamber for an hour or so after completion of the autoclaving cycle in order to purge them of DEPC.

Important

1. Do not add DEPC to any buffer containing mercaptans or 1° amine groups, with which DEPC is reactive (Berger, 1975). Perhaps the most common buffers to which DEPC exposure is to be avoided are the Tris buffers (those containing *tris*(hydroxymethyl)aminomethane). Instead, use DEPC-treated water (add DEPC to water; autoclave to remove DEPC) to make up the Tris-containing solution and then autoclave the solution again.

2. Those buffers consisting of chemicals that demonstrate DEPC incompatibility can be twice filtered through a nitrocellulose membrane to remove RNase activity and other trace proteins.

3. The electrophoresis unit itself, including the gel comb, casting tray, and interior of the gel box can be treated to remove RNase contamination by rinsing and soaking in DEPC-treated water (after autoclaving the water). *Never* expose the unit to DEPC because acrylic is not resistant to DEPC.

Many RNases, including the particularly resilient pancreatic variety, manage to renature following treatment and removal of most denaturing reagents. It is prudent to maintain separate containers of chemicals as well as stock solutions for exclusive use as RNA reagents; chemical solids should be weighed with a RNase-free spatula. This is most easily accomplished by aliquoting sterile reagents into suitable volumes, rather than drawing repeatedly from the stock bottles. Aliquots used once should be discarded. Although such actions may at first seem excessive, they may well preclude the accidental introduction of RNases and facilitate recovery of the highest possible quality RNA.

Alternative: Sterile Water for Irrigation

There remains a fair amount of resistance to using DEPC in the laboratory. It is the opinion of this author that there are three principal reasons for this:

1. DEPC is toxic and not appropriate in a lab where it cannot be handled properly. This objection is with merit, as improper handling may present a serious risk to health.
2. DEPC is too hard to remove from reagents and interferes with PCR, and other, reactions. This is often a direct result of adding too much DEPC (reports range from 0.5 to 5%!) or autoclaving for too short a period.
3. We do not have problems with RNase activity. Such a situation is fortunate indeed, though not widespread.

A suitable alternative to DEPC treatment of H_2O to render it nuclease-free is the use of the water commonly available in hospitals, labeled "sterile water for irrigation." This water, if available, is free of many contaminants commonly found in the laboratory. Purified RNA may be rehydrated in this water and stored at $-80°C$. This water is also excellent for making dilutions of nuclease-free stock solutions. Although the habitual use of this water is generally expensive, its ready availability in hospitals and medical centers makes it an excellent alternative to DEPC.

Hydrogen Peroxide

The need to eliminate RNase from equipment and supplies in the lab, including gel combs, casting trays, chambers, and graduated cylinders, is unquestioned. Soaking such implements in a 3% solution of hydrogen peroxide (H_2O_2) is an extremely effective measure. Such solutions are commonly available in pharmacies and similar stores and are very inexpensive. H_2O_2 is a powerful oxidizing agent that can render a surface nuclease-free by soaking for 20–30 min, followed by rinsing with copious amounts of water that, at the very least, has been autoclaved. *Do not* use the more concentrated forms of H_2O_2, commonly available from standard chemical supply companies as 30% H_2O_2. At this concentration H_2O_2 is extremely dangerous and may well cause irreparable damage to equipment and injury to the investigator. Also, be sure to avoid old solutions of H_2O_2, as they may no longer be solutions of H_2O_2!

NaOH and SDS

An older method for the removal of RNase activity from laboratory surfaces is to soak or wipe surfaces with as much as 0.5 N NaOH and 0.5% SDS. Subsequently, extensive washing is required to remove the alkali. Failure to do so often creates a chemical environment that favors alkaline hydrolysis of the RNA sample under investigation! Even though thorough rinsing with water precludes this potential difficulty, it is not as cost-effective as H_2O_2, especially when NaOH and SDS are purchased premixed.

Other Reagents Used to Control Nuclease Activity

The following is a list of reagents commonly employed to minimize or eliminate RNase activity. These reagents are generally not used alone, but rather constitute popular RNA isolation buffers. For cells enriched in RNase, and even those that are not, homogenization in lysis buffers consisting of guanidinium thiocyanate or guanidinium hydrochloride is widely accepted as the method of choice for consistency from sample to sample. Because of their extremely chaotropic nature, these buffers are quite effective for tissue disruption as well as for cells grown in culture. There are also disadvantages to lysing cells directly in guanidinium-containing buffers, namely the intermixing of nuclear and cytoplasmic message, as described in Chapter 5. Moreover, in the absence of a strongly denaturing lysis buffer, RNase inhibitors, as described above, should be added to the buffer just prior to use (Chapter 5).

Guanidine Hydrochloride

Guanidine hydrochloride (guanidine HCl) is a strong ionic protein denaturant. At a working concentration of about 4–6 M (in water) it is an excellent inhibitor of RNase activity during purification of nucleic acids from cells and whole tissue samples. Solutions containing guanidine HCl or guanidine thiocyanate are often referred to as chaotropic buffers because of their biologically disruptive nature.

Guanidine Thiocyanate

Guanidine thiocyanate (GTC) is a stronger protein denaturant than guanidine hydrochloride and is the denaturant of choice for the preparation of RNA from sources enriched in RNase activity, especially pancreatic tissue (Chirgwin *et al.*, 1979). It is routinely used at a working concentration of 4 M in water and appears repeatedly in the literature in various formulations, usually along with a reducing agent (e.g., β-mercaptoethanol) and an ionic detergent (e.g., sarcosyl).

Sodium Dodecyl Sulfate

Sodium dodecyl sulfate, also known as sodium lauryl sulfate, is an ionic detergent that is useful for the rapid disruption of biological membranes. It is a key component of many reagents used to purify nucleic acids because of its abilities to quickly disrupt the tissue architecture and to inhibit both RNase and deoxyribonuclease (DNase) activity. SDS is prepared as either a 10 or a 20% (w/v) stock solution and is used most often at a working concentration of 0.1–0.5%. The performance of this detergent can be affected significantly by its purity. SDS is easily precipitable in the presence of potassium salts and generally is not added to guanidinium buffers, as it has very low solubility in these high-salt, chaotropic buffers.

N-Laurylsarcosine

N-Laurylsarcosine, also known as sarcosyl, is an ionic detergent, similar to SDS. Unlike SDS, however, sarcosyl manifests excellent solubility in high-salt, chaotropic solutions and is therefore the detergent of choice in guanidinium-based lysis buffers. As such, it is a common component of these buffers. It acts to disrupt tissue and cellular ultrastructure, inhibit nuclease activity, and facilitate the disaggregation of proteins and nucleic acids in a cellular lysate.

Phenol:Chloroform:Isoamyl Alcohol

CAUTION   **Phenol, chloroform, and combinations thereof are caustic, carcinogenic reagents that must be handled with extreme care. Be sure to observe all safety precautions indicated by the manufacturer.**

Phenol and chloroform are organic solvents that very efficiently denature and cause the precipitation of proteins. Molecular biology-grade phenol (redistilled to remove impurities)[3] is itself a good protein denatu-

---

[3]Commercially distilled phenol is readily available, precluding the need by any lab to perform in-house phenol purification. The distillation of phenol in the laboratory is a very dangerous process, which, if absolutely necessary, should be carried out by experienced personnel only. Details of in-lab phenol distillation can be found elsewhere (Wallace, 1987).

rant, although it is very unstable; it oxidizes rapidly into quinones, which impart a pinkish tint to phenol solutions. Quinones form free radicals that break phosphodiester linkages and crosslink nucleic acids. To reduce the rate of phenol oxidation, 8-hydroxyquinoline may be added to a final concentration of 0.1% (w/v) to retard this effect. Phenol reagents are usually prepared by saturating the phenol with aqueous buffer and then adding an equal volume of chloroform. The addition of chloroform stabilizes the phenol, imparts a greater density to this organic extracting material, improves the efficiency of deproteinization of the sample, and also facilitates removal of lipids from the RNA prep. Mixtures of phenol and chloroform have also been shown to increase the yield of poly(A)$^+$ mRNA over phenol alone (Perry *et al.*, 1972). Very often, isoamyl alcohol is used in conjunction with chloroform or with mixtures of phenol and chloroform. Isoamyl alcohol reduces the foaming of proteins that would normally be generated by the mechanics of the extraction procedure. The most common organic extraction buffer formulations consist in part of phenol:chloroform:isoamyl alcohol in a 25:24:1 ratio. In the presence of these solvents, RNase activity is inhibited.

For a more thorough discussion of the preparation and use of phenol, see Appendix A.

8-Hydroxyquinoline

8-Hydroxyquinoline is a partial inhibitor of RNase and is occasionally added to organic extraction buffers that contain phenol (Kirby, 1956). As an antioxidant, 8-hydroxyquinoline stabilizes phenol and retards the formation of quinones (phenol oxidation products). It is usually added to a final concentration of 0.1% (w/v). 8-Hydroxyquinoline imparts a bright yellow color to the organic buffers to which it is added, thereby helping the investigator keep track of the organic and aqueous phases during the nucleic acid purification process.

8-Hydroxyquinoline also chelates heavy materials, making it very useful for removing vanadyl ribonucleoside complexes from cell lysates. Upon binding VDR, the color imparted by the 8-hydroxyquinoline changes from yellow to dark green. When the phenol-containing phase of the extraction buffer remains yellow, all VDR has been removed. The inclusion of 8-hydroxyquinoline in organic extracting buffers may also be advantageous even when VDR is not used because heavy metals can cause RNA degradation when they are present with RNA for extended periods. Of course, all reagents should be prepared in high-purity biochemical quality water.

Cesium Chloride

Cesium chloride (CsCl) has been used in the past as a standard centrifugation medium for both analytical and preparative separation of nucleic acids. CsCl demonstrates a limited ability to inhibit RNase activity. Impure, solid CsCl may be baked at 200°C for 6–8 h to remove residual RNase activity prior to exposure to RNA. The use of CsCl as a menstruum for the purification of nucleic acids in general has declined significantly in favor of newer resins, in kit form, the use of which precludes density gradient centrifugation.

Cesium Trifluoroacetate

Cesium trifluoroacetate (CsTFA) is a highly soluble salt that solubilizes and dissociates proteins from nucleic acids without the use of detergents. In this capacity, CsTFA (available from Pharmacia Biotech) is an excellent inhibitor of RNase; its use precludes removal of proteins from a sample by more traditional methods (e.g., phenol:chloroform, proteinase K). It is more chaotropic than CsCl, inhibits RNase to a greater extent than CsCl, and shows greater solubility in ethanol, which expedites its removal following isopycnic centrifugation of RNA. If desired, RNA can be banded in CsTFA, as well as pelleted, because solutions with densities of up to 2.6 g/ml are possible using highly soluble CsTFA. A major disadvantage of this approach, as with CsCl, is the requirement for ultracentrifugation, a technique used infrequently in the contemporary molecular biology laboratory.

Proteinase K

Proteinase K is a proteolytic enzyme that is purified from the mold *Tritirachium album.* In solution, it is stable over a pH range 4.0–12.5, with an optimum of pH 8.0, and a temperature range 25–65°C (Ebeling *et al.*, 1974). Proteinase K is stable for at least 1 year at –20°C. Although the enzyme has two binding sites for Ca^{2+}, in the absence of this divalent cation sufficient catalytic activity is retained to degrade proteins that are commonly found in nucleic acid preparations. Occasionally, proteinase K digestion is carried out in the presence of EDTA, at 50°C, in order to inhibit labile, Mg^{2+}-dependent nucleases.

Proteinase K is prepared commonly as a 20 mg/ml stock solution in sterile water (stable for 1 year at –20°C) or in a solution of 50 mM Tris, pH 8.0, 1 mM $CaCl_2$ (stable for 1 year at 4°C). It is generally used at a working concentration of up to 50 µg/ml in a buffer of 10 mM Tris–Cl, pH 7.5, 5 mM EDTA, and up to 0.5% SDS. Although the inclusion of 1 mM Ca^{2+} (with no EDTA in the reaction buffer) is optional, maximum proteinase K activity is calcium-dependent.

Pronase (*Streptomyces griseus*) can be used in place of proteinase K, although Pronase self-digestion is required to eliminate contaminating RNase and DNase activity. This is easily accomplished by incubation of the Pronase stock (20 mg/ml in 10 mM Tris–Cl, pH 7.5, 10 mM NaCl) at 37°C for 1 h. Suitable aliquots are then stored at –20°C. Reaction conditions for Pronase are identical to those for proteinase K except that the recommended working concentration for Pronase is about 1 mg/ml.

Protocol: Synthesis of VDR

The protocol described here is a modification of the procedures described by Leinhard *et al.* (1971) and Berger and Birkenmier (1979). In this laboratory, synthesis of VDR is efficiently accomplished by mixing all reagents in a three-necked flask submerged in a hot oil bath (Figure 1). The three-necked configuration permits access of a thermometer and pH meter, while allowing the dropwise addition of NaOH. The volumes and masses indicated may be increased proportionately to accommodate larger scale preparations.

1. Mix 0.5 mmol each of adenosine, guanosine, cytidine, and uridine in a total volume of 8 ml H$_2$O.
2. Heat this mixture in a heating mantle, boiling water bath, or hot oil bath until all solids are dissolved.
3. While flushing or bubbling the solution with a steady stream of nitrogen gas, add 1 ml of 2 M vanadium sulfate (VOSO$_4$).

CAUTION   **Vanadium sulfate is highly poisonous.**

4. Increase the pH of the solution to about 6.0 with the dropwise addition of 10 N NaOH.
5. Increase the pH of the solution to 7.0 with the dropwise addition of 1 N NaOH.

Note: Be sure to change to 1 N NaOH between pH 6.0 and pH 7.0 because the pH changes very rapidly in this interval.

6. At pH 7.0, a heavy precipitate of oxovanadium IV is observed. Accompanying the formation of this precipitate is a radical change of color from bright blue to a characteristic green-black appearance.

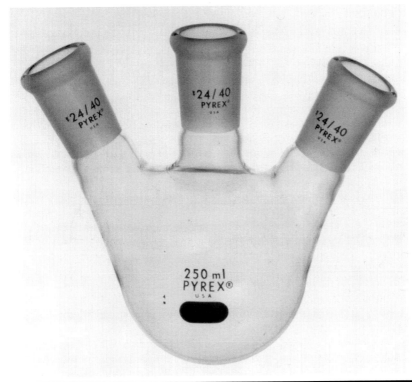

FIGURE 1

The three-neck flask configuration permits easy access to the reaction components and mounting of a pH electrode, thermometer, and N_2 stream. Photograph courtesy of Corning Glass, Corning, New York.

Note: In the absence of a nitrogen atmosphere, rapid oxidation of the oxovanadium IV species to the oxovanadium V species occurs above pH 3.5, the latter being a less efficient inhibitor of RNase.

7. Increase the final volume of the mixture to 10 ml with H_2O.

Note: A by-product of this synthesis reaction, Na_2SO_4, can be removed by brief centrifugation at this point, or just prior to use, though this is not necessary.

8. The product of this synthesis reaction is a 200 mM stock solution of vanadyl ribonucleoside complexes. Store small aliquots of this preparation under N_2 at $-20°C$ for 1–2 months; at $-70°C$ these complexes are stable for at least 1 year.

References

Benecke, B. J., Ben-Ze'ev, A., and Penman, S. (1978). The control of mRNA production, translation, and turnover in suspended and reattached anchorage-dependent fibroblasts. *Cell* **14**, 931.

Berger, S. L. (1975). Diethyl pyrocarbonate: An examination of its properties in buffered solutions with a new assay technique. *Anal. Biochem.* **67**, 428.

Berger, S. L., and Birkenmier, C. S. (1979). Inhibition of intractable nucleases with ribonucleoside-vanadyl complexes: Isolation of messenger ribonucleic acid from resting lymphocytes. *Biochemistry* **18**(23), 5143.

Berger, S. L., Hitchcock, M., Zoon, K. C., Birkenmier, C. S., Friedman, R. M., and Chang, E. H. (1980). Characterization of interferon messenger RNA synthesis in Namalva cells. *J. Biol. Chem.* **255**, 2955.

Blackburn, P. (1979). Ribonuclease inhibitor from human placenta: Rapid purification assay. *J. Biol. Chem.* **254**, 12, 484.

Blackburn, P., and Moore, S. (1982). Pancreatic ribonuclease. *In* "The Enzymes" (P. D. Boyer, Ed.), 3rd ed., pp. 317–433. Academic Press, New York.

Blackburn, P., Wilson, G., and Moore, S. (1977). Ribonuclease inhibitor from placenta. *J. Biol. Chem.* **252**, 5904.

Blade, W. H., and Boulton, R. (1988). Absorption of protein by bentonite in a model wine solution. *Am. J. Vitic.* **39**(3), 193.

Blumberg, D. D. (1987). Creating a ribonuclease-free environment. *Methods Enzymol.* **152**, 20.

Chirgwin, J. M., Przybyla, A. E., MacDonald, R. J., and Rutter, W. J. (1979). Isolation of biologically active ribonucleic acid from sources enriched in ribonuclease. *Biochemistry* **18**, 5294.

Cox, R. (1979). Quantitation of elongating form A and B RNA polymerases on chick oviduct nuclei and effects of estradiol. *Cell* **7**, 455.

Dahle, C. E., and Macfarlane, D. E. (1993). Isolation of RNA from cells in culture using Catrimox-14 cationic surfactant. *BioTechniques* **15**, 1102–1105.

de Martynoff, G., Pays, E., and Vassart, G. (1980). Synthesis of a full length DNA complementary to thyroglobulin 33S mRNA. *Biochem. Biophys. Res. Commun.* **93**, 645.

Ebeling, W., Hennrich, N., Klockow, M., Metz, H., Orth, H. D., and Lang, H. (1974). Proteinase K from *Tritirachium album* Limber. *Eur. J. Biochem.* **47**, 91.

Ehrenberg, L., Fedorcsak, I., and Solymosy, F. (1974). Diethyl pyrocarbonate in nucleic acid research. *Prog. Nucleic Acid Res. Mol. Biol.* **16**, 189.

Fraenkel-Conrat, H., Singer, B., and Tsugita, A. (1961). Purification of viral RNA by means of bentonite. *Virology* **14**, 54.

Henderson, R. E. L., Kirkegaard, L. H., and Leonard, N. J. (1973). Reaction of diethyl pyrocarbonate with nucleic acid components. Adenosine-containing nucleotides and dinucleoside phosphates. *Biochim. Biophys. Acta* **294**, 356.

Kirby, K. S. (1956). A new method for isolation of ribonucleic acids from mammalian tissue. *Biochem. J.* **64**, 405.

Leinhard, G. E., Secemski, I. I., Koehler, K. A., and Lindquist, R. N. (1971). Enzymatic catalysis and the transition state theory of reaction rates: Transition state analogs. *Cold Spring Harbor Symp. Quant. Biol.* **36**, 45.

Nevins, J. R., and Darnell, J. E. (1978). Steps in the processing of Ad2 mRNA: Poly(A+) nuclear sequences are conserved and poly(A+) addition precedes splicing. *Cell* **15**, 1477.

Palmiter, R., Christensen, A., and Schimke, R. (1970). Organization of polysomes from pre-existing ribosomes in chick oviduct by a secondary administration of either estradiol or progesterone. *J. Biol. Chem.* **245,** 833.

Perry, R. P., La Torre, J., Kelley, D. F., and Greenberg, J. R. (1972). On the lability of poly(A) sequence during extraction of messenger RNA from polyribosomes. *Biochim. Biophys. Acta* **262,** 220.

Poulson, R. (1973). Isolation, purification, and fractionation of RNA. *In* "The Ribonucleic Acids" (P. R. Stewart and D. S. Letham, Eds.), pp. 243–261. Springer-Verlag, New York.

Saywell, L. G. (1934). Clarification of wine. *Ind. Eng. Chem.* **26,** 981.

Scheele, G., and Blackburn, P. (1979). Role of mammalian RNase inhibitor in cell-free protein synthesis. *Proc. Natl. Acad. Sci. USA* **76,** 4898.

Scheider, R., Schneider-Scherzer, E., Thurnher, M., Auer, B., and Schweiger, M. (1988). The primary structure of human ribonuclease/angiogenin inhibitor (RAI) discloses a novel, highly diversified protein superfamily with a common repetitive module. *EMBO J.* **7,** 4151.

Sela, M., Anfinsen, C. B., and Harrington, W. F. (1957). The correlation of ribonuclease activity with specific aspects of tertiary structure. *Biochim. Biophys. Acta* **26,** 502.

Wallace, D. M. (1987). Large- and small-scale phenol extractions. *Methods Enzymol.* **152,** 33.

RNA Isolation Strategies

Rationale

Efficient methodologies have been empirically derived to accommodate the expedient isolation of RNA—techniques that should be scrutinized and refined continuously. In general, these methods yield cytoplasmic RNA, nuclear RNA, or mixtures of both. RNA isolated from the nucleus or cytoplasm by direct cell lysis is known as **steady-state** RNA; it represents the final accumulation of the RNA in the cell or a subcellular compartment. Whereas the abundance of specific steady-state RNA species may certainly be interpreted as an indication of how specific genes are modulated, it furnishes information neither about the *rate* at which these RNA molecules are transcribed nor about their stability or half-life. This is the major disadvantage of evaluating the steady-state RNA alone. It is certainly clear that cellular biochemistry responds to environmental change not only by modification of the rate of transcription, but also through the processing efficiency of precursor RNA, the efficiency of nucleocytoplasmic transport, the stability of RNA in the cytoplasm, and the translatability of salient messages. In contrast to the protocols for the analysis of steady-state transcripts presented in this chapter, transcription rate studies require the isolation of intact nuclei that are able to support elongation of, and label incorporation into, RNA molecules whose transcription was initiated at the time of cellular lysis. This family of techniques, known collectively as the nuclear runoff assay (Marzluff and Huang, 1984), has the distinct advantage of maintaining the natural architecture of the nucleus while supporting labeling of initiated transcripts (see Chapter 19 for a discussion of the nuclear runoff assay and protocols).

Protocols for the isolation of RNA begin with cellular lysis mediated by buffers that typically fall into one of two categories: (1) those consisting of harsh chaotropic agents including one of the guanidinium salts, sodium dodecyl sulfate (SDS), *N*-laurylsarcosine (sarcosyl), urea, phenol, or chloroform, which disrupt the plasma membrane and subcellular organelles and which simultaneously inactivate ribonuclease (RNase); and (2) those that gently solubilize the plasma membrane while maintaining nuclear integrity, such as hypotonic Nonidet P-40[1] (NP-40) lysis buffers.[2] Nuclei, other organelles, and cellular debris are then removed

---

[1]Nonidet P-40 is the older name for this nonionic detergent, which is no longer made. An equivalent product that may be substituted is Igepal CA-630 (Sigma Cat. I-3021).

[2]Whereas osmotic lysis is one of the most gentle methods for cell disruptions, it does nothing when carbohydrate-rich cells walls are present, as with certain bacteria, fungi, and plant cells. Appropriate steps must be taken to break through the cell wall and access the cellular contents.

from the lysate by differential centrifugation. The reliability of this approach is often dependent on the inclusion of nuclease inhibitors in the lysis buffer and careful attention to the handling and storage of RNA so purified. It is worth noting that a seemingly endless list of permutations on a few fundamental RNA extraction techniques exist; for example, some techniques support isolation of poly(A)$^+$ material directly from a cellular lysate without prior purification of total RNA.

Goals in the Purification of RNA

The first decision to factor into the design of an RNA isolation strategy is the method of cellular disruption, and it is based on which population of RNA or subcellular compartment the investigator wishes to study. Of equal importance is the method by which RNase activity will be controlled. For example, an investigator may wish to determine whether the observed modulation of gene expression in a model system is regulated transcriptionally or by some posttranscriptional event(s). In such an instance, selection of the method of cellular disruption and subsequent RNA isolation must permit analysis of the nuclear abundance of salient transcripts independently of the cytoplasmic abundance of the same species. Any worthwhile strategy for RNA purification must address and achieve specific goals if data derived from the final RNA preparation are to be meaningful.

Goal 1: Select an appropriate method of membrane solubilization. The method of cell lysis will determine the extent of subcellular disruption of the sample. For example, a lysis buffer that is used successfully with tissue culture cells may be entirely inappropriate for whole tissue samples. Just as importantly, the method by which membrane solubilization is accomplished will dictate whether additional steps will be required to remove DNA from the RNA preparation, and whether compartmentalized nuclear RNA and cytoplasmic RNA species can be purified independently of one another. While DNA can be purged from an RNA preparation with minimal fanfare, it may be difficult if not impossible to determine the relative contribution of RNA from the nucleus and the cytoplasm once RNA from these two subcellular compartments have copurified. A particular lysis procedure must likewise demonstrate compatibility with ensuing protocols once it has been recovered from the lysate. Always think two steps ahead; the proper method of solubilization is dependent on the plans for the RNA after purification and the questions being asked in a particular study.

Goal 2: Ensure total inhibition of nuclease activity.   The imperative for controlling nuclease activity should be abundantly clear, if not from personal experience, then from the discussion in Chapter 4. This includes purging RNase from reagents and equipment and controlling such activity in a cell lysate. Although some lysis reagents inhibit nuclease activity in their own right, other lysis reagents require additional nuclease inhibitors to safeguard the RNA during the isolation procedure. Steps for the inhibition or elimination of RNase activity must, first and foremost, demonstrate compatibility with the lysis buffer.

Goal 3: Select a method for deproteinization of the sample.   The complete removal of protein from a cellular lysate is of paramount importance in the isolation of both DNA and RNA. Meticulous attention to this detail is required for accurate quantification and precision in hybridization. For example, the restriction of double-stranded DNA can be inhibited by carryover protein, especially histone proteins. Further, reverse transcription and amplification by polymerase chain reaction (PCR) (Chapter 15) can be inhibited strongly by the use of "dirty" RNA. Removal of proteins may be accomplished by:

a. Digestion of the sample with the enzyme proteinase K;
b. Repeated extraction with mixtures of organic solvents such as phenol and chloroform;
c. Solubilization in guanidinium buffers;
d. Salting-out of proteins;
e. Any combination of the above.

Any procedure for deproteinization is itself a means of controlling RNase activity. It is important to recognize that the RNA in the sample will be susceptible to nuclease degradation once the protein denaturant is removed and is never more susceptible to nuclease attack than when it is dissolved in simple, nondenaturing aqueous buffer.

Goal 4: Select a method for nucleic acid concentration.   This is the final step in most RNA purification schemes. The most versatile method for concentrating nucleic acids is precipitation using various combinations of salt and alcohol (Table 1). For example, one common method is to add 0.1 volume of 3 M sodium acetate (pH 5.2) to a nucleic acid sample, followed by the addition of 2.5 volumes of 95 or 100% ethanol. Nucleic acids and the salt that drives their precipitation form complexes that have greatly reduced solubility in ethanol and isopropanol. Further, the rate of precipitation using various salt and alcohol combinations is tem-

Salt and Alcohol Combinations for Concentration of Nucleic Acids[a]

Salt	Typical stock concentration	Amount required (volume)	Final concentration
NaOAc	3 M, pH 5.2	0.1	300 mM
NaCl	1 M	0.1	100 mM
LiCl[b]	8 M	0.1	800 mM
NH$_4$OAc[c]	10 M	0.2	2 M
KOAc	2.5 M	0.1	250 mM
Glycogen[d]	10 mg/ml	0.01	100 µg/ml

Alcohol	Volume required after addition of salt
Ethanol (95–100%)	2.2–2.5
Isopropanol (100%)	0.6–1.0

[a]Precipitation thresholds: DNA 50 ng/ml, RNA 100 ng/ml.
[b]Lithium chloride does not coprecipitate with nucleic acids. As with NaCl, it is very soluble in ethanol. Avoid LiCl if RNA is to be reverse transcribed.
[c]Preferred salt for precipitating cDNA after synthesis: ammonium acetate does not precipitate dNTPs.
[d]The judicious use of glycogen is recommended, however, because it inhibits a number of *in vitro* reactions.

perature dependent. Unlike the precipitation of genomic DNA, the precipitation of RNA is much more refined, requiring longer incubation periods at –20°C to ensure complete recovery. This phenomenon is a direct function of genome size and complexity that an organism exhibits, compared to the relatively low complexity of cellular RNA. Other procedures for concentrating nucleic acid samples,[3] including the use of commercially available concentrating devices, dialysis, and centrifugation under vacuum, are not as frequently employed because the degree of success is much more dependent on the skill of the investigator than with simple salt and alcohol concentration.

Goal 5: Select the proper storage conditions for purified samples of RNA. Because of the naturally labile character of RNA, improper storage of excellent RNA samples will often result in degradation in a rela-

[3]An older procedure for concentration of nucleic acids, repeated extraction with butanol, is not recommended for any application because of the inherent difficulties with the complete removal of butanol.

tively short time. There are many opinions as to the proper temperature, buffer, and storage form for both RNA and DNA. A discussion of storage considerations appears at the end of this chapter.

Chaotropic Lysis Buffers

There is probably no better way to deal with seemingly recalcitrant RNases than to disrupt cells in guanidinium lysis buffer (Chirgwin *et al.*, 1979). On contact with guanidinium- and sarcosyl-containing buffers, or other lysis buffers that contain high concentrations of SDS (many different formulations have been described), the tertiary folding of RNases is distorted, resulting in the inactivation of these enzymes. It is not necessary to add additional RNase inhibitors to such lysis buffers, and RNA isolation procedures conducted under these harsh conditions can be performed at room temperature. Because these buffers are so chaotropic (biologically disruptive), organelle lysis also accompanies disruption of the plasma membrane. This occurrence liberates heterogeneous nuclear RNA (hnRNA) and genomic DNA from the nucleus as well as mitochondrial DNA and RNA, all of which copurify with the cytoplasmic RNA. Therefore, further purification is required to remove DNA from the sample. Formerly, the most prevalent of these methods was a technique known as isopycnic centrifugation (see Cooper, 1977, for background information) through a cesium chloride (CsCl) gradient (Glisin *et al.*, 1974; Ullrich *et al.*, 1977) or a cesium trifluoroacetate (CsTFA) gradient (Zarlenga and Gamble, 1987). Isopycnic separation is possible because of the differing buoyant densities of DNA (1.5–1.7 g/ml) and RNA (1.8–2.0 g/ml). For the convenience of the reader, a review of the fundamentals of centrifugation as a tool in molecular biology is presented in Appendix G.

More recently, the differential partitioning of DNA, RNA, and protein by guanidinium–acid–phenol extraction (Chomczynski and Sacchi, 1987) has become one of the most commonly used techniques for rapid purification of RNA. The investigator can fully exploit the salient chemical differences between RNA and DNA by establishing an acidic pH environment and judiciously blending organic solvents. As result of the popularity of this general approach, numerous products and methodologies have been developed for the rapid, efficient purification of both RNA and DNA (and protein) from the same biological source (Majumdar *et al.*, 1991; Chomczynski, 1993). Some of these procedures, and others, are described here.

The principal drawback to these chaotropic procedures is that the investigator is not able to discriminate between cytoplasmic and nuclear RNA, as there is no method for separating hnRNA from spliced messenger RNA (mRNA) once these two populations have been mixed (size fractionation may result in a partial separation, but is not at all definitive).

It is unfortunate that many seasoned investigators begin to show signs of sloppiness with respect to keeping buffers and equipment nuclease-free when routinely working with guanidinium buffers. Although it is true that RNA is safe from nuclease degradation in the presence of these agents, purified RNA is once again susceptible to nuclease degradation when the denaturants have been removed. Therefore, it is necessary to guard against the RNase peril consistently, according to the guidelines described in Chapter 4.

Gentle Lysis Buffers

Cellular lysis mediated by nonionic, hypotonic buffer is not disruptive to most subcellular organelles. The inclusion of NP-40 (Favaloro et al., 1980) and $MgCl_2$ in the lysis buffer facilitates plasma membrane solubilization while maintaining nuclear integrity. Thus, initial lysis of cells with NP-40-containing buffer precludes the mixing of nuclear and cytoplasmic RNA from the onset. In such protocols, intact nuclei, large organelles, and cellular debris are easily removed by differential centrifugation. The resulting cytosolic supernatant is rich in cytoplasmic RNA and proteins, the latter of which are easily removed by a series of extractions with mixtures of organic solvents such as phenol and chloroform. The obvious advantage of this isolation strategy is that the RNA precipitated at the end of the procedure represents only the cytoplasmic population. This may be especially meaningful in the exploration of the level of gene regulation.

A disadvantage of this approach is that the lysis buffer alone is not sufficiently chaotropic to fully inhibit or destroy RNase activity. Keep in mind that upon cellular lysis, normally sequestered RNases are suddenly liberated, and their activity will greatly compromise the integrity of the RNA even as the investigator is working diligently to purify it. Therefore, it is necessary to add some type of RNase inhibitor to the lysis buffer *just prior to use* in order to control nuclease activity. When using this method of RNA isolation, it is extremely helpful to keep the sample on ice at all times unless the protocol specifically dictates otherwise; it is also very helpful to use ice-cold reagents and tubes that have been prechilled. Should one wish to take advantage of the disruptive nature of the guanidinium-based isolation procedures and still maintain the subcellular compartmentalization of RNA, one worthwhile strategy is to begin the isolation procedure with cellular lysis mediated by NP-40-containing buffer, recover intact nuclei, and then lyse the nuclei with guanidinium buffer; one may proceed to purify the nuclear (or cytoplasmic) RNA as if working with intact cells. This approach is particularly suited for the isolation of nuclear RNA.

Note: Protocols describing the use of guanidinium buffers are described here; for a more complete discussion of the isolation of nuclear RNA, see Chapter 19.

Protocol: Isolation of Cytoplasmic RNA by Hypotonic Lysis with NP-40

This procedure is based on nonionic, hypotonic solubilization of the plasma membrane while maintaining nuclear integrity.[4] Cellular lysis is accomplished as quickly as possible, on ice, in the presence of RNase inhibitors. RNase inhibitors must be included until phenol:chloroform mixtures are introduced into the system to remove the protein. An in-house modification of this valuable technique is presented here (Figure 1) and is further modified in Chapter 19 to accommodate isolation of intact nuclei and accompanying purification of nuclear RNA. Following gentle cell lysis, the subsequent phenol:chloroform extractions and alcohol precipitation of the RNA are among the most common preparative techniques. Whereas this method is faster than the older guanidinium-based methods requiring ultracentrifugation, it may appear to be more time consuming than the newer acid–phenol methods (below). This is due, in part, to the requirement for gentle handling of the sample to preclude nuclear breakage, which, to a limited extent, *may* put the RNA molecules at slightly greater risk of RNase degradation. Although better quality RNA may be generated using guanidinium-based lysis buffers, RNA prepared according to this method is adequate for most standard applications. The procedure delineated here works best when starting with cells grown in tissue culture, as the lysis buffer does not disrupt whole tissue samples as efficiently as cells harvested from tissue culture. The protocol may be adapted for whole tissue samples when combined with gentle Dounce homogenization to prevent serious damage to cellular membranes.

---

[4]The investigator may wish to optimize empirically the lysis conditions because cells and their organelles exhibit different sensitivities to detergents, salts, and RNase inhibitors.

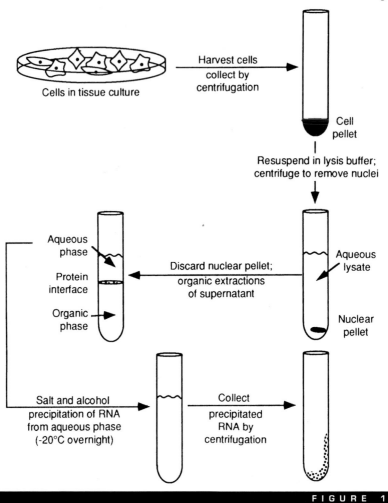

FIGURE   1

Preparation of total cytoplasmic RNA by hypotonic lysis in the presence of NP-40. While this method is well suited for cells grown in culture, tissue samples often require a gentle, nonaggressive homogenization step.

Advance Preparation of Extraction Buffer

An adequate volume of extraction buffer should be freshly prepared before each use, as the oxidation of phenol can greatly compromise the quality of an RNA preparation (Figure 2).

1. Melt redistilled (molecular biology grade) phenol at 60°C and saturate with an equal volume of Tris–SDS buffer (Figure 2A):

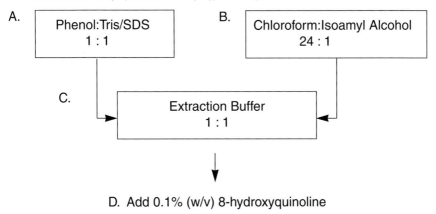

A. Phenol:Tris/SDS 1 : 1

B. Chloroform:Isoamyl Alcohol 24 : 1

C. Extraction Buffer 1 : 1

D. Add 0.1% (w/v) 8-hydroxyquinoline

F I G U R E   2

Method of preparation of the organic extraction mixture used in the NP-40 procedure for the isolation of cytoplasmic RNA. Saturation of phenol with Tris–SDS buffer, rather than Tris alone, and the inclusion of 8-hydroxyquinoline render a superior formulation for the deproteinization of the cytosol. This buffer can be substituted in most applications when the use of phenol:chloroform:isoamyl alcohol is dictated.

100 mM NaCl

1 mM EDTA

10 mM Tris, pH 8.5

0.5% SDS[5]

2. To the resulting Tris–SDS/phenol mixture add an *equal* volume of a mixture of chloroform:isoamyl alcohol (24:1) (Figure 2B). The resulting reagent is known as the extraction buffer (Figure 2C).

3. Finally, add 8-hydroxyquinoline to the extraction buffer (Figure 2D) to a concentration of 0.1%. Allow the phases to separate.

For example: Mix 20 ml of melted, redistilled phenol with 20 ml (an equal volume) of Tris–SDS buffer. To the resulting 40 ml of Tris–SDS-saturated phenol, add 40 ml (an equal volume) of chloroform:isoamyl alcohol, previously mixed in a ratio of 24:1. This is the extracting buffer, the stability of which is enhanced by the addition of 0.08 g of 8-hydrox-

[5]The inclusion of SDS ensures the quantitative recovery of poly(A)$^+$ RNA in the aqueous phase during extraction with organic buffer. Saturation of phenol with Tris–SDS buffer is desirable also because this partitioning phenomenon in the presence of SDS is pH independent. Be sure to autoclave this buffer *before* adding SDS to complete this reagent.

yquinoline (final concentration of 0.1% w/v). This buffer is conveniently stored in a 100-ml glass bottle and has a shelf life of 3–4 weeks at
4°C. Ideally, the bottle should be wrapped in foil when not in use, especially if the refrigerator is opened and closed continuously throughout
the day.

CAUTION   **Be sure to handle all organic solvents with proper skin and
eye protection, in a chemical fume hood, and be sure to dispose of organic
waste according to departmental regulations.**

RNA Isolation

1. **Wear gloves!**
2. Harvest cells from tissue culture and collect by centrifugation in a
suitable tube, prechilled to 4°C.

Note: For trypsinization protocol, see Appendix H.

3. Gently, and completely, resuspend the cell pellet in ice-cold lysis
buffer:

  140 mM NaCl

  1.5 mM MgCl$_2$

  10 mM Tris–Cl, pH 8.5

  0.5% NP-40[6]

  Optional: 10 mM vanadyl ribonucleoside complexes (VDR)[7]

Use at least 100 μl lysis buffer/10^6 cells. Periodic *gentle* inversion of the
tube may facilitate cell lysis. Incubate on ice for 5 min.

*Note 1: VDR oxidizes very rapidly and therefore should be added to the
lysis buffer just before use. It is usually prepared/supplied as a 200
mM stock solution and should be used at a working concentration of
5–20 mM. Alternatively, RNasin (Promega) or RNAguard (Pharmacia
Biotech) can be added to the lysis buffer (250–1000 U/ml) to inhibit
RNase activity. Beyond this point, the tube containing the RNA should
be kept on ice unless otherwise indicated.*

*Note 2: The extent of cell lysis may be assessed by examining a 1-μl
aliquot of the lysate on a microscope slide for the presence of intact
cells versus free nuclei.*

---

[6]Autoclave this lysis buffer *before* adding the NP-40 needed to complete this reagent.
[7]See Chapter 4 for discussion and synthesis protocol.

4. Pellet the nuclei and other cellular debris by differential centrifugation for 5 min at 4°C. Centrifuge at 5000g in a microfuge, or a suitable g-force if another type of tube is used.

Note 1: Although nuclei can be efficiently pelleted with as little as 500g, the added force is useful for the more complete removal of large organelles and the tight packing of the nuclei. This centrifugation will result in the removal of genomic DNA as well as the nuclear hnRNA species and will facilitate the analysis of extranuclear RNA.

Note 2: If the nuclei are to be used in a later application, such as the isolation of hnRNA or genomic DNA, then collect the nuclei at a maximum of 500g and extend the centrifugation time as needed. Additional details may be found in Chapter 19.

5. Transfer the supernatant (cytosol) to a fresh, RNase-free tube and add EDTA (pH 8.0) to a final concentration of 10 mM.

Note: Na$_2$-EDTA is added to prevent the magnesium-mediated aggregation of nucleic acids with proteins and with each other.

6. Add an equal volume of the organic phase (lower phase; yellow) of the extracting buffer to the aqueous cytosol (lysate). Using extreme care, mix thoroughly by inversion for 15 s or longer. Incubate the sample at 55°C for 5 min.

Note 1: If the prep is being done in a microfuge tube, hold the tube between the thumb and index finger and invert the tube several times. This technique will give the most complete intermixing of the phases, the most complete denaturation of proteins, and the least likelihood of accidental spillage.

Note 2: The heating of the sample will facilitate more thorough removal of proteins. Some variations of this procedure direct the investigator to do a series of organic extractions without heating the sample. The approach recommended here reduces the volume of organic waste.

Note 3: Upon contact with VDR, the initially yellow 8-hydroxyquinoline becomes dark green-black. This is because 8-hydroxyquinoline will chelate heavy metals (vanadium). In the absence of 8-hydroxyquinoline (or VDR), no color change is expected.

7. Incubate the sample on ice or in an ice water bath for 5 min.

Note: This rapid cooling will facilitate separation of the organic and aqueous phases.

8. Centrifuge the sample at 500–2000g at 4°C or at top speed in a bench-top microfuge to separate the phases.

Note: The time required to complete the separation will be a direct function of the volume and the size of the tube. A maximum of 5 min is usually adequate. Although a refrigerated microfuge is useful, it is not required.

9. Carefully remove the upper aqueous phase to a fresh, RNase-free tube. Use great care to avoid disturbing the protein interface. It is much better to leave a few microliters of aqueous material behind than to risk transfer of any material at the interface (see Figure 1).

10. Add a fresh aliquot (equal volume) of extracting buffer to the aqueous phase and mix carefully and thoroughly by inversion. Centrifuge at 500–2000g at 4°C or at top speed in a bench-top microfuge to separate the phases. Usually, it is not necessary to heat and cool the sample as in the first organic extraction.

11. Carefully recover the upper aqueous phase and transfer to a new RNase-free tube. If a protein interface is apparent or if the aqueous phase remains cloudy, repeat the organic extraction described in Step 10 until no protein appears at the interface and the aqueous (upper phase) is clear.

Note: Additional phenol extractions are usually necessary when extracting a large number of cells in a relatively small volume.

12. Extract the aqueous material one last time with an equal volume of chloroform or chloroform:isoamyl alcohol (24:1) and mix carefully and thoroughly by inversion. Centrifuge for 30 s to separate the phases.

Note: Because phenol is very soluble in chloroform, a final chloroform extraction will remove any traces of phenol from the RNA-containing aqueous phase. This should be a standard technique whenever phenol is involved. Remember that even trace amounts of phenol will oxidize into quinones and compromise the quality of a nucleic acid sample.

13. Transfer the aqueous phase (upper) to a fresh, RNase-free tube and add 0.1 volume 3 M sodium acetate, pH 5.2, and 2.5 volumes ice-cold 95% ethanol. Mix thoroughly by inversion.

14. Precipitate the RNA overnight at –20°C.

Note: A combination of salt and alcohol is the most versatile method for the precipitation of nucleic acids. A complex consisting of the sample and the salt, which has reduced solubility in high concentrations of alcohol, forms. This precipitate can then be collected by centrifuga-

tion, briefly dried, and resuspended in an appropriate buffer. To ensure complete RNA recovery, samples should be precipitated overnight at –20°C, as the quantity of RNA from a biological source is not always known.

15. Collect the precipitated RNA by centrifugation at 10,000g at 4°C for 30 min if using Corex glass or at top speed in a microfuge for 10 min at 4°C (room temperature is all right, too).

16. Decant the ethanol supernatant. Excess coprecipitated salt should be removed by washing the RNA pellet 2–3 times with 70–75% ethanol, into which the salt will dissolve (sodium acetate is particularly soluble in 70–80% ethanol). This concentration of ethanol is not sufficiently aqueous, however, to redissolve precipitated nucleic acids.

17. Briefly air-dry the tube(s). If desired, a final wash with 95% ethanol will facilitate the drying of the sample. The RNA is best stored at –70°C as an ethanol precipitate or, following quantification, stored in suitable aliquots at –70°C to avoid repeated freezing and thawing. Depending on the cell type, typical yields range from 75 to 100 µg cytoplasmic RNA per 10^7 cells.

Isolation of RNA with Guanidinium Buffers

RNA lysis buffers that contain guanidinium thiocyanate or guanidinium–HCl reproducibly yield very high-quality RNA samples. This is true because of the extremely chaotropic nature that these chemicals exhibit; they are among the most effective protein denaturants (Cox, 1968; Nozaki and Tanford, 1970; Gordon, 1972). This attribute supports the inclusion of guanidinium as the method of choice when planning an RNA extraction from whole tissue samples. The nuclear and organelle disruption that accompanies plasma membrane solubilization liberates nuclear RNA and genomic DNA, both of which copurify with cytoplasmic RNA species. The efficiency of protein denaturation (including disruption of RNases) may be enhanced by the inclusion of β-mercaptoethanol (β-ME) or dithiothreitol (DTT)[8]; these reducing agents act to break intramolecular protein disulfide bonds. The addition of exogenous inhibitors of RNase is not required.

The use of guanidinium-based lysis buffers mandates a procedure for the partitioning of RNA, DNA, and protein in the resulting lysate. There

---

[8]DTT reacts chemically with thiocyanate and is therefore to be avoided when using guanidinium buffers. Use β-ME only as a reducing agent.

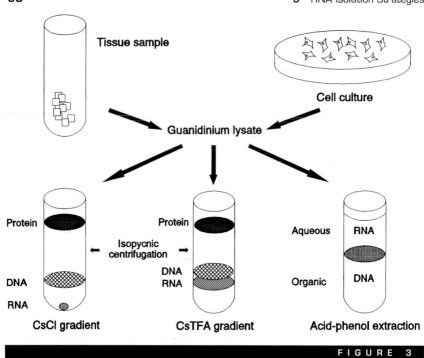

F I G U R E   3

Methods for fractionation of total cellular RNA following cellular disruption with guanidinium buffer: CsCl or CsTFA ultracentrifugation or acid–phenol extraction.

are currently three basic approaches for accomplishing this task; examples of each are presented here. These approaches involve whole cell lysis followed by isopycnic ultracentrifugation[9] or acid–phenol extraction (Figure 3). Although the RNA is most often precipitated directly from a guanidinium-containing lysate, ultracentrifugation through CsCl (or CsTFA), though cumbersome, is useful when ultrapure RNA is required. Gradients in general are time-consuming procedures and usually are no longer required for mainstream molecular biology applications.

In contrast to the gentle NP-40 methods, which facilitate the isolation of total cytoplasmic RNA, procedures involving the use of guanidinium-containing buffers are used in the isolation of total cellular RNA (nuclear and cytoplasmic species). The RNA size distributions from total cytoplasmic and total cellular isolation procedures are compared in Figure 4.

[9]**Caution:** If you are not familiar with the proper operation of an ultracentrifuge or the procedure for derating a rotor be sure to get proper technical assistance. If not used properly, a rotor can be lethal.

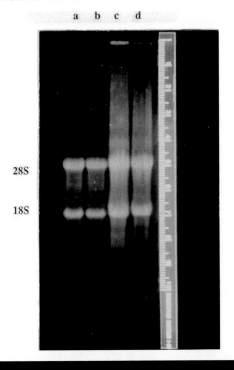

28S

18S

FIGURE 4

Comparison of size distribution of total cellular RNA and total cytoplasmic RNA. Lanes a and b: 20 μg total cytoplasmic RNA isolated with NP-40 lysis buffer and subsequent deproteinization with phenol:chloroform, as described. Lanes c and d: 25 μg of total cellular RNA isolated by guanidinium–acid–phenol extraction. The higher molecular weight species (hnRNA) are clearly visible above the 28S rRNA in lanes c and d. The obvious fluorescence in the wells of lanes c and d suggests a small amount of contaminating DNA. Electrophoretogram made on Polaroid Type 667 film using Model DS-24 camera. Fluorescent ruler courtesy of Diversified Biotech, Inc.

Isopycnic Centrifugation

The original protocols for the isolation of RNA from biological sources enriched in ribonuclease described cell and tissue disruption with guanidinium buffer. The resultant intermixing of subcellular components mandates the separation of these biochemical macromolecules from each other, with particular regard to the removal of DNA from RNA preparations. Subtle though measurable differences in the density of DNA, RNA, and protein (Table 2) allow fractionation by banding them in a density gradient.

Density gradient configuration, known more properly as isopycnic

TABLE 2	
Buoyant Density Ranges of Macromolecules Resolved by Isopycnic Centrifugation	
Molecule	Density (g/ml)
Protein	1.2–1.5
DNA	1.5–1.7
RNA	1.7–2.0

centrifugation, is a technique in which macromolecules move through a density gradient, historically[10] CsCl, CsTFA, or cesium sulfate (Cs_2SO_4), until they find a density equal to their own. These molecules then accumulate at this position in the gradient, floating there until the end of the centrifugation run. In some cases, RNA is of greater density than any position in the gradient and accumulates as a pellet at the bottom of the tube, whereas gradients of greater maximum density may permit the banding of RNA, much as plasmid DNA is banded for purification.

CsCl

Cesium chloride is a dense salt that exhibits the ability to form a linear gradient when a homogeneous suspension (e.g., cell lysate) is subjected to ultracentrifugation g-forces. Thus, there is no need to pre-form the gradient. CsCl gradients are generally steep, and the maximum density within the gradient usually equals or only slightly exceeds that of the most dense material to be sedimented (RNA is collected as a pellet at the bottom of the tube). These gradients are subjected to ultracentrifugation g-forces for a period sufficiently long to complete the migration of all macromolecules into their equilibration positions. The actual time required is a function of the rotor size; traditional RNA preps in floor model ultracentrifuges require overnight centrifugation, while as few as 3 h is needed to accomplish the same type of separation in the newer table-top or "microultracentrifuges," for example, Sorvall RC-M150 (Figure 5).

CsTFA[11]

CsTFA (Pharmacia Biotech, Piscataway, NJ) is an excellent density gradient medium for the separation of RNA and DNA by isopycnic cen-

---

[10]Other materials such as glycerol, ficoll, metrizamide, and sucrose have been used, primarily because they are readily available at low cost and high purity. Nucleic acid resolution by density is best performed using one of the cesium salts.

[11]Data and protocol courtesy of Pharmacia Biotech (Piscataway, NJ).

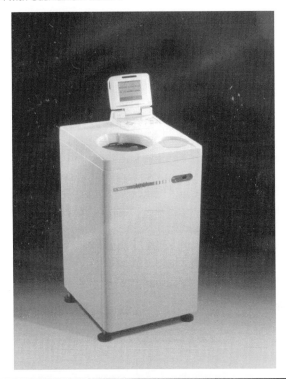

FIGURE 5

Sorvall RC-M150 microultracentrifuge. Smaller radius rotors permit higher g-forces, which, in turn, permit isopycnic preparation of RNA in only 3 h. Photo courtesy of Sorvall, Inc.

trifugation. It is used in a manner similar to CsCl and Cs_2SO_4. The trifluoroacetate anion gives CsTFA properties that result in nucleic acid preparations of extremely high quality, including:

1. High solution density. CsTFA is an extremely soluble salt, capable of forming solutions with densities of up to 2.6 g/ml. In the realm of isopycnic separation of nucleic acids, this is significantly greater than the maximum density achievable with CsCl (1.7 g/ml). The trifluoroacetate ions present in CsTFA promote the hydration and solubilization of nucleic acids and proteins. Consequently, nucleic acids have lower buoyant densities in CsTFA than in CsCl, and proteins are more readily dissociated from nucleic acids.

2. Isopycnic banding of RNA. The lower buoyant densities of nucleic acids and the higher solution density possible with CsTFA allow RNA to be banded in an isopycnic manner, which is not possible in CsCl gradi-

ents nor in Cs_2SO_4, in which precipitation of RNA frequently occurs. If desired, simultaneous banding of DNA and RNA can be achieved in the same CsTFA gradient (Zarlenga and Gamble, 1987). Alternatively, the gradient can be prepared to accommodate pelleting of RNA (Figure 6) for recovery in a manner similar to that of CsCl gradients.

3. Protein denaturation. Since CsTFA solubilizes and dissociates proteins from nucleic acids without the use of detergents, it is possible, in a variety of applications, to reduce or even eliminate hazardous and time-consuming extractions with organic solvents and/or digestion with proteinase K. Reducing the requirement for phenol extractions allows nucleic acids, especially hnRNA and genomic DNA, to be isolated with less risk of shearing or degradation. Moreover, in its capacity as an excellent nuclease inhibitor, CsTFA facilitates the isolation of intact nucleic acids.

4. Improved yields of pure RNA. Procedures using CsTFA give high recoveries of nucleic acids. The salting-in effect of the trifluoroacetate ions, in contrast to the salting-out effect of the chloride and sulfate an-

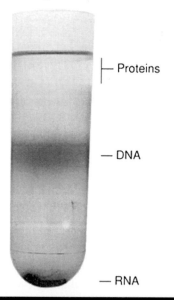

— Proteins

— DNA

— RNA

FIGURE 6

CsTFA gradient. Isopycnic gradient separation of calf liver macromolecules extracted with guanidinium thiocyanate. A guanidinium thiocyanate extract of calf liver (18 ml) was layered on top of a 19-ml cushion of CsTFA (ρ = 1.51 g/ml) and centrifuged at 125,000g for 16 h. Approximately 3.5 mg of total RNA was pelleted at the bottom of the tube ($\rho \approx$ 1.62–1.9 g/ml). This RNA contained less than 0.4% DNA as measured with a DNA probe. DNA forms a band near the original interface of the cushion and the sample solutions ($\rho \approx$ 1.6 g/ml). Protein in the sample collects in the upper portion of the gradient ($\rho \approx$ 1.2–1.5 g/ml). Courtesy of Pharmacia Biotech.

ions, minimizes the incidence of protein precipitation/coprecipitation during centrifugation, which would result in the loss of nucleic acids during isolation.

5. Solubility in polar solvents. The solubility of CsTFA in polar solvents facilitates recovery of nucleic acids from CsTFA solutions. Following gradient fractionation, nucleic acids may be recovered directly from the gradient by precipitation with ethanol and without prior dialysis or extraction to remove the salt.

6. Lower melting temperatures. Nucleic acid duplexes have a lower T_m in CsTFA than in CsCl. Because lower melting points expedite the separation of single-stranded from double-stranded nucleic acids, this method may prove to be useful for recovery of intact single-stranded DNA and RNA from DNA:RNA hybrids.

Disadvantages

Among the disadvantages of such density gradients are the cost of the density gradient material (e.g., CsCl, CsTFA), the requirement for a very expensive ultracentrifuge, the relatively long period of centrifugation, the labor-intensive cleanup of gradient-purified RNA, and the limited ability to prep more than a few samples at a time. For these and other reasons, isopycnic centrifugation has become something of a rarity in the isolation of RNA, in favor of newer reagents and kits that purport to obviate the requirement for ultracentrifugation at all. In some applications, there remains the need to resolve RNA in a density gradient and for this reason, two excellent protocols follow.

Protocol: Cesium Chloride Gradients

The RNA isolation protocol presented here is a modification of the procedures described by Glisin *et al.* (1974), Ullrich *et al.* (1977), and Chirgwin *et al.* (1979).

1. **Wear gloves!**

2a. For tissue culture: Collect harvested cells by centrifugation at 500*g* for 5 min. Decant as much of the supernatant as possible. Wash cells once with phosphate-buffered saline (PBS; per liter: 8.0 g NaCl; 0.2 g KCl, 1.44 g Na_2HPO_4, 0.24 g KH_2PO_4). Decant as much PBS as possible. If the RNase is likely to be a major challenge, perform centrifugations at 4°C and wash cells with ice-cold PBS.

2b. For tissue: *Rapidly* harvest the tissue and mince/dice in ice-cold saline solution. Quickly transfer tissue into a suitable homogenization vessel that already contains an aliquot of lysis buffer (Step 3). If tissue was flash-frozen in liquid nitrogen, it may be helpful to shatter it (while wrapped in aluminum foil) by impact upon the lab bench. In so doing, the sample can be quickly transferred and allowed to thaw in the lysis buffer.

3. Add 5 volumes (the size of the pellet) of a solution consisting of

6 M guanidinium hydrochloride (or 4 M guanidinium thiocyanate)

5 mM sodium citrate (pH 7.0)

0.5% sarcosyl

10 mM β-ME (add just prior to use).

Gentle vortexing may facilitate dissociation of the pellet.

Note 1: Guanidinium hydrochloride and guanidinium thiocyanate may be used interchangeably in the formulation of this buffer.

Note 2: Guanidinium thiocyanate solution may be stored at 4°C for several days, prior to the addition of β-ME. Immediately prior to use, warm a suitable aliquot of the guanidinium thiocyanate solution to 37°C, cool to room temperature, and then add β-ME to a final concentration of 200 mM. If desired, the solution can be clarified by filtration through a 0.2-μm filter.

Note 3: Be sure to check for guanidine and β-ME incompatibility with ultracentrifuge tubes.

4. Disrupt the cells or tissue at room temperature by either manual (repeatedly drawing the lysate through a 19-gauge needle) or mechanical (e.g., polytron or Dounce) homogenization. In either case, shearing forces are desirable to physically break up genomic DNA to preclude the formation of an impenetrable web of high-molecular-weight DNA that would otherwise impede the sedimentation of RNA.

5. If desired, add 1 g of solid CsCl for every 2.5 ml of lysate. This is required only when other molecules are to be copurified from the same gradient along with the RNA.

6. Pipet a cushion of 5.7 M CsCl (dissolved in 100 mM EDTA, pH 7.5) into an ultracentrifuge tube, typically cellulose nitrate or polyallomer, compatible with the swinging bucket rotor to be used. The 5.7 M CsCl cushion should occupy about 25% of the total volume of the tube.

7. Gently layer the lysate on top of the CsCl cushion until the ultracentrifuge tube is filled. Be sure to weigh each centrifuge tube along with its swinging bucket and corresponding top. Each sample (bucket, screw top, and centrifuge tube) should weigh *exactly* the same as the other samples to be loaded onto the rotor, within 0.1 g of each other. Follow the manu-

facturer's instructions for sealing the tubes. It is worth weighing each tube again before loading the rotor.

8. Centrifuge the preparation in a swinging bucket rotor. Floor model ultracentrifuges generally require an overnight run, while microultracentrifuge models require only 3–4 h. Each rotor instruction guide will delineate the correct rcf ($\times$ *g*) needed to purify RNA safely using this type of centrifugation.

Note: Cooling the centrifuge chamber at any time during the run will result in precipitation of the CsCl and ruin the gradient. Perform this type of centrifugation at ambient temperature.

9. At the conclusion of the run, carefully remove the supernatant by aspiration (sterile pasteur pipettes, or sterile 5-ml pipettes with the tip removed, work well). Use great care not to disrupt the pellet at the bottom of the tube. Many investigators prefer to remove 80% of the gradient by pipetting and the remainder by decanting. After the supernatant has been removed, it is useful to cut off the top portion of the ultracentrifuge tube just above the RNA pellet. In this lab, a sterile razor blade is clamped in a hemostat, briefly heated with a Bunsen burner, and used to slice through the emptied centrifuge tube. In so doing, the RNA pellet will appear to be resting in a giant contact lens-shaped cradle (Figure 7).

Note: This technique will help prevent contamination of the RNA pellet with protein and/or DNA that may be clinging to the side of the ultracentrifuge tube, even after removal of the gradient. The translucent RNA pellet is located at the bottom of the tube.

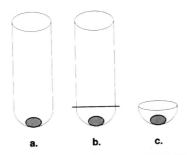

a.   **b.**   **c.**

FIGURE 7

Method for recovery of RNA following isopycnic centrifugation. (a) Appearance of RNA pellet after removal of gradient. (b) Ultracentrifuge tube is cut just above the RNA. (c) RNA rests in contact lens-shaped cradle. This approach favors recovery of RNA free of DNA and protein contamination from the inside walls of the tube.

10. Dissolve the RNA pellet completely in the smallest possible volume of

10 mM Tris, pH 7.4

5 mM EDTA

0.1% SDS (optional)

and then transfer the sample to a sterile microfuge tube. For large-scale RNA extractions, transfer the RNA suspension to a conical 15-ml polypropylene tube.

Note: The RNA pellet is often difficult to dissolve in aqueous buffer after this type of purification, and the larger the sample, the more difficulty will be encountered. Freezing the pellet at $-20°C$ and then thawing usually facilitates dissolving it and removal from the centrifuge tube. Limited vortexing may help.

11. Extract the sample once with 2 volumes of chloroform:isoamyl alcohol (4:1, v/v) or with 2 volumes of chloroform:butanol (4:1, v/v). Pulse centrifuge to separate the two phases. Transfer the aqueous phase to an autoclaved microfuge tube or an acid-washed, autoclaved 15-ml Corex tube.

12. Reextract the organic phase, interface with an equal volume of

10 mM Tris, pH 7.4

5 mM EDTA

0.1% SDS (optional)

and combine the two aqueous phases.

Note: These chloroform extractions are needed to remove residual contaminants from the RNA preparation.

13. Precipitate the RNA for several hours at $-20°C$, with the addition of 0.1 volume 3 M sodium acetate (pH 5.2) and 2.5 volumes of ice-cold 95% ethanol.

14. If the RNA was precipitated in a microfuge tube, collect the precipitate at 12,000g for 10 min, at 4°C if possible. If precipitation was carried out in a Corex glass tube, centrifuge in a suitable adapter for at least 20 min at 7000 rpm.

CAUTION **Be certain not to exceed the maximum g-force rating for the rotor or the type of tube in which the RNA was precipitated.**

15. Carefully decand the supernatant and wash the pellet at least once with 70% ethanol. Decant the 70% ethanol wash and, if desired, wash the

sample once with 95% ethanol to accelerate the drying of the sample. It is wise not to allow the pellet to dry out completely.

16. The RNA is best stored at −70°C as an ethanol precipitate or, following quantification, stored in suitable aliquots at −70°C to avoid repeated freezing and thawing. Depending on the cell type, typical yields range from 50 to 75 μg of very high-quality cellular RNA per 10^6 cells.

Protocol: Cesium Trifluoroacetate Gradients

This protocol is a modification of the procedure of Okayama *et al.* (1987). Following it as presented here will result in the pelleting of RNA. It is recommended for the preparation of total RNA from a variety of tissues and can be modified to accommodate as little as 25 mg or as much as 6 g of starting material.

Advance Preparation of CsTFA. CsTFA (Pharmacia Biotech No. 17-0847-02) is prepared as a working solution with a density of 1.51 ± 0.01 g/ml, prepared in 100 mM EDTA, pH 7.0. To prepare 100 ml of this solution, mix $51/(\rho - 1)$ ml of CsTFA (where ρ = density of product as received) with 40 ml of 250 mM EDTA, pH 7.0, and $60 - [51/(\rho - 1)]$ ml of sterile distilled water. The density of the working solution of CsTFA may be altered to fit a particular experimental protocol.

1. **Wear gloves!**

2a. For tissue culture: Collect harvested cells by centrifugation at 500g for 5 min. Decant as much of the supernatant as possible. Wash cells once with PBS (per liter: 8.0 g NaCl, 0.2 g KCl, 1.44 g Na_2HPO_4, 0.24 g KH_2PO_4). Decant as much PBS as possible. If the RNase is likely to be a major challenge, perform centrifugations at 4°C and wash cells with ice-cold PBS.

2b. For tissue: Rapidly harvest the tissue and mince/dice in ice-cold saline solution. Quickly transfer tissue into a suitable homogenization vessel that already contains an aliquot of lysis buffer (Step 3). If tissue was flash-frozen in liquid nitrogen, it may be helpful to shatter it (while wrapped in aluminum foil) by impact upon the lab bench. In so doing, the sample can be quickly transferred and allowed to thaw in the lysis buffer.

3. For each gram of tissue or cell pellet, add 18 ml of 5.5 M guanidinium thiocyanate solution (5.5 M guanidinium thiocyanate, 25 mM sodium citrate, 0.5% sodium laurylsarcosine; adjust pH to 7.0 with NaOH; add 200 mM β-ME just prior to use).

Note 1: Guanidine thiocyanate and guanidinium hydrochloride may be used interchangeably in the formulation of this buffer.

Note 2: Guanidinium thiocyanate solution may be stored at 4°C for several days, prior to the addition of β-ME. Immediately prior to use, warm a suitable aliquot of the guanidinium thiocyanate solution to 37°C, cool to room temperature, and then add β-ME to a final concentration of 200 mM. If desired, the solution can be clarified by filtration through a 0.2-μm filter.

Note 3: Be sure to check for guanidine and β-ME incompatibility with ultracentrifuge tubes.

4. Disrupt the cells or tissue at room temperature by either manual (repeated shearing through a 19-gauge needle) or mechanical (e.g., polytron or Dounce) homogenization.

Note: For optimal recovery of undergraded sample, avoid generating excess heat or foam during this step.

5. Carefully transfer the lysate to disposable conical-bottom centrifuge tubes (e.g., 50-ml size). Centrifuge at 1500g for 5 min to sediment insoluble materials.
6. Without disturbing the pelleted material, transfer the clarified lysate to a fresh tube.
7. Shear the DNA by drawing the lysate through an 18-gauge needle repeatedly until the viscosity decreases. Between 10 and 12 passes into and out of the syringe is usually adequate.
8. Centrifuge at 5000g for 20 min at 15°C to remove the cell debris. Decant and retain the supernatant, taking great care not to dislodge the pelleted debris.

CAUTION   Do not exceed the maximum *g*-force recommended for any of the centrifuge tubes used in this protocol.

9. If necessary, adjust the volume of the supernatant to the desired sample-loading volume (Table 3) using 5.5 *M* guanidinium thiocyanate solution (see Step 3).
10. Select a rotor suitable for the number and volume of samples. Prepare appropriately sized cushions of CsTFA in these tubes (Tables 4 and 5) and then overlay aliquots of the sample onto these cushions.
11. Centrifuge at 125,000g in a swinging bucket rotor for 16–20 h at 15°C. Under these conditions RNA will pellet at the bottom of the tube and DNA will collect in a band in the lower third of the gradient that forms.
12. After centrifugation, carefully aspirate off most of the liquid in each tube, stopping as soon as the DNA band has been removed. If the DNA band is not visible, discontinue aspiration within 1 cm from the bottom of

Selection of Appropriate Volumes, Tubes, Rotors, and Speeds for CsTFA RNA Purification[a]

	Beckman rotor		
	SW 50.1	SW 28.1	SW 28
Mass of tissue per tube			
Minimum (mg)	25	100	400
Maximum (mg)	140	450	1000
CsTFA per tube (ml)	2.7	8.5	19.0
Sample loading per tube (ml)	2.5	8.0	18.0
RPM required for 125,000g	31,000	25,000	25,000

[a]Data courtesy of Pharmacia Biotech. For physical specifications of additional rotors, see Tables 4 and 5.

the tube. Decant the remaining liquid, taking care not to disturb the RNA pellet. Invert the tubes and allow them to drain on a paper towel for 5 min.

13. Cut off the bottoms of the tubes with a sterile razor blade or scalpel, as shown in Figure 7. Leave a sufficient portion of the sides of the tube to form a small cup. Place tube bottoms on ice.

14. Dissolve RNA pellets directly in the tube bottoms using two

Physical Specifications of Sorvall Preparative Swinging Bucket Ultracentrifuge Rotors[a]

Rotor	Maximum speed (rpm)	Maximum force (g)	k Factor	No. of tubes × nominal volume	Nominal rotor capacity (ml)	Rotor radius (mm) r_{max}
RP55-S	55,000	258,826	43.7	4 × 2.2	8.8	76.6
S55-S	55,000	258,826	43.7	4 × 2.2	8.8	76.6
TH-660	60,000	488,576	44.4	6 × 4.4	26.4	121.5
AH-650	50,000	296,005	53.0	6 × 5.0	30	106.0
TH-641	41,000	287,660	114.0	6 × 13.2	79.2	153.2
SureSpin 630 (17 ml)	30,000	166,880	255	6 × 17	102	166.0
SureSpin 630 (36 ml)	30,000	166,880	216	6 × 36	216	166.0
AH-629 (17 ml)	29,000	155,846	284.0	6 × 17	107	165.0
AH-629 (20 ml)	29,000	121,464	176.0	6 × 20	120	129.3
AH-629 (36 ml)	29,000	151,243	242.0	6 × 36	216	161.0

[a]Data courtesy of Sorvall, Inc.

T A B L E   5

Physical Specifications of Beckman Preparative Swinging Bucket
Ultracentrifuge Rotors[a]

Rotor	Maximum speed (rpm)	Maximum force (g)	k Factor	No. of tubes × nominal volume	Nominal rotor capacity (ml)	Rotor radius (mm) r_{min}	Rotor radius (mm) r_{max}
SW 65Ti	65,000	421,000	46	3 × 5.0	15	41.2	89.0
SW 60Ti	60,000	485,000	45	6 × 4.4	26.4	63.1	120.3
SW 55Ti	55,000	369,000	49	6 × 5.0	30	61.0	109.0
SW 50.1	50,000	300,000	59	6 × 5.0	30	59.7	107.3
SW 41Ti	41,000	286,000	125	6 × 13.2	79.2	67.0	152.0
SW 40Ti	40,000	285,000	137	6 × 14	84	66.7	158.8
SW 30.1	30,000	124,000	138	6 × 8	48	75.3	123.0
SW 30	30,000	124,000	138	6 × 20	120	75.3	123.0
SW 28.1	28,000	150,000	276	6 × 17	102	72.9	171.3
SW 28	28,000	141,000	245	6 × 38.5	231	75.3	161.0
SW 25.2	25,000	107,000	335	3 × 60	180	66.7	152.3
SW 25.1	25,000	90,400	338	3 × 34	102	56.2	129.2

[a]Data courtesy of Beckman Instruments, Inc.

aliquots of an appropriate buffer (DEPC-treated H_2O or TE, pH 7.4 (10
mM Tris–Cl, pH 7.4; 1 mM EDTA).

*Note: At least two small aliquots (100–200 μl) of buffer are recom-
mended to recover all of the RNA sample from the tube bottoms. Repeated pipetting may be necessary for complete recovery. As always
keep the sample as concentrated as possible so that all manipulations,
including reprecipitation (if necessary), can be carried out in a microfuge tube.*

15. Combine both aliquots of RNA and vortex. Heat tubes to 65°C for
10 min and vortex again.

16. Spin tubes briefly and at low speed to remove any insoluble material.

17. Determine RNA concentration (this chapter). If desired, RNA can
be precipitated to concentrate the RNA sample and/or change buffers.

18. Store RNA in suitable aliquots at –80°C until ready for use.

Guanidinium–Acid–Phenol Extraction Techniques

The highest quality RNA indisputably results from the extraction of
RNA mediated by chaotropic lysis buffers, and guanidinium-containing

buffers are among the most effective. As described above, these isolation procedures have the advantage of disrupting cells grown in culture or whole tissue samples rapidly and completely, simultaneously inactivating RNase activity, even when it is present in great abundance. The original procedures are labor-intensive because of the inclusion of an ultracentrifugation step to separate RNA from DNA based on differences in buoyant density. In contrast, Chomczynski and Sacchi (1987) describe the isolation and purification of undegraded RNA by treatment of cells with guanidinium thiocyanate-containing lysis buffers, but without the need for subsequent CsCl ultracentrifugation of the sample. In this and related procedures, the RNA is isolated in a very short time by extraction of a guanidinium cell or tissue lysate with an acidic phenol solution; chloroform is added to facilitate partitioning of the aqueous and organic material. Although such extraction buffers are easily prepared in the molecular biology laboratory by mixing water-saturated phenol with an acidic solution of sodium acetate, premixed monophasic formulations of phenol and guanidinium thiocyanate (e.g., Trizol and RNAzol) are readily available from a variety of suppliers of molecular biology reagents; these and others have improved on the original published protocol. Upon phase separation, RNA is retained in the aqueous phase, while DNA and proteins partition into the protein interface and organic phase. Then RNA is recovered by precipitation with isopropanol and collected by centrifugation. In this procedure, RNA can be efficiently isolated from as little as 1 mg of tissue or 10^6 cells, usually in less than 1 h.

Protocol: Guanidinium–Acid–Phenol Extraction

1. **Wear gloves!**

2a. For tissue: Mince freshly isolated tissue (up to 100 mg) on ice. For each 0.1 mg of tissue add 1 ml of solution D:

> 4 M guanidinium thiocyanate
>
> 25 mM sodium citrate, pH 7.0
>
> 0.5% sarcosyl
>
> 100 mM 2-mercaptoethanol

and homogenize at room temperature in a glass–Teflon homogenizer.

2b. For tissue culture: Collect harvested cells by centrifugation and resuspend cell pellets in 100 μl solution D per 10^6 cells.

Note: Cells may be lysed directly by the addition of 1.0–1.5 ml of solution D per 100 mm tissue culture dish. Lysis in the culture dishes and flasks usually mandates the use of larger reagent volumes, and it may prove difficult to recover all of the lysate due to its tremendous viscosity. In such an event, a sterile cell scraper (Corning Costar) may be helpful. Scaled down, the entire procedure can be performed in a 2.2-ml microfuge tube.

3. Transfer the homogenate to a polypropylene tube. For each 1 ml of solution D lysis buffer used in Step 2, add

0.1 ml 2 *M* sodium acetate, pH 5.2

1.0 ml water-saturated phenol (molecular biology grade)

0.2 ml chloroform:isoamyl alcohol (49:1)

Cap tube and mix carefully and thoroughly by inversion following the addition of each reagent and invert vigorously for an additional 30 s after all reagents have been added.

4. Cool sample on ice for 15 min; centrifuge at 4°C to separate the phases.

5. Transfer aqueous phase (containing RNA) to a fresh tube and mix with 0.75 volume of ice-cold isopropanol. Store at –20°C for at least 1 h to precipitate RNA.

Note: For larger scale preparations, Corex glass tubes, placed in the correct adapters, can be used for precipitation and recovery of RNA.

6. Collect precipitate by centrifugation at 10,000*g* for 20 min at 4°C. Carefully decant and discard supernatant.

CAUTION   Do not exceed the maximum *g*-force for any of the tubes used in this protocol.

7. Completely dissolve RNA pellet in 300 µl of solution D (see Step 2a) and then transfer to a RNase-free 1.7 ml microfuge tube.

8. Reprecipitate the RNA by the addition of 0.75 volume of ice-cold isopropanol and store at –20°C for 1 h.

9. Collect precipitate at top speed in a microcentrifuge for 10 min at 4°C. Carefully decant and discard supernatant.

10. Wash pellet 3–4 times in 500 µl 70% ethanol. If the RNA does not dislodge during these washes, there is no need to recentrifuge. Allow tubes to air-dry to remove residual ethanol.

11. Redissolve RNA in the smallest possible volume of TE buffer or DEPC-treated H$_2$O. Incubation at 65°C for 10 min may facilitate solubilization, though this is unnecessary if the RNA did not dry out completely following the ethanol washes. Store RNA as an ethanol precipi-

tate until ready to use. Following determination of concentration, store the RNA in suitable aliquots at –70°C. Avoid repeated freezing and thawing.

Protocol: RNA Isolation from Adipose Tissue

The isolation of RNA from adipose tissue is a simple matter, being accomplished most easily through the use of guanidinium buffers to disrupt the tissue architecture. Perhaps the most cumbersome issue related to this method is removal of the preponderance of lipids stored in this type of tissue. In general, cells are lysed with guanidinium buffer, followed by a brief, low-speed centrifugation during which the lipid layer floats to the top of the tube. The lipid layer is then aspirated away. This low-speed centrifugation is then repeated again, as needed. Alternatively, one may begin with the addition of chloroform directly to the lysate, in order to draw the lipids into the organic phase. RNA isolation then continues with the addition of acidic phenol and chloroform, followed by the purification of the RNA.

1. **Wear gloves!**
2a. For tissue: Mince freshly isolated tissue (up to 100 mg) on ice. For each 0.1 mg of tissue add 1 ml of solution D:

$4\ M$ guanidinium thiocyanate

25 mM sodium citrate, pH 7.0

0.5% sarcosyl

100 mM 2-mercaptoethanol

and homogenize at room temperature in a glass–Teflon homogenizer. Add an equal volume of chloroform and then pulse centrifuge to separate the phases. Remove and discard the chloroform (lower) phase. Add a fresh volume of chloroform, pulse centrifuge to separate the phases, and then transfer the upper aqueous phase to a fresh tube.

2b. For tissue culture: Collect harvested cells by centrifugation and resuspend cell pellets in 100 μl solution D per 10^6 cells.

Note: Cells may be lysed directly by the addition of 1.0–1.5 ml of solution D per 100-mm tissue culture dish. Lysis in the culture dishes and flasks usually mandates the use of larger reagent volumes, and it may prove difficult to recover all of the lysate due to its tremendous viscosity. In such an event, a sterile cell scraper (Corning Costar) may be helpful. Scaled down, the entire procedure can be performed in a 2.2-ml microfuge tube.

3. Transfer the homogenate to a polypropylene tube. For each 1 ml of solution D lysis buffer used in Step 2, add:

0.1 ml 2 M sodium acetate, pH 5.2

1.0 ml water-saturated phenol (molecular biology grade)

0.2 ml chloroform:isoamyl alcohol (49:1)

Cap tube and mix carefully and thoroughly by inversion following the addition of each reagent and invert vigorously for an additional 30 s after all reagents have been added.

4. Cool sample on ice for 15 min.

5. Centrifuge to separate the phases; for microfuge tubes use 12,000g and do so at 4°C, if possible.

6. Transfer aqueous phase (containing RNA) to a fresh tube and mix with 0.75 volume of ice-cold isopropanol. Store at –20°C for at least 1 h to precipitate RNA.

Note: For larger scale preparations, Corex glass tubes, placed in the correct adapters, can be used for precipitation and recovery of RNA.

7. Collect precipitate by centrifugation at 10,000g for 10 min at 4°C. Carefully decant and discard supernatant.

CAUTION   **Do not exceed the maximum g-force for any of the tubes used in this protocol.**

8. Completely dissolve RNA pellet in 300 μl of solution D (see Step 2a) and then transfer to a RNase-free 1.7-ml microfuge tube.

9. Reprecipitate the RNA by the addition of 0.75 volume of ice-cold isopropanol and store at –20°C for 1 h.

10. Collect precipitate at top speed in a microcentrifuge for 10 min at 4°C. Carefully decant and discard supernatant.

11. Wash pellet 3–4 times in 500 μl 70% ethanol. If the RNA does not dislodge during these washes, there is no need to recentrifuge. Allow tubes to air-dry to remove residual ethanol. Optional: wash sample once more with 95% ethanol to accelerate the drying process.

12. Redissolve RNA in the smallest possible volume of TE buffer or DEPC-treated H_2O. Incubation at 65°C for 10 min may facilitate solubilization, though this is unnecessary if the RNA did not dry out completely following the ethanol washes. Store RNA as an ethanol precipitate until ready to use. Following determination of concentration, store the RNA in suitable aliquots at –70°C. Avoid repeated freezing and thawing.

Protocol: Rapid Isolation of RNA with SDS and Potassium Acetate Reagents

This procedure (Peppel and Baglioni, 1990; Salvatori *et al.,* 1992; Zolfaghari *et al.,* 1993) exploits the ability of the ionic detergent SDS to inhibit RNase activity. It is a very rapid method for isolating total cellular RNA without guanidinium buffers or isopycnic centrifugation, and it is suitable for the isolation of both eukaryotic and prokaryotic RNA. When RNA is to be isolated from tissue, it is best to couple this method with some form of mechanical disruption.

Most of the reagents required are commonly found in labs performing even the most rudimentary molecular biology procedures. The original procedure, a modification of which is presented here, describes the isolation of RNA from cells plated into six-well cluster dishes (9.3 cm^2). The volumes can be scaled up to accommodate larger extractions, and one may expect to harvest approximately 100 μg RNA per 10^6 cells.

1. **Wear gloves!**

2a. For tissue culture: Remove growth medium from the tissue culture vessel containing cells of interest and rinse with PBS (per liter: 8.0 g NaCl, 0.2 g KCl, 1.44 g Na_2HPO_4, 0.24 g KH_2PO_4). Remove PBS by aspiration and discard. Lyse cells directly on tissue culture plastic with lysis buffer (2% SDS; 200 mM Tris–Cl, pH 7.5, 0.5 mM EDTA), using 50 μl lysis buffer/cm^2. Gently agitate the culture vessel on the bench top for 2 min to ensure complete lysis. This lysis buffer can be prepared ahead of time and stored at room temperature.

2b. For tissue: *Rapidly* harvest the tissue and mince/dice in ice-cold saline solution. Quickly transfer tissue into a suitable homogenization vessel that already contains an aliquot of lysis buffer (Step 2a). If tissue was flash-frozen in liquid nitrogen, it may be helpful to shatter it (while wrapped in aluminum foil) by impact upon the lab bench. In so doing, the sample can be quickly transferred and allowed to thaw in the lysis buffer. Use 1 ml lysis buffer for each milligram of tissue. Disrupt the cells or tissue at room temperature by either manual (repeated shearing through a 19-gauge needle) or mechanical (e.g., polytron or Dounce) homogenization and avoid generating excess heat or foam during this step.

3. Transfer the lysate to a sterile microfuge tube (1.7- or 2.2-ml size). Cap and invert tube sharply 15–20 times.

4. For each 500 μl lysis buffer used in Step 2, add 150 μl of potassium acetate solution (50 g potassium acetate; 11 ml glacial acetic acid; H_2O to 100 ml). This solution can be prepared ahead of time and stored at $-20°C$.

5. Carefully invert the tube sharply 15–20 times to ensure thorough mixing.

6. Incubate on ice for at least 3 min but no longer than 5 min.

7. Centrifuge sample for 5 min at room temperature at maximum speed in a microcentrifuge.

8. Carefully recover the supernatant and transfer to a fresh tube.

9. Extract the supernatant with 300 μl of a mixture of chloroform:isoamyl alcohol (24:1). Centrifuge for 30 s to separate the phases. Transfer the aqueous (upper) phase to a fresh tube without disturbing the protein interface and then repeat the extraction.

10. Transfer the aqueous (upper) phase to a fresh microfuge tube prechilled on ice.

11. Precipitate the RNA with the addition of an equal volume (approximately 650 μl) ice-cold isopropanol. Incubate on ice or at –20°C for 20–30 min.

12. Collect the RNA precipitate by centrifugation for 5 min at room temperature.

13. Carefully decant the supernatant; wash the cell pellet 2–3 times with 70% ethanol. If the pellet becomes dislodged, centrifuge again. Decant and discard the ethanol, and allow the sample to air-dry. If desired, a final wash with 95% ethanol may facilitate air-drying.

14. Dissolve RNA in the smallest possible volume of TE buffer or DEPC-treated H_2O. Determine RNA concentration and store sample in suitable aliquots at –70°C until further use.

Protocol: Isolation of Prokaryotic RNA

This rapid and efficient protocol for the isolation of bacterial RNA is a modification of the procedures of Peppel and Baglioni (1990), Salvatori *et al.* (1992), and Zolfaghari *et al.* (1993). Be sure to use the lysis buffer described below; *never* use the standard NaOH/SDS lysis buffer used to disrupt bacterial cells for plasmid preps, as this buffer will cause rapid and extensive hydrolysis of the RNA!

1. **Wear gloves!**

2. In advance: Prepare the NP-40 lysis and potassium acetate buffers, and store them at 4°C until ready to use. Although these buffers have a shelf life of several months, they should be inspected prior to use for the growth of opportunitistic prokaryotes.

NP-40 buffer:

250 mM sucrose

20 mM EDTA, pH 8.0

autoclave

0.75% NP-40

Potassium acetate buffer:

3 M potasium acetate (KCH$_3$O$_2$) in 80% of final volume

pH to 5.5 with glacial acetic acid

add remaining H$_2$O

add 100 g guanidinium thiocyanate (GTC) per 100 ml

mix carefully and thoroughly

3. Harvest bacterial cells from a 100 ml culture by centrifugation. Completely remove supernatant.

4. Combine 1.5 ml of ice-cold lysis buffer with 2 ml ice-cold potassium acetate buffer (Step 2) and then gently resuspend the cell pellet in this mixture. Allow lysate to incubate on ice for 10 min.

5. Centrifuge at 10,000g for 10 min at 4°C, if possible. Be sure to use centrifuge tubes rated for this g-force. It may be helpful to divide the lysate into several microcentrifuge tubes so that the investigator always has access to several aliquots of RNA made from the same culture on the same day.

6. Transfer the supernatant to a fresh tube, taking care not to transfer any of the precipitated solids. It is far better to leave a small amount of supernatant behind than to risk carryover of cell debris. Incubate supernatant on ice for an additional 15 min.

7. Centrifuge at 10000g for 10 min at 4°C, if possible.

8. Transfer the supernatant to a fresh polypropylene tube, and the extract with an equal volume of chloroform:isoamyl alcohol (24:1). Mix *carefully* and thoroughly. Centrifuge for 3 min to separate the phases. If a significant band appears between the upper and lower phases, repeat the extraction with a fresh aliquot of chloroform:isoamyl alcohol (24:1).

9. Transfer the upper aqueous phase to a fresh tube, and then add 0.6 volume of ice-cold isopropanol. Store sample at –20° for 20 min to precipitate the RNA.

10. Collect RNA by centrifugation at 10,000g for 10 min at 4°C, if possible.

11. *Carefully* decant supernatant and then wash pellet with several aliquots of 70% ethanol (diluted in sterile H$_2$O).

12. Allow sample to air-dry and then resuspend in TE buffer or sterile H$_2$O. After determination of RNA concentration and purity, store in suitable aliquots at –80°C. If RNA is not to be used right away, store as an ethanol precipitate at –80°C.

Protocol: Simultaneous Isolation of RNA and DNA

Many noteworthy procedures have been described for the isolation of RNA from cells and tissues representing varying degrees of complexity. Many companies now offer package-type devices, reagents, solutions, or resins for the *"painless"* isolation of RNA from numerous biological sources. These products vary widely in their simplicity, speed, cost, and effectiveness. Several procedures have also been described for the simultanous recovery of RNA, DNA, and protein from the same biological source (Majumdar *et al.,* 1991; Chomczynski, 1993) and these, too, are now available commercially.

The method presented here has the advantage of requiring no specialized equipment other than what is found even in the most modest of molecular biology labs. At the heart of this technique is the selective partitioning of RNA and DNA by adjusting the pH of a phenolic lysate as described previously (Chomczynski and Sacchi, 1987). Succinctly, the RNA is recovered in the aqueous phase by first extracting the sample at acidic pH; DNA in the sample is retained in the interface and organic phase. Following transfer of the RNA-containing aqueous phase to a fresh tube, the DNA is eluted from the interface and organic phase by establishing an alkaline pH. This is known as reverse extraction. In this procedure, the investigator is afforded the option of making the phenolic lysate acidic by the addition of sodium acetate, pH 4.9, or simply extracting the SDS/EDTA lysate (pH 8.7) with an equal volume of phenol:chloroform:isoamyl alcohol (25:24:1), the overall pH of which is approximately 7.6.

Cells may be lysed directly in the tissue culture vessel or harvested first. Cell pellets should be gently dispersed prior to addition of the lysis buffer because shearing of the DNA is likely to occur when an aggressive attempt is made to resuspend a cell pellet *following* addition of the lysis buffer. In this regard the investigator is strongly urged not to succumb to the temptation to vortex this or any lysate from which genomic DNA is to be purified, as the enormous shearing forces created will damage DNA in a very short time. The size distribution of the resulting "fragments" is likely to be unacceptably low.

1. **Wear gloves!**

2. In advance: Prepare phenol:chloroform:isoamyl alcohol extraction buffer as described in this chapter, with the exception that the phenol should be *water*-saturated 2–3 times prior to the addition of other reagents.

3a. For tissue culture: Remove growth medium by aspiration and wash the monolayer twice with 5-ml aliquots of ice-cold PBS (per liter: 8.0 g NaCl, 0.2 g KCl, 1.44 g Na_2HPO_4, 0.24 g KH_2PO_4). Maintain culture dishes on ice throughout. Remove, by aspiration, as much PBS as possible. Optional: Prepare cell cultures for lysis by cooling them directly on a bed of ice for 30 min.

Note: Cooling reagents and cells on ice is an excellent method for controlling RNase activity. It is the responsibility of the investigator, however, to empirically determine whether a 30-min incubation on ice prior to lysis influences the cellular biochemistry and accompanying interpretation of the data.

3b. Add 2 ml of ice-cold lysis buffer (10 mM EDTA, pH 8.0; 0.5% SDS) per 100-mm tissue culture dish (4–5×10^6 cells). Tilt the dish gently from side to side to ensure lysis of all cells. Incubate on ice for 3 min. If desired, the lysate can be transferred to a second dish, followed by incubation for an additional 3 min. Gentle agitation throughout may help. Do not use an aliquot of lysis buffer for more than two 100-mm tissue culture dishes or the equivalent. Cells that were harvested from tissue culture should be lysed using 200 µl lysis buffer per 10^6 cells.

3c. For tissue: Mince freshly isolated tissue (up to 100 mg) on ice. For each milligram of tissue, add 250 µl (10 mM EDTA, pH 8.0; 0.5% SDS) lysis buffer and gently Dounce homogenize at room temperature. Do not use a polytron to homogenize, as this will cause unacceptable shearing of the genomic DNA to be recovered later in this protocol.

4. Transfer the lysate into a prechilled 15 ml polypropylene tube on ice. When scaling down, a polypropylene microfuge may be satisfactory.

Note: Due to the viscous nature of liberated chromatin, a sterile cell scraper (e.g., Corning Costar) may be useful for the quantitative recovery of the lysate.

5. Optional: Rather than scraping the cell lysate to transfer it from plate to plate, the investigator has the option of rinsing each plate with the same 2-ml aliquot of a solution of 10 mM EDTA, 100 mM sodium acetate, pH 4.9, or, to keep the lysate more concentrated, with a 1-ml aliquot of a solution of 10 mM EDTA, 0.15 M sodium acetate, pH 4.9.

6. Add an equal volume of cold organic extraction buffer, prepared as described in Step 2. Be sure to check for centrifuge tube compatibility with phenol and chloroform.

7. Carefully mix the tube(s) by gentle inversion several times. Do not vortex! The yield will improve dramatically if the tubes are maintained on ice at all times when not mixing.

8. Centrifuge samples in an appropriate rotor in order to separate the aqueous material from the organic material.

CAUTION   Be careful not to exceed the recommended rotor speed or the maximum g-force recommended for any of the tubes used in this protocol.

Recovery of RNA

9. Collect the aqueous (upper) phase and transfer to a fresh tube, prechilled and resting on ice. Take care to avoid disrupting the protein interface. Recovery of DNA, which must be eluted from the interface and organic phase, can be delayed until all steps pertaining to RNA recovery have been completed. Leave tubes on ice until then.

10. Reextract the aqueous phase with a fresh, equal-volume aliquot of the cold organic mixture described in Step 2.

11. Centrifuge as described above. Carefully recover the aqueous phase and transfer to a fresh tube, taking care to avoid disrupting the interface or organic material.

12. Add 500 μl of ice-cold 1 M Tris–Cl, pH 8.0, and 200 μl of 5 M NaCl for every 4 ml of RNA-containing aqueous material. Mix carefully and thoroughly. Add 12 ml (or 2.5 volumes) of ice-cold 95% ethanol and mix thoroughly. Store at –20°C for at least 1 h.

13. Collect RNA precipitate by centrifugation at 10,000g for 10 min, assuming that the tube in which the RNA is precipitated can withstand such speeds. Centrifugation at 4°C is preferred, though not essential. Carefully decant and discard the supernatant.

14. Wash pellet 2–3 times with 70% ethanol. Centrifuge again to recover the precipitate. Carefully decant and discard supernatant.

15. Allow sample to air-dry. If desired, a final wash with 95% ethanol may accelerate air-drying.

16. Dissolve the RNA pellet in 200 μl of ice-cold TE buffer (10 mM Tris–Cl, pH 8.0; 1 mM EDTA). Incubate on ice for 1 h and then transfer to a microfuge tube. A second aliquot of ice-cold TE buffer may be used to rinse the centrifuge tube and should be combined with the first aliquot.

17. Add 6 μl of 5 M NaCl and 0.8 ml of ice-cold ethanol for every 300 μl RNA solution. Incubate on ice for 5–10 min, after which the solution should appear milky, manifesting the precipitation of RNA. The sample can now be stored as an ethanol precipitate at –80°C or processed immediately.

18. Collect RNA precipitate by centrifugation in a microfuge for 10 min, preferably at 2–4°C. Carefully decant the supernatant and dry to remove excess ethanol as described in Step 14.

19. Resuspend the RNA pellet(s) in a minimum volume of the desired

buffer (DEPC-treated water, for example) and allow 15 min on ice for the sample to redissolve. Depending on the degree to which the sample has dried, longer incubation on ice may be necessary to completely redissolve the sample.

20. Calculate the concentration of RNA. Store hydrated samples in suitable aliquots at –80°C.

Recovery of DNA

21. Return to the tubes containing organic material saved from Step 9. Remove any remaining aqueous material as completely as possible. If desired, remove a small amount of the interface to facilitate removal of any contaminating RNA. This is usually unnecessary because residual RNA is likely to be hydrolyzed in the steps that follow. Alternatively, the purified DNA may be incubated with RNase (Appendix D) to remove all contaminating RNA.

22. To the remaining organic/interface material add an equal volume of 1 M Tris base solution. This solution should not be pH adjusted (pH ≈ 10.5). Mix thoroughly by gentle shaking. Do not vortex.

Note: This step is known as reverse extraction; creating a strongly alkaline environment will draw the DNA into the aqueous phase.

23. Centrifuge samples at 2000g for 15 min, at 4°C if possible, in order to separate the aqueous material from the organic material. Carefully transfer the aqueous (upper) phase to a fresh tube.

CAUTION   Be careful not to exceed the recommended rotor speed or the maximum g-force recommended for any of the tubes used in this protocol.

24. Repeat Steps 22 and 23, but do not mix the two aqueous phases. Keep them in separate centrifuge tubes.

25. To each Tris:DNA aqueous sample, add an equal volume of chloroform:isoamyl alcohol (24:1). Mix carefully and thoroughly. Pulse centrifuge samples to separate the phases (recall that chloroform, in the absence of phenol, settles out very quickly). Carefully transfer the aqueous (upper) phase to a fresh tube.

26. Precipitate the DNA with the addition of 0.1 volume of 3 M sodium acetate, pH 7.0, and 2.5 volumes of 95% ethanol. Mix thoroughly, first by gentle swirling and then by thorough inversion. Do not vortex.

Note: Adding ethanol should cause the immediate precipitation of the genomic material, which, characteristically, should appear quite

stringy. DNA so precipitated may be collected by low-speed centrifuga-
tion (500g) for 2–3 min. Alternatively, genomic DNA can be "fished"
out of the ethanol solution using a silanized Pasteur pipet (Appendix F)
such that there is minimal sticking of the DNA to the glass.

27. Transfer the precipitated DNA to a fresh microfuge tube. Wash the
DNA with 2–3 500-µl aliquots of 70% ethanol to remove excess salt. Allow DNA to redissolve in a suitable buffer (TE buffer, pH 8.0) and then
determine DNA concentration.

Protocol: Isolation of RNA from Yeast

Yeast product a cell wall that acts as a formidable impediment to rapid
recovery of RNA from these eukaryotic cells. Many of the classical methods for the isolation of RNA from yeast involve vortexing yeast cultures
in the presence of glass beads, a technique intended to compromise severely the ultrastructure of these cells in a nonenzymatic manner.[12] This
approach is tedious and, when scaled down to the miniprep level, often
results in poor yields.

The RNA isolation procedure given here exploits the chaotropic nature
of phenol and SDS, enhanced by heating, freezing, and then thawing the
sample. As the sample freezes, the phenol crystals that form pierce the
yeast cell, thereby liberating its contents. DNA is separated from RNA by
phenol extraction under acidic conditions, a tactic frequently exploited
for RNA purification under chaotropic conditions. The procedure presented here is a modification of the method of Schmitt *et al.* (1990). It
has the advantages of being a rapid isolation procedure and having minimum cell culturing requirements.

1. In advance: Inoculate 0.5 ml of yeast culture medium (YPD)[13] in a
sterile 15-ml tube. Grow culture overnight at 30°C with gentle shaking.

Note: Ensure that the tubes are properly covered, not only to prevent
evaporation but also to prevent infection by opportunistic prokaryotes.

2. The following morning dilute the culture 1:40 with fresh YPD

---

[12]The purification of yeast genomic DNA often involves incubation of yeast cells with the
enzyme zymolyase, the result of which is degradation of the cell wall and concomitant
spheroplast formation. This method of gentle cellular disruption favors the recovery of
high-molecular-weight DNA, and is not necessary for RNA isolation.

[13]YPD medium (per liter) = Mix 10 g yeast extract, 20 g peptone, 20 g dextrose; autoclave. After autoclaving, YPD medium acquires a dark brown appearance due to the
"caramelizing" effect on the dextrose at high temperatures.

medium. Continue to incubate for 4 h or until the culture is in exponential growth phase.

3. Remove a 5-ml aliquot of the culture and collect cells by centrifugation (500 g for 5 min). Decant and discard supernatant.

4. Resuspend cell pellet in 400 μl of AE buffer (50 mM sodium acetate, 10 mM EDTA, adjust pH to 5.2). If the cells do not resuspend quickly and thoroughly, add more AE buffer.

Note: If desired, an appropriate RNase inhibitor may be added to the solution just prior to use, though this is usually not necessary.

5. Transfer the cell suspension to a microfuge tube and add 40 μl of 10% SDS. Invert the tube several times to mix.

Note: If the volume of AE buffer was increased in Step 4, be sure to increase the volumes of the subsequently used reagents proportionately.

6. Add an equal volume (approx. 1 ml) of molecular biology-grade (redistilled) phenol that has been equilibrated with AE buffer.

Important: Do not adjust the pH of the equilibrated phenol. See Appendix A for tips on phenol preparation.

7. Mix the sample carefully and thoroughly and then incubate at 65°C for 4 min.

8. Rapidly cool the lysate in an ethanol/dry ice bath or on dry ice until phenol crystals appear. If an ethanol/dry ice bath is not available, the sample may be incubated at −20°C for 20–30 min.

9. Centrifuge the sample at 12,000g (top speed) in a microfuge for 2 min to separate the phases.

Note: It is not unusual to observe a large amount of precipitate, consisting of cellular debris, at the bottom of the tube.

10. Carefully recover the upper (aqueous) phase, taking care not to disturb either the protein interface or the organic phase. Transfer the aqueous material to a fresh microfuge tube and then add an equal volume of room temperature phenol:chloroform (1:1) or phenol:chloroform:isoamyl alcohol (25:24:1). Mix carefully and thoroughly by inversion for 2–3 min.

11. Centrifuge the sample at top speed in a microfuge for 2 min to separate the phases.

12. Carefully recover the upper (aqueous) phase, taking care not to disrupt either the protein interphase (if present) or the organic phase. Transfer the aqueous material to a new microfuge tube.

13. Extract the aqueous material once with an equal volume of chloro-

form or chloroform:isoamyl alcohol (24:1). Centrifuge briefly to separate phases and then carefully transfer the aqueous phase to a fresh tube.

14. Add 0.1 volume 3 M sodium acetate, pH 5.2, and 2.5 volumes of 95% ethanol. Store sample overnight (or at least for a few hours) at –20°C to precipitate RNA.

15. Collect the precipitate by centrifugation (at 4°C if possible).

16. Decant ethanol. Prepare a solution of 70% ethanol/30% DEPC-H_2O and then wash the RNA pellet 2–3 times with 500 μl per wash. Briefly dry the pellet, and *do not* allow the pellet to dry completely. For long-term storage, maintain sample at –70°C. If desired, a final wash with 95% ethanol may accelerate air-drying the RNA pellet.

17. Resuspend the pellet in 25–50 μl sterile (RNase-free) water or TE (10 mM Tris, 1 mM EDTA, pH 7.5).

18. Determine RNA concentration and purity. Store the sample in aliquots at –70°C. The expected yield is about 150 μg total RNA from an exponential 10-ml yeast culture.

19. As always, be sure to electrophorese an aliquot of the newly isolated RNA to ascertain the integrity of the sample.

Determination of Nucleic Acid Concentration and Purity

The efficiency of manipulations involving nucleic acids is heavily dependent on the mass and purity of nucleic acid in a reaction. Not only are precise enzyme:primer:RNA ratios important for efficient conversion of RNA into cDNA, but it is also necessary to have some basis for normalization of samples to each other in studies involving the up- or down-regulation of specific transcripts. A minimal profile of a purified sample of RNA or DNA should include information about its concentration, purity, and integrity.

Spectrophotometric Methods

The concentration and purity of DNA and RNA samples are determined readily by taking advantage of the ability of nucleic acids to absorb ultraviolet light, maximally at 260 nm. Absorbance measurements at this wavelength permit the direct calculation of nucleic acid concentration in a sample:

$$[RNA]μg/ml = A_{260} \times dilution \times 40.0,$$

where

A_{260} = absorbance (in optical densities) at 260 nm

dilution = dilution factor (usually 200–500)

40.0 = extinction coefficient of RNA.

Likewise, the concentration of a sample of DNA can be determined in a similar manner:

$$[DNA]\mu g/ml = A_{260} \times dilution \times 50.0,$$

where

A_{260} = absorbance (in optical densities) at 260 nm

dilution = dilution factor (usually 200–500)

50.0 = extinction coefficient of DNA.

The ultraviolet (UV) measurement is best taken by diluting the sample 1:200–1:500 in an acid-washed quartz cuvette;[14] the disposable plastic variety is not suitable for measurements in this portion of the UV spectrum. Dilutions may be made in water or TE buffer, and the spectrophotometer should be zeroed in the presence of the same. Note that the above equations for RNA and DNA concentration render units of $\mu g/ml$; for the molecular biologist, the mathematics are often greatly simplified when concentration is expressed in $\mu g/\mu l$, so remember to divide by 1000!

Note 1: For greater precision in determining nucleic acid concentration, the extinction coefficient for a particular (unique) RNA or DNA sample can be calculated, using Beer's Law,[15] if the G+C content of the organism is known. The average extinction coefficients of 40 and 50 for RNA and DNA, respectively, are more or less standard for most molecular biology applications.

Note 2: Because the extinction coefficients of thymine and uracil differ, DNA contaminating RNA samples and RNA contaminating DNA sam-

---

[14]Quartz cuvettes can be efficiently cleaned and purged of ribonuclease activity by soaking them in chromic acid and then rinsing them thoroughly with sterile (nuclease-free) water. Exercise extreme caution when handling chromic acid.

[15]According to the Bouguer-Beer Law, commonly known as Beer's Law, $A = \epsilon bc$ where A is absorbance, ϵ is the molar extinction coefficient, b is the light path length, and c is the concentration of the absorbing material, such as adenosine, cytidine, guanosine, thymidine, and uridine. This equation can be rearranged such that $\epsilon = A/bc$. In a mixture of polynucleotides, the extinction coefficient can be determined because $A_{TOT} = A_1 + A_2 + A_3 \ldots$. For example, a precise extinction coefficient for human RNA is 44.19.

ples can falsely elevate or depress spectrophotometric readings. Treatment of a DNA sample with 40 µg/ml RNase A for 30 min at 37°C, followed by reprecipitation, should remove virtually all RNA from the sample. In addition, DNA may be carried forward electrophoretically when RNA is present. The increased mobility may well result in an underestimation of DNA fragment size.

When determining nucleic acid concentration in this manner, one should be certain to use a sufficient mass in order to obtain at least the minimum absorbance recommended for measurement by the manufacturer of the spectrophotometer. For example, spectrophotometric measurements of less than 0.1 optical density (OD) are often unreliable in some of the older model instrumentation. For smaller scale nucleic acid samples, where dilution into the minimum readable volume in the cuvette renders the sample too dilute to be measured, the investigator is urged to invest in smaller volume, though more expensive, quartz cuvettes.

It is also important to realize that these measurements reflect the total mass of nucleic acid in a sample, and not the relative contribution of any single component, such as mRNA in a mixture of total cellular RNA, or contaminating DNA in an RNA prep. Calculations based on absorbance at 260 nm provide little information about the quality and purity of the sample; in the presence of excess salt, contaminating proteins, and/or carryover organic solvents, the absorbance value can be skewed significantly. For this reason, calculation of the so-called "260 to 280 ratio" (A_{260}/A_{280}) provides a reasonable estimate of the purity of the preparation. A pure sample of RNA has an A_{260}/A_{280} ratio of 2.0 ± 0.15, and a pure sample of DNA has an A_{260}/A_{280} ratio of 1.8 ± 0.15. Variations outside this range generally indicate contaminants, and investigators attuned to this detail may wish to take corrective action including, but not limited to, additional phenol:chloroform extractions, 70% ethanol washes, dialysis, and/or double precipitation.

Although the value obtained for a particular sample is a good indicator of the quality of the sample, it is also of paramount importance to examine the absorbance spectrum of the sample, usually between 240 and 320 nm. Pure samples of nucleic acids will produce a characteristic skewed bell curve (Figure 8), which exhibits an absorbance maximum close to 260 nm. The absorbance of ultraviolet light by nucleic acids is 0 at 320 nm, meaning that the curve should return to baseline, and below 240 nm the absorbance profile has little bearing on computation of the concentration and determination of the purity of a sample. This is because high-quality nu-

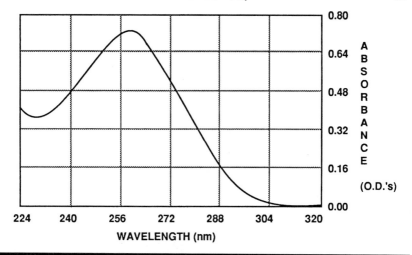

F I G U R E   8

Typical UV absorbance spectrum of purified RNA. The spectrophotometric profile for purified DNA is similar. Note the positive slope of the curve below 260 nm and the negative slope above 260 nm.

cleic acid samples manifest an absorbtion profile with a positive slope below 260 nm and a negative slope above 260 nm. This is not immediately obvious when measuring absorbance only at 260 and 280 nm.

The rationale for doing these types of preliminary analyses to gain information about the quality of a sample should be obvious. In the absence of UV quantitation, the amount of nucleic acid in a sample would be largely unknown, thereby making normalization among samples a virtual impossibility. Without computation of the A_{260}/A_{280} ratio, it may not be immediately obvious if contaminants have copurified with the sample. In the absence of a wavelength scan between 240 and 320 nm, there will be no indication of the shape of the curve. It is frequently noted in the laboratory that contaminants change the characteristic shape of the curve at points other than 260 or 280 nm. For example, excessive quantities of carryover salt (nucleic acid coprecipitate) can mask the positive slope of the curve that is usually associated with the 240–260 nm portion of the spectrum. Washing a nucleic acid coprecipitate with 70–75% ethanol immediately following centrifugation (as suggested in several protocols) will remove most of the salt[16] that was used to drive the precipitation of the nucleic acid molecules in the first place. In the absence of a wave-

[16]Sodium acetate (NaOAc) is particularly easy to remove with 70% ethanol washes.

length scan, the presence of excess salt, and probable negative influence, would not be immediately obvious.

Salt that has coprecipitated with the RNA is not the only concern when quantifying the amount of RNA in a sample. One problem associated with the isolation of RNA by nonionic NP- 40 lysis, for example, is the incomplete removal of protein from the cell lysate (by repeated extraction with phenol:chloroform). Because proteins increase the absorbance at 280 nm, the A_{260}/A_{280} ratio generally decreases, indicating a problem with purity. In the event of incomplete protein extraction, the sample (RNA or DNA) should be dissolved 100–200μl sterile (nuclease-free) buffer, extracted once with an equal volume of phenol:chloroform (1:1) or phenol:chloroform:isoamyl alcohol (25:24:1) and then once with an equal volume of chloroform:isoamyl alcohol (24:1). The appearance of a protein interphase, no matter how small, between the lower organic phase and the upper aqueous phase clearly indicates protein contamination. RNA samples in large volumes may then be reprecipitated to increase the concentration of the sample. This type of manipulation should not be used gratuitously, however, as the stability and recovery of the sample is compromised by repeated precipitation, not to mention the tremendous opportunity to introduce nuclease activity into the sample. It is definitely worthwhile to perform an extra phenol:chloroform extraction to remove contaminating protein during the isolation procedure if an additional extraction appears to be warranted. Precipitated samples of RNA should be washed with 70–75% ethanol and the concentration determined as described earlier.

Nonspectrophotometric Methods

In the absence of fairly expensive spectrophotometric equipment, or if the amount of nucleic acid in a sample is very small, there are several methods through which a reasonable attempt can be made to assess the concentration of nucleic acid in a sample.

Option 1

The prevalence of image analysis software in the molecular biology laboratory affords an opportunity for quantifying nucleic acid concentration by digitizing a fluorescence image; for example, a known mass standard, such as dilutions of the commonly used molecular weight standard ϕX174, and a sample of unknown concentration can be electrophoresed side by side on a gel (Figure 9). Assuming that the software is available, this approach for determining nucleic acid concentration is rapid and usually quite accurate, especially in view of the ready availability of super-

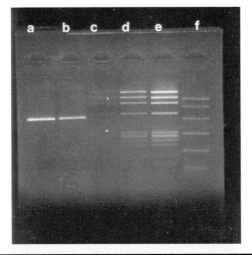

FIGURE 9

Determination of sample mass and molecular weight by comparison with dilutions of a DNA standard. Electrophoresis was performed on a 2.5% agarose gel: (a) 2 µl sample of unknown concentration; (b) 1 µl sample of unknown concentration; (c,d,e) 100, 200, and 400 ng, respectively, of φX174 DNA; (f) PCR Marker (Promega). Molecular weights are 1000, 750, 500, 300, 150, and 50 bp. Calculation method and further details are found in Chapter 9.

sensitive stains such as SYBR Green I and II. The subject of image analysis is discussed and critiqued in detail in Chapter 10.

Option 2

Another method for determining nucleic acid concentration involves immobilization of a small aliquot of a sample onto a proprietary membrane and subsequent chromogenic (colorimetric) assay of sample mass. Two such products, DNA Dipsticks (Invitrogen, San Diego, CA) and Nucleic Acid QuickSticks (Clontech, Palo Alto, CA) permit efficient quantitation of very small quantities of single- or double-stranded DNA, RNA, or oligonucleotides, typically in the 0.1–10 ng/µl range, in aqueous buffer. Briefly, dilutions of nucleic acid samples are spotted onto the membrane, followed by a series of washes and incubation in substrate. Nucleic acid concentration is determined by comparing the resulting color intensity to standards provided with the kit. Because of the mechanics involved and the fact that these membranes do not register individual nucleotides, Dipsticks and QuickSticks are quite handy for quantitating dilute samples in preparation for PCR amplification as well as for monitoring the progress of the reaction.

Option 3

As a third alternative to spectrophotometric determination of nucleic acid concentration, a reasonable guess at the amount of nucleic acid in a sample can be made by spotting known amounts of nucleic acid onto an ethidium bromide/agarose plate.[17] After the standards are spotted onto the plate, dilutions of test material can be added in 5- μl aliquots. Ethidium bromide/agarose plates can be observed directly on the surface of a UV transilluminator or irradiated from above with a handheld UV monitor.[18] Be sure to allow samples to be absorbed into the agarose before assessing the fluorescence intensity of the ethidium bromide. Alternatively, dilutions of a nucleic acid sample of unknown concentration can be added to an equal volume (5 μl or less) of a 1 μg/ml solution of ethidium bromide. These samples can then be applied directly onto plastic wrap and irradiated with UV light, and the resulting fluorescence can be compared to DNA standards of known concentration (Wienand *et al.*, 1978). Although this method is barbaric, it may be useful teaching tool when funds are limited.

Short- and Long-Term Storage of Purified RNA

The correct storage method and form of purified RNA samples is often a hotly discussed topic within the laboratory. Improper storage, over a period of hours or months, is likely to have a profound negative impact on the probable utility of a sample of purified RNA. The key issues are:

a. Has the RNA been purified from the cellular/tissue architecture?
b. Is –80°C storage available in the lab?
c. Will the RNA be used within the next 7–10 days?
d. Has the RNA been redissolved in aqueous buffer (sterile water or TE buffer?

---

[17]To make ethidium bromide/agarose plates, prepare a solution of 1% agarose in water, cool to 55°C, and then add ethidium bromide stock solution (10 mg/ml) to a final concentration of 0.5–1.0 μg/ml. Carefully swirl flask to mix, and avoid inhaling vapors. The molten agarose/ethidium bromide mixture can then be poured into 100-mm petri dishes and allowed to solidify.

[18]**Caution:** Handheld UV irradiating devices can inflict injuries as serious as those from transilluminators. Be sure to wear proper eye protection and cover exposed skin when working with UV light.

Tissues and cell pellets may be harvested and then flash-frozen for storage in liquid nitrogen or at $-80°C$. When stored as such, RNA can be purified up to a year later. In general, the frozen material is thawed in lysis buffer, accompanied by mechanical disruption. If the investigator has access to a freezer capable of maintaining $-80°C$, the RNA may be purified and confidently stored as an ethanol precipitate for several months, the maximum storage time being a direct function of the biological source.[19] If the sample will surely be used within the next 1–2 weeks, the RNA may be stored stably at $-20°C$, again as an ethanol precipitate. The "clock begins ticking" when a purified sample of RNA is dissolved in aqueous buffer, either sterile water or TE buffer. At this point, the RNA should be maintained on ice during handling, unless a protocol specifically dictates otherwise. The most prudent course of action would be to determine the RNA concentration (above, this chapter) and then store the RNA in suitable aliquots at $-80°C$. This will preclude repeated freezing and thawing of the sample. Some investigators add RNase inhibitors to RNA samples during long-term storage, an action that is usually unnecessary and may even be counterproductive. It is incumbent upon the investigator to ensure that any RNase inhibitors added to the prep at any time will not interfere with any subsequent manipulations and/or reactions involving the RNA.[20]

If ever in doubt, store purified RNA samples, in whatever form, at $-80°C$.

References

Andersson, K., and Hjorth, R. (1985). Isolation of bacterial plasmids by density gradient centrifugation in cesium trifluoracetate (CsTFA) without the use of ethidium bromide. *Plasmid* **13**, 78.

Bailey, J. M., and Davidson, N. (1976). Methylmercury as a reversible denaturing agent for agarose gel electrophoresis. *Anal. Biochem.* **70**, 75.

Benecke, B. J., Ben-Ze'ev, A., and Penman, S. (1978). The control of mRNA production, translation, and turnover in suspended and reattached anchorage-dependent fibroblasts. *Cell* **14**, 931–939.

Berger, S. L., and Birkenmier, C. S. (1979). Inhibition of intractable nucleases with ribonucleoside-vanadyl complexes: Isolation of messenger ribonucleic acid from resting lymphocytes. *Biochemistry* **18**(23), 5143–5149.

[19]RNA purified from certain sources (plants, for example) is remarkably stable for years, when maintained as an ethanol precipitate at $-80°C$; this RNA also shows enhanced stability when stored for months in hydrated form at $-20°C$.

[20]Historically, RNA has been stored in widely varying concentrations of SDS, VDR, and even formamide!

Chirgwin, J. M., Przybyla, A. E., MacDonald, R. J., and Rutter, W. J. (1979). Isolation of biologically active ribonucleic acid from sources enriched in ribonuclease. *Biochemistry* **18**, 5294.

Chomczynski, P. (1993). A reagent for the single-step simultaneous isolation of RNA, DNA, and proteins from cell and tissue samples. *BioTechniques* **15**(3), 532.

Chomczynski, P., and Sacchi, N. (1987). Single-step method of RNA isolation by acid guanidinium thiocyanate-phenol-chloroform extraction. *Anal. Biochem.* **162**, 156.

Church, G. M., and Gilbert, W. (1984). Genomic sequencing. *Proc. Natl. Acad. Sci. USA* **81**, 1991–1995.

Cooper, T. G. (1977). "The Tools of Biochemistry," pp. 309–354. Wiley, New York.

Cox, R. A. (1968). The use of guanidinium chloride in the isolation of nucleic acids. *Methods Enzymol.* **12**, 120–129.

Favaloro, J., Triesman, R., and Kamen, R. (1980). Transcriptional maps of polyoma virus-specific RNA: Analysis by two-dimensional S1 gel mapping. *Methods Enzymol.* **65**, 718.

Glisin, V., Crkvenjakov, R., and Byus, C. (1974). Ribonucleic acid purified by cesium chloride centrifugation. *Biochemistry* **13**, 2633.

Gordon, J. A. (1972). Denaturation of globular proteins. Interaction of guanidinium salts with three proteins. *Biochemistry* **11**, 1862.

Leinhard, G. E., Secemski, I. I., Koehler, K. A., and Lindquist, R. N. (1971). Enzymatic catalysis and the transition state theory of reaction rates: Transition state analogues. *Cold Spring Harbor Symp. Quant. Biol.* **36**, 45.

Majumdar, D., Avissar, Y. J., and Wyche, J. H. (1991). Simultaneous and rapid isolation of bacterial and eukaryotic DNA and RNA: A new approach for isolating DNA. *BioTechniques* **11**(1), 94–101.

Marzluff, W. F., and Huang, R. C. C. (1984). Transcription of RNA in isolated nuclei. *In* "Transcription and Translation: A Practical Approach" (B. O. Hames and S. J. Higgins, Eds.), pp. 89–129. IRL Press, Washington, DC.

McMaster, G. K., and Carmichael, G. C. (1977). Analysis of single- and double-stranded nucleic acids on polyacrylamide and agarose gels by using glyoxal and acridine orange. *Proc. Natl. Acad. Sci. USA* **74**, 4835–4838.

Meselson, M., and Stahl, F. W. (1958). The replication of DNA in *E. coli. Proc. Natl. Acad. Sci. USA* **44**, 671.

Meselson, M., Stahl, F. W., and Vinograd, J. (1957). Equilibrium sedimentation of macromolecules in density gradients. *Proc. Natl. Acad. Sci. USA* **43**, 581.

Nevins, J. R., and Darnell, J. E. (1978). Steps in the processing of Ad2 mRNA: Poly(A$^+$) nuclear sequences are conserved and poly(A$^+$) addition precedes splicing. *Cell* **15**, 1477.

Nozaki, Y., and Tanford, C. (1970). The solubility of amino acids, diglycine, and triglycine in aqueous guanidinium hydrochloride solutions. *J. Biol. Chem.* **245**, 1648.

Okayama, H., Kawaichi, M., Brownstein, M., Lee, F., Yokata, T., and Arai, K. (1987). High efficiency cloning of full length cDNA: Construction and screening of cDNA expression libraries from mammalian cells. *Methods Enzymol.* **154**, 3.

Peppel, K., and Baglioni, C. (1990). A simple and fast method to extract RNA from tissue and other cells. *BioTechniques* **9**(6), 711.

Salvatori, R., Bockman, R. S., and Guidon, P. T., Jr. (1992). A simple modification of the Peppel/Baglioni method for RNA isolation. *BioTechniques* **13**, 510–511.

Schmitt, M. E., Brown, T. A., and Trumpower, B. L. (1990). A rapid and simple method for preparation of RNA from *Saccharomyces cerevisiae*. *Nucleic Acids Res.* **18**(10), 3091.

Thomas, P. S. (1980). Hybridization of denatured RNA and small DNA fragments transferred to nitrocellulose. *Proc. Natl. Acad. Sci. USA* **77,** 5201–5205.

Ullrich, A., Shine, J., Chirgwin, J., Pictet, R., Tisher, E., Rutter, W. J., and Goodman, H. M. (1977). Rat insulin genes; construction of plasmids containing the coding sequences. *Science* **196,** 1313.

Wienand, U., Schwarz, Z., and Felix, G. (1978). Electrophoretic elution of nucleic acids from gels adapted for subsequent biological tests: Application for analysis of mRNAs from maize endosperm. *FEBS Lett.* **98,** 319–323.

Wilkens, T. A., and Smart, L. B. (1996). Isolation of RNA from plant tissue. *In* "A Laboratory Guide to RNA Isolation, Analysis, and Synthesis" (P. A. Krieg, Ed.). Wiley-Liss, New York.

Zarlenga, D. S., and Gamble, H. R. (1987). Simultaneous isolation of preparative amounts of RNA and DNA from *Trichinella spiralis* by cesium trifluoroacetate isopycnic centrifugation. *Anal. Biochem.* **162,** 569.

Zolfaghari, R., Chen, X., and Fisher, E. A. (1993). Simple method for extracting RNA from cultured cells and tissues with guanidine salts. *Clin. Chem.* **39,** 1408–1411.

Isolation of Polyadenylated RNA

Rationale

Why purify mRNA? To answer this question fully, statistical consideration must be given to the number and variety of products of transcription. Eukaryotic mRNAs are categorized as falling into an abundance class based on prevalence (i.e., the average number of copies of an RNA transcript) in the cell. The three officially recognized classifications are referred to as high abundance, medium abundance, and low abundance mRNAs.

High abundance mRNAs are those that are present in hundreds of copies per cytoplasm. If a cell is highly specialized to perform a particular function, or produces large amounts of a particular protein, then it is reasonable to expect that the cell transcribes a correspondingly elevated mass of the message. For example, one begins with reticulocytes and oviduct cells to clone β-globin (Efstratiadis *et al.,* 1976) and ovalbumin (Buell *et al.,* 1978), respectively. Medium abundance mRNAs are those present in dozens of copies per cytoplasm. Many of the standard housekeeping genes (e.g., β-actin, GAPDH), the expression of which are assayed frequently as control or reference genes in RNA-based analyses, produce transcripts of sufficient number to be in this class. Low abundance mRNAs are those transcripts that are present in 14 or fewer copies per cytoplasm (Williams, 1981); these are the so-called "rare" mRNAs, the detection of which has been enhanced greatly through the application of such supersensitive assays as the polymerase chain reaction. It has been estimated that fully 30% of cellular mRNAs fall in the low abundance category. Further, this author's definition of *very low abundance* transcripts are those present in fewer than 1 copy per cell, meaning that an investigator would have to look at a mixed population of cells in order to assay mRNA of the very low abundance variety.

The underlying rationale for purifying mRNA is to increase the statistical representation of a particular mRNA in a sample and the concomitant increase in the statistical representation of all mRNAs as a function of the total mass of purified RNA (polyadenylated and nonpolyadenylated). Any approach of this nature is known as an enrichment strategy. Common enrichment techniques include the manipulation of biological material prior to cellular disruption to superinduce, or accumulate, certain types of transcripts in the cell; enrichment may also involve the manipulation of a previously purified RNA sample, the expressed goal of which is to increase the prevalence of one, several, or all mRNAs in a sample. Reasons for enriching samples in favor of mRNA may include, but are certainly not limited to, (1) reducing the amount of RNA necessary to observe specific transcripts by Northern analysis, S1 nuclease analysis, or ribonuclease protection analysis (and in so doing increasing the sensitivity of the assay), or (2) conversion of mRNA into cDNA.[1] By increasing statistical representation one increases the likelihood of being able to identify all elusive, rare mRNA molecules either on a blot or in a library.

The Poly(A) Caveat

Perhaps the most important point of this entire chapter pertains to the significance of mRNA that is purified by some type of affinity chromatographic selection directed toward the poly(A) tail. Although most eukaryotic mRNAs are polyadenylated, a small percentage are not. Thus, populations of poly(A)-selected mRNAs molecules lack nonpolyadenylated mRNAs, an element that may or may not be meaningful in the context of a particular model system.

Example 1: Northern analysis of poly(A)$^+$ mRNA could not be utilized to derive qualitative or quantitative information about cellular or cytoplasmic poly(A)$^-$ species, such as the histone mRNAs. The situation becomes more complicated when one considers the likelihood that alternative posttranscriptional processing pathways might well yield a collection of polyadenylated and nonpolyadenylated transcripts from the same locus.

The reader is cautioned that kits that purport to facilitate the selection of "total mRNA" through poly(A) selection will fail to yield poly(A)$^-$ mRNA. Of equal gravity is the fact that the selection of poly(A)$^+$ RNA from a whole cell lysate will unquestionably render adenylated nuclear RNA as well as adenylated cytoplasmic RNA; the mixture of the two cannot be construed as representing functional (translated) mRNA. Recall that a great deal of hnRNA never matures into cytoplasmic mRNA and is eventually degraded in the nucleus. Since the poly(A) tail is added in the nucleus, selection of adenylated species from total cellular RNA will contain a mixture of both. In order to focus more closely on the mass of poly(A)$^+$ mRNA being produced by the cell, one could lyse the cells gently with hypotonic, NP-40 buffer (Chapter 5), remove the intact nuclei, and then proceed to select the adenylated RNA from this cytosol. Alternatively, to give even clearer definition to those specific mRNAs destined

---

[1]cDNA synthesis may be performed for traditional library synthesis, transcript measurement by PCR, or other relevant purposes.

to be translated, one could prepare the polysome fraction to determine which RNAs are actually engaged by the ribosomes. In one version of this approach, a technique known as the EDTA-release assay, the chelation of Mg^{2+} by EDTA causes dissociation of ribosomes into their individual subunits and release of the RNA that was undergoing translation at that very moment (Aziz and Munro, 1986; Meyuhas *et al.*, 1987; reviewed by Pierandrei-Amaldi and Amaldi, 1994). This method is very useful for discerning the active mRNA population. In addition, a more recent protocol directed toward the same experimental end has been published (Meyuhas, *et al.*, 1996).

Example 2: The classical approach to the synthesis of cDNA and subsequent propagation of these molecules as libraries begins with the selection of poly(A)$^+$ mRNA. To initiate conversion of mRNA into first-strand cDNA, an oligo(dT) primer is annealed to the mRNA template to provide the 3'-OH group in the proper orientation; primers such as these are essential for supporting all types of polymerase activity, in this case by utilizing reverse transcriptase[2] for the synthesis of first-strand cDNA.

The net effect of this approach is cDNA synthesis primed from the 3' end of the mRNA. Therefore, cDNA libraries prepared in this fashion do not contain clones corresponding to poly(A)$^-$ mRNA. Scientists should be aware that such libraries persist in circulation. In order to circumvent the potential shortcomings of oligo(dT)-primed cDNA libraries, many newer libraries are prepared by random-primed synthesis of the first strand cDNA, and sometimes generated through the use of random primers *and* oligo(dT) in the same reaction tube. Briefly, a number of nucleotide hexamers or nonomers of random nucleotide sequence are allowed to anneal to heat-denatured RNA. This annealing effectively produces more than one 3'-OH primer upon which reverse transcriptase can act simultaneously. This will occur regardless of 3' terminal structure of the molecule [poly(A)$^+$ or poly(A)$^-$]. cDNAs synthesized in this fashion are unlikely to include sequence corresponding to the poly(A) tail unless an oligo(dT) random primer had been included in the primer mixture. When screening a library that may have been synthesized some years ago, knowledge of the method of cDNA priming may influence interpretation of the outcome of the experiment.[3]

---

[2] RNA-dependent DNA polymerase. The role of reverse transcriptase is discussed in depth in Chapter 15.

[3] In some extreme cases, total RNA has been treated with poly(A) polymerase to polyadenylate all RNA prior to cDNA synthesis! This is neither necessary nor recommended.

Polyadenylation

The existence of the poly(A) tail has been documented for several years. This structure is a tract of adenosine nucleotides that characterizes the 3' end of most, but not all, eukaryotic mRNA molecules. Polyadenylation, the process by which this structure is added to mRNA, is not coupled directly to transcription termination (Darnell, 1979; Manley, 1988); rather it is catalyzed by a family of nuclear enzymes (Moore and Sharp, 1985), known collectively as poly(A) polymerase. Thus, polyadenylation *in vivo* is not an integral part of transcription, but rather a very rapid (Nevins and Darnell, 1978) posttranscriptional regulatory event.

The length of the poly(A) tail is variable among mRNAs, even among those RNA molecules that were derived by transcription of the same locus. The typical mammalian poly(A) tail is 200–250 nucleotides long, though these tracts often range from 50 to 300 nucleotides. Collectively, all RNA molecules that possess a poly(A) tail are referred to as poly(A)$^+$ mRNA and are correctly said to be polyadenylated. Although of unknown function, the poly(A) tail is believed to somehow facilitate transport of the mature mRNA out of the nucleus (nucleocytoplasmic transport): the adenosine analog 3'-deoxyadenosine (cordycepin) inhibits polyadenylation of nuclear RNA and curtails the appearance of these transcribed, nonadenylated sequences in the cytoplasm. Further, its presence may well regulate stability in the cytoplasm (reviewed by Ross, 1995), where the mRNA may or may not be translated.

The generation of the 3' poly(A) tract is actually the product of two sequential enzymatic activities, namely a cleavage reaction of the precursor RNA (hnRNA) followed by the actual polyadenylation event. It is clear that primary RNA transcripts may extend up to 20 kilobases beyond the polyadenylation signal (Nevins and Darnell, 1978; Ziff, 1980). The precise site at which the poly(A) tail is added is formed by an endonucleolytic cleavage of the precursor RNA molecule (Nevins and Darnell, 1978), and not by exonucleolytic degradation from the 3' end of the molecule generated by the termination of transcription. This cleavage event is then followed by the sequential addition of the adenylate residues that ultimately constitute the poly(A) tail.

The splicing that precedes polyadenylation is not a haphazard occurrence; the efficiency and location of this event are precisely regulated by the aptly named polyadenylation signal. Among the higher eukaryotes, virtually all polyadenylated mRNA molecules exhibit the highly conserved hexanucleotide motif 5' . . . AAUAAA . . . 3', or a closely related variant thereof, which can typically be identified fewer than 30 bases upstream from the polyadenylation start site (Proudfoot and Brownlee, 1976). For efficient polyadenylation to occur, it is also necessary that a second element, a GU-rich cluster, be present (Gil and Proudfoot, 1984; McDevitt *et al.*, 1984; Sadofsky and Alwine, 1984; Cole and Stacy, 1985; Conway and Wickens, 1985). This GU-rich sequence is less highly conserved than the AAUAAA motif and is found approximately 20–40 nucleotides downstream from the actual cleavage site. *In vivo*, the poly(A) tract associates with poly(A)-binding protein (PABP), which binds stoichiometrically to the poly(A) tail every 10–20 bases (reviewed by Kramer, 1996). Of course this, and other proteins, dissociate from the RNA during the course of most RNA isolation procedures.

Selection of Polyadenylated Molecules

The extremely heterogenous nature of mRNA, the fact that it represents only a small fraction of the total RNA mass, and the availability of minute amounts of tissue sample frequently preclude the chromatography of mRNA by more traditional approaches such as agarose gel electrophoresis or sucrose gradient size fractionation. At the other extreme, gel purification of RNAs becomes limiting when large quantities of material are required for assays and biophysical studies. Column chromatography, on the other hand, can resolve milligram quantities of material.

In order to resolve adenylated from nonadenylated RNA species, the naturally occurring hydrogen bonding between the bases adenine and thymine can be exploited to isolate the relatively small poly(A)$^+$ mRNA fraction from total cytoplasmic or total cellular RNA. Fractionation in this manner is a fine example of affinity chromatography, that is, a separation or chromatography based on a biological activity or structure. In this case, the biological structure is the poly(A) tail. Historically, this has been accomplished using various oligo(dT)-celluloses, consisting of oligodeoxythymidylate residues covalently bound by their 5' phosphate to cellulose beads (Aviv and Leder, 1972). While no longer the most commonly used matrix, oligo(dT)-celluloses remain commercially available from a number of suppliers and in a variety of grades, each of which has a characteristic level of purity and binding capacity. Alternatively, oligo(dT)-celluloses may be synthesized as originally described by Gilham (1964). Although oligo(dT)-cellulose has proven to be fairly effective for the resolution and detection of poly(A)$^+$ mRNA, newer methods, described below, which require minimal amounts of starting material are now favored strongly.

Over the past 5 years, it has become rather stylish to employ one of a number of variations on the theme of affinity separation of poly(A)$^+$ mRNA. In all cases, the same principles apply to the affinity selection of the poly(A) tail with oligodeoxythymidylate. These approaches have emerged to streamline the somewhat cumbersome mechanics of "running an oligo(dT) column" (discussed later). These newer approaches include the following:

1. Mixing oligo(dT)-cellulose and total RNA together directly in a microfuge tube without any prior formation of some type of matrix geometry.

2. Using paramagnetic beads to which oligo(dT) has been covalently linked. Following solution hybridization, the hybrids are sequestered using a magnet.

3. Exploiting the natural affinity between biotin and avidin. Biotin can be conjugated to a variety of molecules, including nucleic acids, for the most part without changing their biological activities. The tight binding of biotin by streptavidin (K_d $10^{-15}M$) makes many applications possible; in this case biotinylated oligo(dT) molecules, following base pairing between oligo(dT) and poly(A), can be precipitated as a biotin/streptavidin complex. The hybrids may also be recovered using streptavidin linked to a paramagnetic bead. Many permutations of this general approach are available as kits for RNA purification.

In all cases, the most important aspect of poly(A)$^+$ selection is keeping the volumes involved as small as possible and, in so doing, keeping the sample as concentrated as possible.

Oligo(dT)-Cellulose Column Chromatography

A variety of techniques have been developed for the purification of poly(A)$^+$ mRNA. In general, each of these involves annealing poly(A)$^+$ material to the matrix in a buffer of high ionic strength. This is necessary to eliminate the electrostatic repulsion that naturally occurs between the target RNA and the thymidylate residues, which are linked to the cellulose beads. Under high-salt conditions, sodium acting as a counterion neutralizes the net negative charge intrinsic to the phosphodiester backbone of polynucleotides. In the absence of a high-salt milieu, the electrostatic repulsion would be of sufficient magnitude to prevent base pairing and the purification of poly(A)$^+$ mRNA in this manner.

The binding capacity of oligo(dT)-cellulose increases with salt concentration up to about 500 mM NaCl, KCl, or LiCl,[4] though the binding capacity is greater when KCl or LiCl is used in place of an equivalent amount of NaCl (Mercer and Naora, 1975). It is important to realize, however, that this increase in binding capacity is accompanied by an increase in nonspecific hybridization to the matrix. Once the nonhybridized, non-poly(A)$^+$ material is removed in the eluate, hybridized species of poly(A)$^+$ RNA are recovered from the matrix by changing to a very low-salt buffer (0–5 mM salt). Removal of the monovalent cation from the system reestablishes electrostatic repulsion, favoring rapid dissociation of the hybridized RNA from the matrix. Once chromatographed in this fashion both the poly(A)$^+$ and non-poly(A)$^+$ material can be concentrated by salt and alcohol precipitation. The non-poly(A) RNA fraction should likewise be precipitated; poly(A)$^-$ RNA makes an excellent negative control in all types of hybridization analyses as well as in the synthesis of cDNA (Chapters 13 and 15, respectively).

In one older variation of this technique, the oligo(dT) matrix is packed in a column configuration in a 1-ml tuberculin syringe. The eluate is directed through an ultraviolet absorbance detection monitor, such as the ISCO UA-6 model (ISCO, Inc., Lincoln, NE); this instrument is quite popular for a number of protein chromatography applications (Figure 1). As RNA flows from the column and through the optical unit of the instrument (Figure 2), the absorbance of UV light, and hence the concentration of RNA flowing, is proportional to the observed deflection of the pen on the monitor chart recorder. While this approach is quite a bit more time-consuming than some of the other techniques described here, it has the advantage of allowing the investigator to observe directly the magnitude of the absorbance spectra of the poly(A)$^+$ and poly(A)$^-$ material. In addition to quantity, the *shape* of the elution profile (Figure 3) is an important indicator of the quality/integrity of the sample and the efficiency of elution from the matrix. There are two principal drawbacks to this approach, however: (1) the RNA [both poly(A)$^+$ and poly(A)$^-$] elutes from the column in a relatively large volume, even at slow elution flow rates, making subsequent concentration of the RNA somewhat cumbersome,[5] and (2) due to the lack of sensitivity below 0.1 optical density (OD), this

[4]See manufacturer's package insert for the particulars of each grade and lot of oligo(dT)-cellulose.
[5]If the concentration of RNA at any time falls below the level of precipitability, it will be necessary to add carrier RNA to accomplish precipitation. Having to resort to this approach defeats the entire purpose of purifying the mRNA in the first place. It is always to one's distinct advantage to keep nucleic acid samples as concentrated as possible.

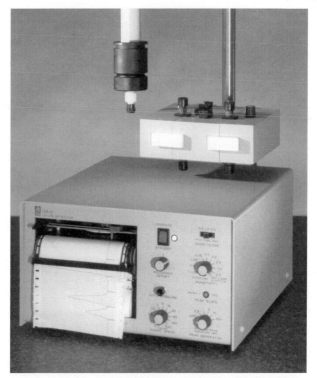

FIGURE 1

UA-6 ultraviolet absorbance detection monitor. Courtesy of ISCO, Inc.

approach is unsuitable for small quantities of starting material (less than 100–150 μg total RNA).

Protocol: Purification of Biophysical Quantities of Poly(A)⁺ RNA

1. **Wear gloves!**

2. Prepare oligo(dT)-cellulose according to the instructions of the manufacturer. Typically, 10 mg of oligo(dT)-cellulose is suspended in 3–5 ml of elution buffer (10 mM Tris, pH 7.4; 0.05% SDS[6]) or DEPC-treated water.

[6]The inclusion of SDS in elution buffers and binding buffers may not be desirable in some instances because the SDS may precipitate (especially in air-conditioned labs!) and clog the column. Further, residual SDS may inactivate subsequent enzyme activities. Excellent RNase-free technique precludes the requirement for SDS.

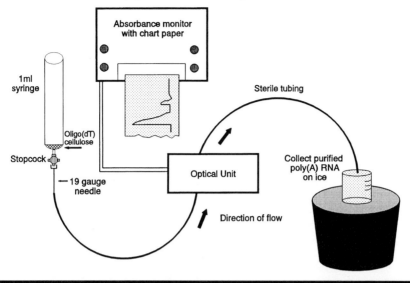

FIGURE 2

Method for purification of large quantities of poly(A)⁺ material. Oligo(dT) is prepared and packed in a 1-ml syringe, the flow rate through which is regulated by an autoclavable (RNase-free) stopcock placed beneath. The needle below the stopcock channels the eluate into the tubing, which leads to the optical unit of the UV detection monitor. The optical unit is outfitted with a 254-nm absorbance filter for nucleic acid chromatography. The concentration of nucleic acid flowing through the optical unit is interpreted by the magnitude of the deflection of the pen mounted on the monitor, thereby providing the user with a chart paper record of the binding and elution profiles for each sample. The eluate is then directed out of the optical unit and is collected in a sterile tube, usually resting in an ice bucket. The movement of the pen on the chart recorder indicates the precise moment for the investigator to change collecting tubes.

Note: Depending on the degree of refinement, 1 g of oligo(dT)-cellulose is capable of binding 20–50 OD of poly(A)⁺ RNA. Assuming that 5% of the total RNA sample is polyadenylated, use 3–5 times more oligo(dT)-cellulose material than is required to bind the estimated mass of poly(A)⁺ material in the sample. It is wise to remain within these parameters as excessive matrix bulk may distort the elution characteristics of the RNA as well as exert pressure on departmental resources: oligo(dT)-celluloses cost as much as $100 per gram! The oligo(dT)-cellulose material is insoluble and should neither be shaken vigorously nor vortexed in an attempt to dissolve it; rather it may be gently resuspended by tapping or gently inverting the tube. For much larger starting quantities of total RNA, as much as 10 mg of RNA

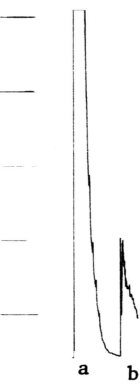

a   b

FIGURE   3

Binding and elution profiles of poly(A)⁺ RNA from oligo(dT)-cellulose matrix. The eluate was directed through the optical unit of an ISCO UA-6 ultraviolet absorbance detection monitor. Sensitivity setting was 0.1 OD, which is necessary to observe the elution of
less than 100 μg poly(A)⁺ mRNA. (a) Application of 150 μg of total cytoplasmic RNA in
binding (high salt) buffer. The column was washed continuously until the monitor re-
turned to baseline, indicating that all non-poly(A) RNA had passed through the column.
(b) Elution of poly(A)⁺ RNA with the application of elution (low salt) buffer. Degraded
samples of RNA tend to produce a wider elution profile without the spiked appearance
obvious in this example.

*can be passed over a larger bore column containing 1 ml of packed
oligo(dT)-cellulose.*

3. Plug the bottom of a sterile 1-cc tuberculin syringe with autoclaved,
shredded polyester fill. Be sure to use only the *smallest amount* necessary
to act as a trap for the oligo(dT)-cellulose. Avoid plugging the neck of the
syringe; instead, push the polyester down only as far as the bottom of the
barrel of the syringe.

Note: Shredded polyester fill is available at most hobby shops and is much easier to manipulate than glass wool. It can be autoclaved briefly (15 min) and should, of course, be handled with gloves.

4. Apply the oligo(dT)-cellulose to the syringe using a ribonuclease-free 9-in. Pasteur pipet. Allow the matrix to pack itself on top of the polyester. Together, the bed volume of the matrix and the polyester plug should not exceed 0.1 cc.

5. After the matrix has been set in place, wash with 2–3 ml binding buffer (500 mM NaCl; 10 mM Tris, pH 7.4; 0.05% SDS) to equilibrate the column in high salt and ensure that the matrix is firmly packed in the column.

6. Attach a sterile stopcock to the bottom of the syringe to regulate the flow rate. Wash the column once again with binding buffer to make sure that no air bubbles have been trapped in or above the stopcock.

Note: Polypropylene stopcocks are quite useful because they can be autoclaved and reused. When assembling the column, it is best to position the stopcock on the bottom of the syringe with the lever in the "open" position so as not to disturb the matrix.

7. Carefully attach a sterile 19-gauge needle or blunted needle to the bottom of the stopcock. Wash the column with binding buffer to make sure that no air bubbles have been trapped in or above the stopcock.

8. Connect the column to an absorbance monitor outfitted with a 254- or 280-nm filter, such as the ISCO UA-6 model. The connection between the column and the absorbance monitor is made by attaching RNase-free tubing that will accommodate the 19-gauge needle and fit inside the flow cell of the optical unit. The sample enters the flow cell from the bottom and exits from the top: this way, the flow rate of the system is dependent only on hydrostatic pressure.

Note: At this point it is very important to make sure that the column does not run dry. Adjust the flow rate to about 2 ml per minute. This relatively slow rate is needed to permit base pairing between the oligo(dT) matrix and the poly(A) tail to occur. It is not advisable to open and close the stopcock repeatedly because this usually dislodges some of the matrix. Oligo(dT)-cellulose can easily clog the flow cell and result in the loss of valuable material. As with any chromatography, separation is optimal when the column is flowing uniformly.

9. Continue to wash the column with binding buffer until a baseline has been established on the monitor. In this laboratory, the sensitivity is

routinely set at 0.1 OD. This sensitivity is necessary to observe the elution of low to moderate quantities of polyadenylated RNA (10–200 μg). Of course, the required sensitivity setting is a function of the RNA mass in the sample.

Note: Setting the baseline at the 10% tick is suggested, as differences in the refractive indices of the binding and elution buffers may cause baseline shift.

10. Dissolve the RNA in 500–1000 μl of binding buffer and apply the sample [up to 50 mg RNA per gram of oligo(dT)-cellulose] to the column. Draw the sample into a sterile Pasteur pipet and follow the buffer meniscus down the barrel of the syringe until the column *almost* runs dry. Carefully release the sample directly onto the matrix and allow it to enter the bed volume. Immediately follow with a copious volume of binding buffer. Do not let the column run dry.

Suggested: before loading the sample onto the column, heat it to 65°C for 5 min and then rapidly cool it on ice. This heating step will disrupt secondary structures that, if present, could make the poly(A) tail inaccessible to the matrix.

11. Observe the enormous deflection of the baseline as the non-poly(A) component of the sample passes through the column (Figure 3a). Continue to wash the column with binding buffer.

Suggested: Collect and load the eluate onto the column a second time to maximize the capture of the poly(A)$^+$ fraction on the matrix. This may be helpful because some poly(A)$^+$ RNAs manage to slip through the matrix during the first passage.

12. Continue to wash the column with binding buffer until the monitor pen returns to the baseline. If the poly(A)$^-$ fraction is to be saved, it should be collected in a tube on ice as it flows from the optical unit.

Note 1: The poly(A)$^-$ fraction may be reprecipitated for future use as RNA size standards or as a negative control lane when performing Northern analysis or a nuclease protection assay.

Note 2: Occasionally it becomes necessary to unclog the column as the flow rate diminishes dramatically when samples loaded onto the column contain large amounts of salt and/or genomic DNA carried over in the previous precipitation of the sample. The flow rate may also be diminished by the unexpected appearance of air bubbles. These problems can often be corrected by simply opening the stopcock all the way, to expel air or debris, and then readjusting the flow rate. In more extreme cases, it may be necessary to disrupt the matrix bed directly to unclog it. This approach is acceptable on an as-needed basis. It will usually not cause a loss of resolution because efficient affinity chromatography, unlike gel filtration, is not totally dependent on the geometry of the matrix. Once the oligo(dT)-cellulose particles have settled back in place, one may resume the chromatography. Poly(A)+ material will not be eluted from the matrix until the counterion is removed upon application of elution buffer.

13. Recover the bound poly(A)+ RNA with elution buffer (Step 2) prewarmed to 45°C. The eluted RNA should be recovered in an RNase-free Corex glass tube, or other appropriate tube, resting in an ice bucket.

Note: Prewarming the elution buffer will greatly facilitate the rapid elution of poly(A)+ material in a minimal volume, as it is always desirable to keep nucleic acid samples as concentrated as possible.

14. Reprecipitate the RNA at –20°C with 0.1 volume of 3 M sodium acetate, pH 5.2, and 2.5 volumes of ice-cold 95% ethanol. The average recovery per 10^7 cells is 1–5 μg poly(A)+ mRNA.

Note: If the yield is expected to be low or if the eluted RNA is believed to be dilute, reprecipitation of the RNA may be facilitated with the addition of glycogen. Add 10 μl of a 10 mg/ml glycogen solution for every milliliter of poly(A)+ eluate collected, followed by the addition of 0.1 volume 3 M sodium acetate, pH 5.2, and 2.5 volumes ice-cold 95% ethanol. Store at –20°C for at least 4 h.

15. The oligo(dT)-cellulose column can be regenerated by washing it with 2–3 ml of 0.1 N NaOH followed by reequilibration with binding buffer. The column matrix can also be stored in the syringe at 4°C under 100–200 μl of buffer; do not forget that bacterial cell growth is favored in celluloses stored at room temperature.

Protocol: Rapid, Noncolumn Poly(A)+ Purification

The selection of poly(A)+ RNA can be accomplished without traditional column chromatography. In fact, relatively small masses of starting material (typically less than 500 μg total RNA) mandate a method of separation other than an affinity column. This is mainly because of the dilution of RNA that inevitably occurs and the very low percentage of the total mass that the poly(A)+ species represent, not to mention the labor-intensive nature of preparing and running the column. If the yield is extremely small or the sample too dilute, it will be impossible to precipitate the purified RNA without the addition of carrier molecules such as tRNA or glycogen. Of course, such a strategy defeats the purpose of poly(A)$^+$ selection in the first place; further, glycogen is incompatible with several subsequent RNA applications.

An alternative matrix configuration that supports the affinity separation of polyadenylated from nonpolyadenylated RNA is known as microcrystalline oligo(dT)-cellulose (New England Biolabs, Beverly, MA). Microcrystalline oligo(dT)-cellulose generally can bind in excess of 100 OD per gram of matrix. By comparison, the typical binding capacity of the oligo(dT)-cellulose under standard assay conditions generally ranges from 50 to 80 OD per gram of matrix, depending on the manufacturer, the grade of refinement, and the monovalent cation used to prepare the binding buffer (Li>>K>>Na). The selection of poly(A)$^+$ RNA is best accomplished by repeated suspension of the matrix in traditional high salt buffers, followed by centrifugation. Using this approach, one avoids the excessively slow flow rates commonly associated with column chromatography, and it should be noted that microcrystalline oligo(dT)-cellulose is not recommended for use in column format. If scaled down, the entire procedure can be carried out in a microfuge tube. The eluted RNA thus remains in a relatively small volume and at a concentration that permits precipitation with salt and alcohol alone, without an added carrier.

The following protocol is adapted from the recommended procedure for microcrystalline oligo(dT)-cellulose. Although similar protocols have been used with standard oligo(dT)-celluloses, the microcrystalline grade maximizes quantitative recovery. Note that all requirements for controlling RNase activity must be satisfied for quantitative and qualitative recovery.

1. **Wear gloves!**
2. Weigh out an appropriate mass of matrix material. For microcrystalline oligo(dT)-cellulose, use 0.3 g (dry weight) for each 0.5 mg of total RNA starting material. Place matrix in an appropriate centrifuge/microcentrifuge tube.

Note: Oligo (dT)-cellulose should be weighed out under RNase-free conditions. While autoclaved or baked metal spatulas may be used (see Chapter 4 for details), it is better to gently tap the bottle of oligo(dT)-cellulose to remove a suitable mass.

3. In a separate tube, dissolve RNA in DEPC-treated H_2O and heat to 65°C for 5 min.

Note: The heating step in this and related protocols is necessary to denature RNA molecules, thereby making the poly(A) structure more available for base pairing to the matrix.

4. Add an equal volume of binding buffer (500 mM LiCl; 20 mM Tris–Cl, pH 7.5; 1 mM EDTA; optional 0.05% SDS) and cool the sample to room temperature. Add the mixture to the oligo(dT)-cellulose.

5. Allow the RNA sample to stand at room temperature or place in a 37°C water bath for 5 min. Occasionally agitate the tube gently by hand during this brief incubation.

6. Centrifuge at 1500g for 5 min, at 4°C if possible. *Carefully* decant the supernatant to a fresh tube and set aside on ice.

Note 1: If SDS is added to the binding buffer, carry out the centrifugation at room temperature or only as low as 15°C. Otherwise, SDS will begin to precipitate in the tube.

Note 2: The poly(A)$^-$ fraction can be extremely useful as a negative control in a variety of applications or can itself be used as carrier RNA in future experiments.

7. Wash the oligo(dT)-cellulose up to 5 times with 0.5- to 1-ml aliquots of binding buffer. Centrifuge at 1500g after each wash and then *carefully* remove the supernatant each time by gentle aspiration.

8. Elute the poly(A)$^+$ RNA from the matrix with as many as five 500-µl aliquots of elution buffer (10 mM Tris–Cl, pH 7.5; 1 mM EDTA; optional 0.05% SDS). Alternatively, RNA may be eluted in sterile DEPC-H_2O.

Note: The number of washes needed to recover all poly(A)$^+$ material is a function of the mass of the bound poly(A)$^+$ and the volume of elution buffer used. The first aliquot of elution buffer will contain a great majority of the poly(A)$^+$ material. Logistically speaking, it may be useful to elute the poly(A)$^+$ in empirically determined smaller volumes. RNA should be reprecipitated in one or more microfuge tubes. This is easily accomplished by adding 0.1 volume 3 M NaOAc (pH 5.2) and 2.5 volumes ice-cold 95% ethanol. Store sample(s) on dry ice for 20 min or at –20°C for at least 4 h.

CAUTION Be certain not to exceed the recommended g-force for the tubes used in this application.

9. Following precipitation and centrifugation, carefully decant the ethanol supernatant. Wash the RNA pellet two or three times with 500-μl aliquots of 70% ethanol (prepared in sterile H_2O). A final wash with 95% ethanol may facilitate air-drying. Resuspend RNA in a minimum volume of TE buffer, pH 7.5, or sterile H_2O.

10. Determine poly(A)$^+$ concentration by measuring A_{260}, as described in Chapter 5.

11. Aliquot hydrated RNA and store at $-80°C$ until further use.

Magnetic Bead Technology for Poly(A)$^+$ Purification

The covalent linkage of a variety of ligands to what are essentially iron beads has become an invaluable tool for the resolution of a variety of compounds from complex mixtures. These ligands include antibodies, which can be used to recognize and bind cell surface antigens, and, in the context of this resource, oligothymidylate, for the efficient recovery of adenylated transcripts. Several companies now produce magnetic bead-based products; Dynabeads oligo(dT) (Dynal, Inc., Great Neck, NY) is one example. Dynabeads are superparamagnetic beads, 2.8 μm in diameter, with a polystyrene coating that encloses iron oxide (Fe_2O_3) particles. These particles form a magnetic dipole when exposed to a magnetic field. Chains of deoxythymidylate, 25 nucleotides long, have been covalently attached to the surface of the beads via a 5$'$ linker group. A mixture of total RNA added to the beads under conditions of high ionic strength will favor hybridization between polyadenylated RNA and the oligo(dT) tract linked to the beads (Figure 4). The hybridization kinetics are similar to those found in free solution, with complete hybridization being observed within 5 min. The magnetic beads and hybridized poly(A)$^+$ RNA are then concentrated by placing the sample tube in the vicinity of a magnet (Figure 5). Washing and elution are performed by a standard combination of temperature and very low salt buffer. Isolation of polyadenylated RNA by magnetic capture precludes the requirement for traditional column affinity chromatography and has achieved a new level of sensitivity in the chromatography of nucleic acids. The principal advantage of this technique is the rapid isolation of pure, intact poly(A)$^+$ RNA in concentrated form. This precludes the requirement for reprecipitation, which often results in the further loss of unique or low abundance transcripts. The average binding capacity of these beads is 2 μg poly(A)$^+$ RNA per milligram of beads. Transcripts affinity purified in this fashion may then be used directly for the construction of cDNA libraries, nuclease protec-

15 MINUTE PROTOCOL

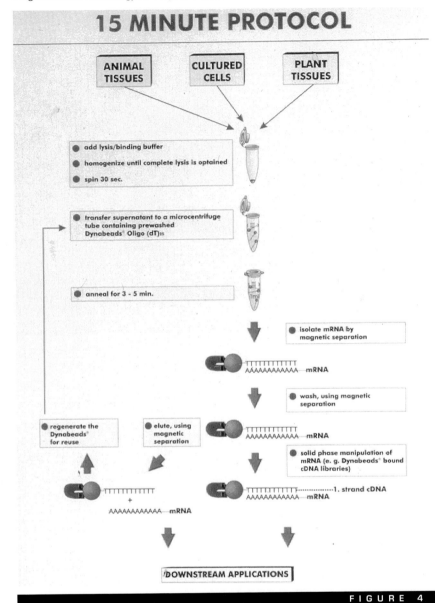

FIGURE 4

Basic strategy for the selection of poly(A)⁺ RNA using magnetic bead technology. Courtesy of Dynal, Inc.

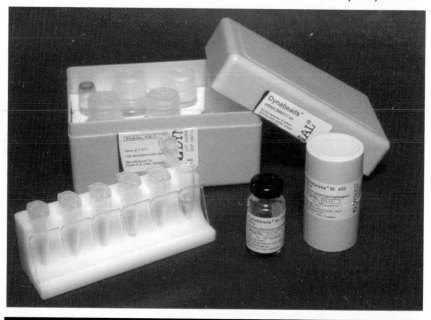

FIGURE 5

Magnetic capture of poly(A)$^+$ RNA after base pairing with oligo(dT) linked to superpara-magnetic particles. Courtesy of Dynal, Inc.

tion analysis, primer extension, PCR amplification, *in vitro* translation, RNA blotting procedures, and so forth.

The protocol that follows describes the selection of polyadenylated sequences using Dynabeads, and begins by offering a selection of starting materials: (a) previously purified total RNA (total cellular or total cytoplasmic), (b) cells growing in culture, or (c) solid tissue. It is very important to realize that in cases where there is a limited amount of starting material it may be prudent to add the beads directly to a whole cell lysate or tissue homogenate for direct harvest of poly(A)$^+$ RNA. In the context of assigning gene regulation as either transcriptional or posttranscriptional, the investigator is also cautioned that thymidylate residues recognize and capture *any* transcript with a poly(A) tract, both mRNA *and* hnRNA, and will fail to capture non-poly(A) mRNA. The two most important points to remember that will influence the success of this technique are (1) be sure to completely remove all of the buffer from the tube containing the magnetic beads before moving on to next step, and (2) *never* remove anything from the tube containing the magnetic beads unless the tube is in the magnet.

Finally, note that other formats for poly(A)$^+$ selection are available from a variety of vendors, including magnetic-based separation products and spin column configurations, all of which in some way target the poly(A) tail.

Protocol: Purification of Poly(A)$^+$ RNA with Magnetic Beads[7]

a. Poly(A)$^+$ Isolation from Total RNA

1a*i*. Begin with 75 μg total RNA, from which 1–4 μg of poly(A)$^+$ mRNA is expected. Dissolve RNA in a total of 100 μl of RNase-free H$_2$O or TE buffer. If the RNA sample is more dilute than 75 μg/100 μl, add an equal volume of 2× binding buffer (1 *M* LiCl; 20 m*M* Tris–Cl, pH 7.5; 2 m*M* EDTA).

1a*ii*. Heat sample to 65°C for 2 min.

Note: Optimal conditions for hybridization are promoted by disrupting any secondary structures, thereby making the poly(A) tail more available for base pairing to the matrix.

1a*iii*. While the sample is being heat denatured, remove 1 mg Dynabeads Oligo(dT)$_{25}$ (usually a 200-μl aliquot) from the resuspended stock[8] and place in an RNase-free microfuge tube. Then place this tube in the accompanying magnet. After 30 s, remove the supernatant completely with a micropipettor while the tube remains in the magnet.

1a*iv*. Add a 100-μl aliquot of 2× binding buffer to the tube containing the beads. Again, remove the supernatant while this tube remains in the magnet.

1a*v*. Remove the tube from the magnet, and place in an Eppendorf tube rack. Add 100 μl 2× binding buffer to the beads and resuspend by gentle tapping. If the RNA sample has already been diluted in binding buffer (Step 1a*i*), omit addition of an aliquot of the binding buffer in this step.

1a*vi*. Add the heat-denatured RNA to the bead suspension and mix gently. Allow hybridization to proceed for 5 min at room temperature. Proceed to Step 2.

---

[7]Adapted, in part, from Dynal, Inc. (Great Neck, NY). Details on the isolation of total cytoplasmic RNA and total cellular RNA can be found in Chapter 5.

[8]Resuspend Dynabeads stock solution by thorough inversion and tapping of the tube. Avoid pipetting repeatedly, and never vortex, as these actions may strip the oligo(dT) from the beads, severely diminishing their utility.

b. Poly (A)$^+$ Selection Directly from Cell Cultures

1bi. Harvest cells as usual (10^6–10^7 is sufficient) and wash with ice-cold PBS (137 mM NaCl; 2.7 mM KCl; 4.3 mM Na$_2$HPO$_4$·7H$_2$O; 1.4 mM KH$_2$PO$_4$). Transfer cells to a microfuge tube and pellet gently (200g).

1bii. Resuspend the cell pellet in 100 μl lysis buffer (10 mM Tris–Cl, pH 7.5; 0.14 M NaCl; 5 mM KCl; 1% Triton X-100 or NP-40) and place on ice for 1 min. If desired, RNase inhibitors may be added to the lysis buffer.

Note: Other popular RNA lysis buffers may be used instead. See Section 1c for an example.

1biii. Centrifuge sample for 30 s.

1biv. Transfer the supernatant to a microfuge tube and store on ice.

1bv. Remove a suitable aliquot of Dynabeads Oligo(dT) from the re-suspended stock and place in an RNase-free microfuge tube. Plan to use 400 μg beads per 10^6 cells. Then place this tube in the accompanying magnet. After 30 s, remove the supernatant completely with a micropipettor while the tube remains in the magnet.

1bvi. Add a 100-μl aliquot of 2× binding buffer to the tube containing the beads. Again, remove the supernatant while this tube remains in the magnet.

1bvii. Remove the tube from the magnet and resuspend the beads in 100 μl of 2X binding buffer (20 mM Tris–Cl, pH 7.5; 1 M LiCl, 2 mM EDTA; 0.4–1% SDS). Next, transfer these resuspended beads to the microfuge tube containing the cell lysate (from Step 1biv).

1b$viii$. Mix gently and allow hybridization to proceed for 5 min at room temperature. Proceed to Step 2.

c. Poly(A)$^+$ Selection Directly from Solid Tissues

This method is designed to facilitate the isolation of poly(A)$^+$ RNA directly from tissues, and has been used successfully with a variety of plant and animal tissues.

1ci. Grind frozen tissue (0.1 g) in liquid nitrogen.

1cii. Transfer the frozen powder to a standard homogenizer containing 1 ml lysis/binding buffer [100 mM Tris–Cl, pH 8; 500 mM LiCl; 10 mM EDTA, pH 8; 1% LiDS or SDS; 5 mM dithiothreitol (DTT)].

1ciii. Homogenize sample thoroughly, taking care not to generate excessive heat or foaming.

1civ. Spin the lysate for 30 s. Transfer the supernatant to a microfuge tube and set aside while preparing Dynabeads.

1c*v*. Remove 1.5 mg Dynabeads Oligo(dT)$_{25}$, usually a 300-µl aliquot, from the resuspended stock, and place in an RNase-free tube. Then place this tube in the accompanying magnet. After 30 s, remove the supernatant completely with a micropipettor while the tube remains in the magnet.

1c*vi*. Add a 150 µl aliquot of lysis/binding buffer (100 m*M* Tris–Cl, pH 8; 500 m*M* LiCl; 10 m*M* EDTA, pH 8; 1% LiDS or SDS; 5 m*M* DTT) to the tube containing the beads. Again, remove the supernatant while the tube is in the magnet.

1c*vii*. Remove the tube from the magnet and place in an Eppendorf rack. Resuspend the beads in 150 µl of lysis/binding buffer and then transfer the beads into the tube containing the tissue homogenate from Step 1c*iv*.

1c*viii*. Mix gently and allow hybridization to proceed for 5 min at room temperature. Proceed to Step 2.

2. Place the tube containing the Dynabeads and RNA sample into the magnet. After a minimum of 30 s, pipet off the supernatant as completely as possible.

3. Remove the tube from the magnet. Wash beads with 200 µl wash buffer (0.15 *M* LiCl; 10 m*M* Tris–Cl, pH 7.5; 1 m*M* EDTA). Place the tube back in the magnet and be sure to remove all of the wash buffer supernatant. Repeat this wash step and again remove the supernatant.

Note: It is essential to pipet away all of the supernatant, especially when working in small volumes, in order to maximize the recovery of enriched poly(A)$^+$ RNA.

4. Remove the tube from the magnet, and add the desired amount of elution buffer (TE buffer, pH 7.5, or sterile H$_2$O) to the Dynabeads. One may use as little as 5 µl elution buffer per milligram of Dynabeads (from Step 1). Heat to 65°C for 2 min to favor rapid and complete separation of RNA from the Dynabeads. Without delay, place the tube into the magnet and after 30 s, transfer the supernatant to a microfuge tube and label as poly(A)$^+$ RNA.

5. The eluted RNA may be used at once. Alternatively, store the mRNA frozen in suitable aliquots at −70°C. Although some investigators add RNase inhibitors to samples that are to be stored, excellent RNase-free technique generally precludes this approach.

References

Aviv, H., and Leder, P. (1972). Purification of biologically active globin messenger RNA on oligothymidylic acid-cellulose. *Proc. Natl. Acad. Sci. USA* **69**, 1408.

Aziz, N., and Munro, H. N. (1986). Both subunits of rat liver ferritin are regulated at a translational level by iron induction. *Nucleic Acids Res.* **14,** 915.

Buell, G. N., Wickens, M. P., Payvar, F., and Schimke, R. T. (1978). Synthesis of full length cDNAs from four partially purified oviduct mRNAs. *J. Biol. Chem.* **253,** 2471.

Cole, C. N., and Stacy, T. P. (1985). Identification of sequences in the herpes simplex virus thymidine kinase gene required for efficient processing and polyadenylation. *Mol. Cell. Biol.* **5,** 2104.

Conway, L., and Wickens, M. (1985). A sequence downstream of A-A-U-A-A-A is required for formation of simian virus 40 late mRNA 3′ termini in frog oocytes. *Proc. Natl. Acad. Sci. USA* **82,** 3949.

Darnell, Jr., J. E. (1979). Transcription units for mRNA production in eukaryotic cells and their DNA viruses. *Prog. Nucleic Acids Res.* **22,** 327.

Efstratiadis, A., Kafayos, F. C., Maxam, A. M., and Maniatis, T. (1976). Enzymatic *in vitro* synthesis of globin genes. *Cell* **7,** 279.

Gibbs, R. A., Nguyen, P. N., Edwards, A., Civitello, A. B., and Caskey, C. T. (1990). Multiplex DNA deletion detection and exon sequencing of the hypoxanthine phosphoribosyltransferase gene in Lesch-Nyhan families. *Genomics* **7,** 235.

Gil, A., and Proudfoot, N. J. (1984). A sequence downstream of AAUAAA is required for rabbit beta-globin mRNA 3′-end formation. *Nature* **312,** 473.

Gilham, P. T. (1964). Synthesis of polynucleotide celluloses and their use in fractionation of polynucleotides. *J. Am. Chem. Soc.* **86,** 4982.

Green, A., Roopra, A., and Vaudin, M. (1990). Direct single stranded sequencing from agarose of polymerase chain reaction products. *Nucleic Acids Res.* **18**(20), 6163.

Hultman, T., Murby, M., Stahl, S., Hornes, E., and Uhlén, M. (1989). Solid phase *in vitro* mutagenesis using plasmid DNA template. *Nucleic Acids Res.* **18**(17), 5107.

Kramer, A. (1996). The structure and function of proteins involved in mammalian pre-mRNA splicing. *Annu. Rev. Biochem.* **65,** 367–410.

Manley, J. L. (1988). Polyadenylation of mRNA precursors. *Biochim. Biophys. Acta* **950,** 1.

McDevitt, M. A., Imperiale, M. J., Ali, H., and Nevins, J. R. (1984). Requirement for a downstream sequence for generation of a poly(A) addition site. *Cell* **37,** 993.

Mercer, J. F. B., and Naora, H. (1975). A comparison of the chromatographic properties of various polyadenylate binding materials. *J. Chromatogr.* **114,** 115.

Meyuhas, O., Bibrerman, Y., Pierandrei-Amaldi, P., and Amaldi, F. (1996). Isolation and analysis of polysomal RNA. *In* "A Laboratory Guide to RNA: Isolation, Analysis, and Synthesis" (P. A. Krieg, Ed.). Wiley Liss, New York.

Meyuhas, O., Thompson, A. E., and Perry, R. P. (1987). Glucocorticoids selectively inhibit the translation of ribosomal protein mRNAs in P1798 lymphosarcoma cells. *Mol. Cell. Biol.* **7,** 2691.

Moore, C. L., and Sharp, P. A. (1985). Accurate cleavage and polyadenylation of exogenous RNA substrate. *Cell* **41,** 845.

Nevins, J. R., and Darnell, J. E. (1978). Steps in the processing of Ad2 mRNA: Poly(A)$^+$ nuclear sequences are conserved and polyadenylation precedes splicing. *Cell* **15,** 1477.

Pierandrei-Amaldi, P., and Amaldi, F. (1994). Aspects of regulation of ribosomal protein synthesis in *Xenopus laevis. Genetica* **94,** 181.

Proudfoot, N. J., and Brownlee, G. G. (1976). 3′ Non-coding region sequences in eukaryotic mRNA. *Nature* **263,** 211.

Rosenthal, A., and Jones, D. S. C. (1990). Genomic walking and sequencing by oligo-cassette mediated polymerase chain reaction. *Nucleic Acids Res.* **18**(10), 3095.

Ross, J. (1995). mRNA stability in mammalian cells. *Microbiol. Rev.* **59**, 423–450.

Sadofsky, M., and Alwine, J. C. (1984). Sequences on the 3′ side of hexanucleotide AAUAAA affect efficiency of cleavage at the polyadenylation site. *Mol. Cell Biol.* **4**, 1460.

Williams, J. G. (1981). The preparation and screening of a cDNA clone bank. *In* "Genetic Engineering" (R. Williamson, Ed.), Vol. 1, pp. 1–59. Academic press, New York.

Ziff, E. B. (1980). Transcription and RNA processing in the DNA tumor viruses. *Nature* **287**, 491.

Quality Control for RNA Preparations

Rationale

Every application involving purified RNA requires an assessment of the quality of the sample. It is of paramount importance to demonstrate the integrity of the mRNA component. Because of the labor-intensive and costly nature of most subsequent applications, it is prudent to show that the template material is intact. One would certainly not begin with degraded or undetectable quantities and simply "hope for the best" for the same reasons that one would not attempt to build a house with wood that had rotted.

Checking a purified sample of RNA for integrity and overall quality may be accomplished in a variety of ways, some of which, in certain applications, are also very good positive control techniques. Of the methods delineated below, Quality Control Technique 1 should always be performed immediately after purification and, if the RNA has been stored for a while, a second aliquot should be tested.

Quality Control Technique 1

In general, running the RNA out on an agarose gel is the single best diagnostic that the investigator has at his disposal; parameters and protocols for electrophoresis are described in Appendix K and Chapter 9, respectively. Chemically intact, biologically competent RNA always produces a characteristic banding profile when denatured, accomplished simply by heating the sample briefly before loading it onto an agarose gel. The inclusion of formaldehyde and formamide, necessary when performing the complete Northern analysis, are optional at this stage. Following a relatively brief period of nondenaturing electrophoresis, visual inspection of the predominant 28S and 18S ribosomal RNA (rRNA) species confirms that the RNA is intact (Figure 1). Moreover, the ribosomal RNAs are usually quite evident when electrophoresed under nondenaturing conditions.

Excellent RNA (eRNA) samples show a minimum of smearing above, between, and below the 28S and 18S rRNAs. Transfer RNA (tRNA), and the low-molecular-weight 5S and 5.8S rRNA species all comigrate at the leading edge of the gel and usually appear as an indiscrete splotch. Lack of definition to the 28S and 18S rRNAs usually means that the sample has been subjected to nuclease assault, especially if the smearing is confined to the lower portion of the gel, below the level of the 18S rRNA as observed in adjacent lanes on the same gel containing intact RNA. Heavy smearing along the length of the gel suggests limited degradation

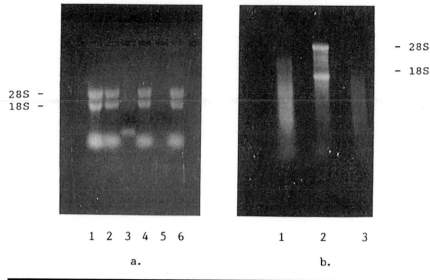

28S -
18S -

- 28S
- 18S

1 2 3 4 5 6   1   2   3

a.   b.

FIGURE 1

Assessment of the integrity of RNA samples. (a) Minigel (7 × 8 cm) electrophoresis of RNA isolated from exponential culture of *Saccharomyces uvarum.* Total RNA (5 μg) was prepared from a log-phase culture, using the yeast RNA isolation protocol found in Chapter 5. Purified yeast RNA was then electrophoresed in a 1% agarose–formaldehyde gel. The clear definition of the ribosomal 28S and 18S rRNA species in lanes 1, 2, 4, and 6 demonstrates the integrity of the sample. RNA Ladder (3 μg; Gibco BRL) was loaded into lane 3 to provide size standards. Lane 5 was left empty. (b) Medium-size gel (12 × 14 cm) electrophoresis of RNA. Samples were formaldehyde denatured and electrophoresed in a 1% agarose gel for 4 h at 50 V. Samples in lanes 1 and 3 are degraded, most likely due to RNase contamination. The sample in lane 2 is high-quality RNA, with minimum smearing above, between, and below the 28S and 18S rRNAs.

of the sample, though it is quite possible that the RNA was not denatured properly prior to electrophoresis: The resultant smear may very well reflect persistent secondary RNA structure. Detergents and excess salt in the sample can also cause smearing of the RNA and, in some cases, inhibit the entry of the RNA into the gel itself. If this is a chronic difficulty, the use of *N*-laurylsarcosine (sarcosyl) may alleviate detergent-related symptoms. Moreover, if care was not taken to remove DNA from an RNA prep, as described Chapter 5 and Appendix C,[1] contaminating

[1]Techniques for the removal of DNA from RNA preparations include differential centrifugation of intact nuclei, treatment of the sample with RNase-free DNase I, acid–phenol extraction, isopycnic partitioning of the RNA, or any combination thereof.

DNA will coprecipitate with the RNA in the final stages of RNA purification.

Genomic DNA on an EtBr- or SYBR Green-stained RNA gel appears as an area of fluorescence within and just below the well into which the sample was loaded (Figure 2). If this phenomenon is observed, it is wise to ascertain the cause of the DNA carryover before using the RNA for any type of quantitative analyses. Of course, it is also critical to understand that failure to observe high-molecular-weight fluorescence does not mean that the RNA sample is devoid of DNA, only that if present it is below the level of detection. Contaminating DNA has a profound negative influence on the outcome of most RNA-based PCR assays in its capacity as an alternative template for primers (see Chapter 15). The only reliable methods to ensure the absence of DNA from RNA samples are to incubate the sample with RNase-free DNase *and* run a control reaction that will produce a result only if genomic DNA is present in the sample, for example, a "no reverse transcriptase" control, as described in Chapter 15.

Protocol

The integrity of experimental RNA can be determined by minigel (gel dimensions: 7 × 8 cm) electrophoresis through a standard, nondenaturing 1% agarose gel. The RNA sample itself is briefly heat denatured just prior to electrophoresis in order to disrupt secondary structure.

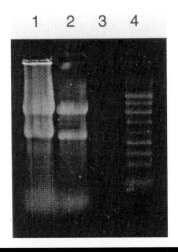

FIGURE 2

SYBR Green II staining of RNA samples. Fluorescence confined to a well into which a sample was loaded strongly suggests genomic DNA contamination.

1. Prepare a 1% solution of molten agarose in 1× TAE[2] buffer. Cast gel when agarose has cooled to 65°C.

Note: If desired, EtBr can be added directly to the molten agarose to a final concentration of 0.5 µg/ml. In so doing, the gel can be examined as soon as the electrophoresis has been completed. Do not add SYBR Green to gels prior to electrophoresis because it will profoundly interfere with the formation of sharp bands. Always stain gels with SYBR green after electrophoresis.

2. While the gel is solidifying, dilute 1–5 µg RNA in 10 µl sterile H$_2$O or TE buffer.

Note: If the RNA proves difficult to denature by heating alone, then add 2.5 µl formaldehyde (37% stock solution), and 5.0 µl formamide to the 10-µl aliquot of RNA. Otherwise, omit these denaturants.

3. Heat to 65°C for 5 min. Pulse centrifuge to collect sample at bottom of tube.

4. Add 1 µl of 10× loading buffer (50% glycerol, 1 mM EDTA, 0.25% bromophenol blue) and then load the gel.

Note 1: Use 2 µl loading buffer if formaldehyde and formamide were added to the RNA sample.

Note 2: A molecular weight marker is not needed; this is a quick and dirty assessment of the integrity of the sample.

5. Electrophorese at 75–100 Volts until dye front has migrated 2–3 cm into the gel. This is all that is necessary to visualize the sample; however, allowing the gel to run longer will certainly favor greater resolution.

6. Stain gel for 5 min (unless the stain was added directly to the agarose) in a solution of EtBr (0.5 µg/ml in H$_2$O) or SYBR Green II (1× in 1× TAE). Inspect gel as usual for indications of degradation and/or DNA contamination.

CAUTION   **When working with a transilluminator or handheld UV light source, be sure to protect eyes and skin from ultraviolet light.**

7. When performing quality control checks on RNA preps, photodocumentation (Chapter 10) is extremely important for future reference.

---

[2]50× TAE (per liter) = 242 g Tris base; 100 ml 0.5 M Na$_2$-EDTA, pH 8.0; 57.1 ml glacial acetic acid; autoclave. Melt agarose in 1× TAE buffer, and also use as the electrophoresis running buffer at a concentration of 1×.

Quality Control Technique 2

Nucleic acid concentration can be determined with minimum fanfare by the spectrophotometric quantification of the mass of RNA in a sample. The absorbance of UV light provides a direct means of determining both nucleic acid concentration and purity, as well as an indirect means of calculating the total yield from the biological source and even an estimation of the mass of RNA per cell, assuming that approximate cell number is known at the onset of the isolation. In addition, pure samples of nucleic acid have a characteristic absorbance profile between 240 and 320 nm (see "Determination of Nucleic Acid Concentration and Purity," Chapter 5). Deviations from the standard shape of the curve indicate the presence of contaminants, the nature of which is suggested by the location of the skewing of the curve. Because most downstream applications are heavily influenced by the input mass of RNA, the importance of the UV absorbance profile of a sample cannot be overstated.

Quality Control Technique 3

The polymerase chain reaction (PCR) has become a mainstream tool for the molecular biologist, and its use is widespread. Without a doubt, it has revolutionized the way most problems pertaining to cell biology are approached. As such, it is likely that in any given laboratory, PCR is being used to amplify one or another transcript,[3] using primers and conditions that have long since been optimized. The frustration index rises when one is unable to generate a product using a freshly prepared sample of RNA, or an RNA sample that "used to work." As such, one standard protocol in this laboratory is to use primers, the products from which may not be of direct experimental interest, but are at least known to work all the time when RNA is isolated, intact and clean, from a particular biological source. Testing a sample in this way can be performed using very boring primers, such as those for β-actin or GAPDH, just to show that one can at least amplify *something* from the RNA in question. If a PCR product is evident with control primers, then the reason for failure to generate a product from experimental primers can be investigated more intelligently. In summary: no signal may mean no transcription *or* that the RNA sample is poor and unable to support amplification of anything. This latter possibility will have been validated, or ruled out, through the use of these "control" primers.

---

[3]Techniques pertaining to RNA-based PCR applications are described in detail in Chapters 15–17.

Quality Control Technique 4

Northern analysis is a most useful experimental tool by which to obtain both a qualitative and a semiquantitative profile of any RNA sample. In the context of using Northern analysis as a quality control device, the stringent hybridization of a labeled oligo(dT) or poly(T) probe to total RNA or poly(A)$^+$ purified material will produce an image, by autoradiography or by chemiluminescence, which describes the size distribution of the polyadenylated component. Subsequently, the poly(T) probe can be removed from the blot for hybridization with gene-specific probes. Alternatively, aliquots of purified RNA can be probed with labeled poly(T) *before* electrophoresis to normalize RNA samples with respect to poly(A) content when the mass of the RNA is so small that it precludes normalization by conventional methodologies, such as affinity selection (Chapter 6). A detailed protocol for using a labeled poly(T) probe to normalize samples prior to electrophoresis is presented in Chapter 9.

Quality Control Technique 5

The biological integrity of mRNA is reflected by its ability to direct the translation of its encoded polypeptide. This ability may be assayed either by microinjection into a suitable cell type (*Xenopus* oocytes, for example) or in a cell-free translation system, such as a reticulocyte lysate or wheat germ extract. Biologically competent mRNAs will, when translated, generate a defined and reproducible pattern of protein bands. This technique, commonly known as *in vitro* translation, is often coupled to immunoprecipitation, predicated upon the availability of a suitable antibody. Whereas this technique is valuable as a quality control method in certain specialized applications, it is the most technically challenging and infrequently used technique. For the most part, it is not an essential quality control technique for most routine applications.

Dot Blot Analysis

Rationale

The isolation of high-quality RNA from tissue culture cells and tissue samples is merely the first (although the most critical) step in the evaluation of a model system. The subsequent analysis of RNA by Northern blot analysis, nuclease protection analysis, conversion into cDNA, and so forth can be a time-consuming and expensive road to travel. When evaluating a new model system, cell, or tissue type, or experimental regimen for the first time, it may be worthwhile to assess the system for mRNAs of interest by dot blot analysis. This simple technique allows the investigator to make "quick and dirty" statements about the biochemical composition of a sample. In addition, the dot blot approach may be useful for simply showing that the purified RNA (or DNA) is capable of hybridizing to something (anything), and that the sample warrants further characterization.

Succinctly, dot blots, and the closely related variant known as slot blots, permit rapid detection of the relative amount of a particular RNA sequence both in purified samples of RNA and in cell lysates. Salient information can be obtained without electrophoresing the sample or performing any sophisticated enzymatic manipulations, either traditional cDNA synthesis or amplification mediated by the polymerase chain reaction, although these methods will undoubtedly factor into the experimental design at a later stage. In dot-blotting, RNA samples are applied directly onto a membrane under vacuum through a multiwell dot blot (Figure 1) or slot blot (Figure 2) filtration manifold, after which the samples are immobilized onto the surface of the filter membrane. Dilutions of the samples are arranged either vertically or horizontally (Figure 3); samples arranged in this geometry are quite simple to quantify by image analysis software (Chapter 10). The degree of hybridization is assessed by measurement of each "dot" or "slot" signal from the filter, as recorded on X-ray film.

Advantages and Disadvantages

Among the most obvious advantages of the dot-blotting technique is the speed with which samples are prepped for nucleic acid hybridization: A denatured sample is applied directly to a membrane and there is no gel to run. Both previously purified RNA (from any extraction method) and partially purified cell lysates, the latter of which are occasionally referred to as cytodots or quick blots (Costanzi and Gillespie, 1987), can be utilized successfully with this type of assay. The method is rapid and facili-

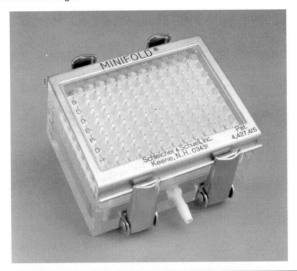

FIGURE 1

Minifold I dot blot apparatus. Sample dilutions are applied under vacuum directly to the surface of the filter membrane resting beneath the faceplate. Courtesy of Schleicher & Schuell.

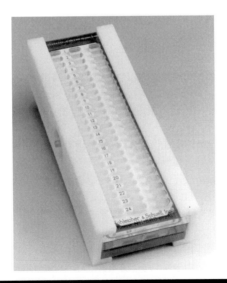

FIGURE 2

Minifold II slot blot apparatus. Sample dilutions are applied under vacuum directly to the surface of the filter membrane resting beneath the faceplate. Courtesy of Schleicher & Schuell.

Slot blot autoradiogram of the temporal pattern of ACTH-mediated increases in P-450$_{SCC}$ mRNA in bovine adrenocortical cells. Cells were harvested and the RNA extracted at the intervals indicated. Twenty micrograms of RNA was applied per sample to nitrocellulose using the Manifold II slot blotter (Schleicher & Schuell) and hybridized to nick-translated pBSCC-2 insert. The three lower samples are duplicates of those above them. Photograph courtesy of Dr. Maliyakal John. From *Proc. Natl. Acad. Sci. USA* **81,** 5628 (1984).

tates the handling of many samples simultaneously; as such, it may be used to generate a quick and dirty profile of the cellular biochemistry. That which is observed by dot-blotting may serve as a basis for the next set of experiments. Moreover, the dot blot format is ideal for making one or more dilutions of a sample. The method of dot blotting is also very easy and does not require any particular skills, other than the requirement for maintaining good RNase-free technique throughout the protocol. Assuming that the samples are applied to a nylon membrane, they can be probed repeatedly by stripping off the old probe once the data have been generated. In some cases, when an X-ray film image is not satisfactory (and when the probes have been radiolabeled), the dots themselves can be cut from the filter and dropped directly into scintillation cocktail; Cerenkov counting will generate a number (cpm) to complement the visual data generated by autoradiography.

The disadvantages of performing dot blots are closely related to the mechanics of how the method is performed. As there is no gel involved, the sample being applied directly to a membrane, dot-blotting renders quantitative data only; signal strength correlates with the abundance of a transcript and there is no way to determine the molecular weight of the transcript. Further, using this method alone, one cannot discern how many different sized transcripts have mananged to hybridize to the probe. This is especially problematic when the probe manifests even a minimal affinity for the enormously abundant ribosomal RNAs. The potential for high background hybridization thus mandates the inclusion of excellent positive and negative controls.

Appropriate Positive and Negative Controls

Dot blot and slot blot analyses yield purely quantitative data. The main drawback of this approach is that it lacks the qualitative component that accompanies electrophoresis. To be truly reliable, dot blot analysis must include excellent positive and negative controls, in order to demonstrate hybridization specificity and to gauge nonspecific probe binding to the filter membrane. For example, applying dilutions of rRNA or tRNA to unused wells would show the degree of cross-hybridization to the non-poly(A)$^+$ component of the sample, especially the ribosomal RNAs. Further, application of the popular λ-*Hind*III digested- and ϕX174-*Hae*III-digested bacteriophage genomes (used as DNA molecular weight standards on agarose gels) should yield no observable hybridization to probe molecules if the hybridization and posthybridization washes are conducted with adequate stringency. It may also be useful to apply nothing but buffer to at least one well to demonstrate lack of buffer-associated signal. Positive controls might include dilutions of cDNA complementary to the transcripts of interest (the probe itself), which, depending on the dilution prepared, may well yield the strongest signal on the filter.[1] Moreover, good internal dot blot controls are always in order: One should observe equally intense signals from wells into which equal amounts of positive control target were applied. When attempting this type of blot analysis for the first time or with a new system, it is strongly suggested that dilutions of the positive control target material be made in order to determine the linear range of the assay. For example, it would be useless, quantitatively speaking, if the hybridiza-

---

[1]Because of the very strong signals usually generated by positive controls, it is best to prepare a substantial dilution of the positive control material before applying it to the filter. Moreover, because strong signals from dot blots tend to bleed over and obscure signals from proximal wells, it is also best to avoid positioning experimental samples into those wells adjacent to the positive control.

tion signals were too intense to be accurately measured on X-ray film (recall that all films have a defined linear range).

When choosing between dot blots or slot blots, it may be useful to consider the area into which samples are concentrated by the required manifold. The slot configuration often generates a higher signal to noise ratio because of the smaller surface area into which the sample is concentrated (6 mm^2 for slots compared to 32 mm^2 for dots). Thus, the more quantitative slot blot configuration may be helpful when working with low mass quantities or in the assay of very low abundance transcripts. Practically speaking, however, it really does not make a great deal of difference how the sample is configured, as this technique, like many others, is only semiquantitative at best; the data from it are generally used as a stepping stone to a higher level of quantitativeness.

Protocol: RNA Dot Blots[2]

1. **Wear gloves!** Be sure that all reagents are purged of nuclease activity prior to contact with RNA samples.

2. Purify RNA according to any of the protocols presented in this volume or elsewhere. RNA should be dissolved in 50 µl of TE buffer, pH 7.5 (10 mM Tris–Cl, 1 mM EDTA, pH 7.5). Plan to apply 1–10 µg of RNA per dot/slot in a volume of 100–200 µl per well.

3. For each 100 µl RNA in TE buffer, add 60 µl 20× saline sodium citrate (SSC) and 40 µl 37% formaldehyde stock solution to the RNA sample. Incubate at 60°C for 15 min to denature the sample.

Note: The manifold should be prepared at this point so that the sample may be applied to the membrane immediately following denaturation.

4. While the RNA is denaturing, pre-wet the nylon or other filter in autoclaved water for 5 min. Be sure to handle the filter as little as possible and only with gloves. Equilibrate the filter in 6× SSC just prior to use and saturate two sheets of absorbant blotting paper [Schleicher & Schuell (S&S) GB003 or the equivalent] in 6× SSC as well.

Note: The ionic strength required is dependent on the chemical composition and surface charge of the filter. As always, it is best to follow the manufacturer's instructions for equilibrating any filter prior to any application.

---

[2]Protocol adapted, in part, from Schleicher & Schuell (1995).

5. Place the saturated sheets of GB003 blotting paper on the filter support plate of the filtration manifold. Place the nylon filter on top of the blotting paper, and clamp the sample well faceplate into position.

6. Apply a low vacuum to the dot blot device, ensure that the residual buffer from the pre-wetting step is being drawn through the membrane, and wash individual wells with 500 μl of 6× SSC. Apply sample to the wells in a volume of 100–400 μl per well. Dilutions, if required, are made in 6× SSC. When all of the sample has been pulled through the nylon filter, wash each well with an additional 300 μl of 6× SSC. When this wash aliquot has been pulled through the membrane, disconnect the vacuum source and remove the membrane from the manifold.

7. Immobilize the RNA onto the filter membrane according to the instructions provided by the manufacturer, usually by UV crosslinking (see Chapter 11 for immobilization strategies). If the filter will not be probed right away, store it in a cool, dry place, out of direct light. Otherwise proceed to the next step.

8. Perform prehybridization blocking, probe hybridization, and posthybridization stringency washes (parameters described in Chapter 13).

9. Perform detection protocols appropriate for the method by which the probe was labeled.

Protocol: DNA Dot Blots[3]

The ability to perform rapid nucleic acid analysis by dot-blotting also extends into the realm of DNA characterization; the method is identical to that prescribed for RNA dot blots, the only exception being the method for double-standed DNA denaturation. The same concerns addressed above for RNA also apply to the design of DNA dot blots.

1. **Wear gloves!** Be sure that all reagents are purged of nuclease activity prior to contact with DNA samples.

2. Purify DNA according to any of the standard protocols for genomic DNA or cDNA preparation in use today. A convenient DNA stock concentration is 0.5 mg/ml. DNA should be dissolved in 50 μl of TE buffer, pH 8.0 (10 mM Tris–Cl, 1 mM EDTA, pH 8.0). Plan to apply 5–25 μg of DNA per dot/slot in a volume of 100–200 μl per well, and dilute the DNA sample as needed in TE buffer or sterile H_2O.

[3]Protocol adapted, in part, from Schleicher & Schuell (1995).

Note: If genomic DNA is to be probed for the presence of specific genes, or parts thereof, it may be necessary to add even greater amounts of DNA per well to be able to detect single copy sequences.

3. For each 100 μl DNA in TE buffer, add 10 μl 2 N NaOH and incubate at 37°C for 10 min. Add 40 μl 20× SSC and then place on ice if not prepared to apply to the membrane immediately. As an alternative to the addition of 20× SSC, one may add an equal volume of 2 M NH₄OAc, pH 7.0, and place the sample on ice.

Note: The manifold should be prepared at this point so that the sample may be applied to the membrane immediately following denaturation.

4. While the DNA is denaturing, pre-wet the nylon or other filter in autoclaved water for 5 min. Be sure to handle the filter as little as possible and only with gloves. Equilibrate the filter in 6× SSC just prior to use, and saturate two sheets of absorbant blotting paper (S&S GB003 or the equivalent) in 6× SSC as well.

Note: The ionic strength required is dependent on the chemical composition and surface charge of the filter. As always, it is best to follow the manufacturer's instructions for equilibrating any filter prior to any application.

5. Place the saturated sheets of GB003 blotting paper on the filter support plate of the filtration manifold. Place the nylon filter on top of the blotting paper, and clamp the sample well faceplate into position.

6. Apply a low vacuum to the dot blot device and ensure that the residual buffer from the pre-wetting step is being drawn through the membrane. Wash individual wells with 500 μl of 6× SSC. Apply sample to the wells in a volume of 100–400 μl per well. Dilutions, if required, are made in 6× SSC. When all of the sample has been pulled through the nylon filter, wash each well with an additional 300 μl of 6× SSC. When this wash aliquot has been pulled through the membrane, disconnect the vacuum source and remove the membrane from the manifold.

7. Immobilize the DNA onto the filter membrane according to the instructions provided by the manufacturer, usually by UV crosslinking (see Chapter 11 for immobilization strategies). If the filter will not be probed right away, store it in a cool, dry place, out of direct light. Otherwise proceed to the next step.

8. Perform prehybridization blocking, probe hybridization, and posthybridization stringency washes (parameters described in Chapter 13).

9. Perform detection protocols appropriate for the method by which the probe was labeled.

Limitations of the Data

The greatest limitations of data generated by dot-blotting are the lack of a qualitative aspect because there is no gel and, because of the mechanics of the assay, the achievable sensitivity is compromised, compared to other standard techniques for the assay of gene expression. One should also be acutely aware that highly concentrated nucleic acid samples, including those suspended in relatively high salt buffers, are notorious for clogging filters when the sample is applied to the well of the dot blot manifold. Further, overloading the wells can result in "spot saturation," meaning that the investigator has applied more material than the filter can bind, or has diluted the sample in a reagent that reduces the binding capability of the membrane.

The dot blot assay, while semiquantitative, has a relatively narrow linear range, due to the ease with which one can exceed the binding capacity of the filter as well as the narrow linear range of X-ray films. Samples that generate enormous signals will literally burn out the film so that the true differences between the dots are completely obscured. Never forget that image analysis software can only analyze that which the user provides, and it is incumbent upon the investigator to at least try to minimize ambiguity wherever possible. For these reasons, dilutions of each sample are strongly recommended, at least when performing this assay for the first time with uncharacterized material, in order to assess the dynamic range of the sample. Following autoradiography, the individual dots may be cut out and the cpm measured if a number is needed to go along with the picture. While this additional numerical data may be quite helpful in this regard, it should also be obvious that cut up filters cannot be used again.

Finally, as suggested above, there really is no method by which to assess cross-hybridization within a particular spot on the filter, though well-thought-out negative and positive controls can provide fairly convincing, though indirect, evidence of the fidelity of the assay.

References

Costanzi, C., and Gillespie, D. (1987). Fast blots: Immobilization of DNA and RNA from cells. *In* Guide to Molecular Cloning Techniques (S. L. Berger, and A. R. Kimmel, Eds.). Academic Press, San Diego, CA.

Schleicher & Schuell (1995). *In* "Blotting, Hybridization, and Detection," 6th ed. Schleicher & Schuell, Keene, NH.

Electrophoresis of RNA

Electrophoresis is accomplished with the use of high-voltage power supplies that have the potential to cause life-threatening injuries. Always use extreme caution when working with such equipment.

Rationale

The expression and differential modulation of discrete genes and complex gene relays result in the precise orchestration of cellular biochemistry. One fundamental approach to the study of salient transcription events involves the analysis of messenger RNA (mRNA) and precursor heterogeneous nuclear (hnRNA) as a parameter of gene expression. Such an approach is predicated on the fact that the abundance of a particular RNA species may well represent a significant biochemical event in the cell cycle.

The sensitivity of this approach is due, in part, to the ability of nucleic acid probes to discriminate between individual members of an extremely heterogeneous mixture of RNA molecules, and do so with high fidelity. Conditions that promote hybridization between target mRNA and a complementary nucleic acid probe may be established for the qualitative and quantitative assessment of those RNA species of immediate interest to the investigator. The Northern analysis, S1 and RNase protection assays, and all forms of quantitative RNA PCR are standard techniques by which steady-state[1] RNA molecules are evaluated by size and abundance, as manifested by electrophoresis of the sample. RNA samples may be purified by any of a variety of RNA extraction procedures described in Chapter 5 and elsewhere. In each of these cases, results are evaluated by electrophoretic separation of the purified products. For example, in the Northern analysis, colloquially known as Northern blotting, electrophoretically resolved samples of total cellular, total cytoplasmic, poly(A)$^+$, and/or poly(A)$^-$ RNA are transferred to (blotted), and immobilized on, the surface of a suitable membrane (Alwine et al., 1977, 1979), such as nylon or nitrocellulose. The blot is then subjected to hybridization conditions in an attempt to base-pair specific RNA molecules on the filter to a complementary probe provided by the investigator. This is a variation on the theme of Southern analysis (Southern, 1975), also known as Southern blotting, in which electrophoretically separated DNA fragments

---

[1]The final accumulation of RNA in the cell or a subcellular compartment such as the nucleus or the cytoplasm. Analysis of steady-state RNA alone renders little information about the rate of transcription or turnover of transcripts. See Chapter 19 for details.

are blotted onto a membrane for nucleic acid hybridization analysis. Electrophoresis remains a preferred method of size-fractionating RNAs; the technique does not become limiting unless very large quantities of RNA are needed, as in biophysical studies.

Messenger RNA molecules *in vivo* are essentially single-stranded polyribonucleotides complexed to cytoplasmic proteins, which together constitute messenger ribonucleoprotein (mRNP) particles. These mRNPs presumably represent the natural form of mRNA in the cell and exhibit considerable secondary structure. This is due in part to RNA:protein interactions and in part to formidible intramolecular base pairing between complementary sequences. Virtually all RNA molecules are believed to contain short double-helical regions consisting not only of the standard A::U and G:::C base pairs but also weaker G::U base pairs. The extent of these inter- and intramolecular interactions is commonly believed to regulate the translatability of the transcripts in the cytoplasm (for review, see Lewin, 1997). Moreover, a rapidly growing body of evidence supports the notion that the transient formation of double-stranded regions of mRNA, particularly in the vicinity of the 5′ end of the molecule, may interfere tremendously with its translatability. This is essentially the basis of antisense technology, both the experimental and the clinical, in which one attempts to influence gene expression by influencing the "strandedness" of certain nucleic acid molecules in the cell.

One of the parameters that governs the migration of nucleic acids in an electric field is the molecular conformation: Secondary structures or "RNA hairpins" impede electrophoretic separation based solely on molecular weight. Some RNA molecules, of which transfer RNA (tRNA) is the premier example, manifest so much self-complementarity that the molecule as a whole assumes its very characteristic, easily recognized, and amazingly stable tertiary structure. Thus, identical species of RNA molecules exhibiting varying degrees of intramolecular base pairing will migrate toward the anode (the positive electrode) at different rates, resulting in the smearing of distinct RNA transcripts. To circumvent this problem, as it is obviously desirable for like species of RNA to comigrate, the electrophoresis of RNA is routinely carried out under what are known as **denaturing conditions**, primarily because heating alone does not solve the problem of secondary structure. The successful electrophoresis of RNA is accomplished in two parts: (1) RNA is denatured prior to loading the gel, and (2) during the electrophoresis, conditions that support and maintain the denatured state are established. If nothing else, remember that the best possible resolution is achieved when gels are poured as thin as possible and run at low voltage.

RNA Denaturing Systems for Agarose Gel Electrophoresis

The choice of denaturing system is determined primarily by the objectives of the experiment. In general, the choice of gel matrix and denaturant is a function of the size range of the RNAs to be separated and also whether the characterization of RNA is to be preparative or strictly analytical in nature. For optimal resolution of very small nucleic acids (fewer than 500 bases or base pairs) the matrix of choice is polyacrylamide (3–20%) (Maniatis et al., 1975). Such applications routinely include the S1 nuclease assay (Berk and Sharp, 1977; Favaloro et al., 1980; Sharp et al., 1980), the RNase protection assay (Zinn et al., 1983; Melton et al., 1984), and DNA sequencing. However, for procedures such as the Northern analysis, the investigator is working with a sample containing thousands of different transcripts, all of which are of different sizes: The optimal balance between electrophoretic resolution and efficiency of transfer (blotting) from the gel to a membrane for hybridization is achieved with a 1.0–1.2% denaturing agarose gel.

The most commonly used RNA denaturants for agarose gel electrophoresis include formaldehyde[2] (Boedtker, 1971; Lehrach et al., 1977; Rave et al., 1979) and glyoxal/dimethyl sulfoxide (DMSO) (McMaster and Carmichael, 1977; Bantle et al., 1976; Thomas, 1980). As an historical note, the methylmercuric hydroxide system (Bailey and Davidson, 1976; Alwine et al., 1977, 1979) was the first to be used to denature RNA for agarose electrophoresis. Ironically, while it is the most efficient RNA denaturant (Gruenwedel and Davidson, 1966), it is also by far the most toxic. Due to the great toxicity of this chemical (for toxicity review, see Junghans, 1983) it is not in use today for this application. For polyacrylamide gels, formamide[3] (Spohr et al., 1976) and urea (Reijnders et al., 1973) have been used successfully as RNA denaturants. Each of these denaturing systems has unique characteristics and safety requirements as well as distinct advantages and limitations. Methodologies pertaining to the use of formaldehyde and the glyoxal/DMSO system are presented in this chapter. Because of the intrinsic difficulties and great health hazards

[2]**Caution:** Formaldehyde is carcinogenic and a teratogen.
[3]**Caution:** Formamide is carcinogenic and a teratogen.

associated with methylmercuric hydroxide it is not recommended as a denaturant, nor is its use described herein.

Normalization of Nucleic Acids

In the preparation of RNA samples for electrophoresis, it is incumbent
upon the investigator to ensure that equal quantities of RNA are "normalized" and loaded in each lane of the gel. But what, exactly, constitutes a
battery of normalized samples? Do the normalizations of samples of total
cellular RNA, total cytoplasmic RNA, and poly(A)$^+$ mRNA have the
same meaning? Normalization must constitute a logical and proper basis
for the comparison and interpretation of hybridization signal intensity observed in each experimental sample; it must also give meaning to signals,
if any, generated from positive and negative controls. The ramifications
of normalization are far-reaching indeed.

When purified poly(A)$^+$ mRNA is to be electrophoresed, loading
equal microgram quantities into each lane is a widely accepted normalization protocol (usually 2–5 μg per lane is sufficient to detect many low
abundance mRNAs). In other cases, however, it may be desirable or even
necessary to electrophorese total cellular RNA or total cytoplasmic RNA
without prior selection of the poly(A)$^+$ RNA. For example, the yield of
poly(A)$^+$ RNA that can be generated in a particular model system may
be so small as to preclude its isolation by conventional oligo(dT) chromatography. Moreover, RNA samples enriched for mRNA by poly(A)
selection are often contaminated with varying amounts of non-poly(A)
material.

The most prevalent component of a purified sample of total cellular or
total cytoplasmic RNA is ribosomal RNA (rRNA), which constitutes
80–85% of the sample by mass. This is clearly an enormous excess compared to the mass of the mRNA transcripts, which is usually no more than
3–4% of the total. Thus, changes in the transcription of discrete mRNAs,
even significant changes, are virtually undetectable by standard UV spectrophotometry. Yet, accurate normalization of experimental samples mandates a rationale that will ensure that any observed variations in the abundance of specific mRNAs truly indicate differential transcription or posttranscriptional regulation of these species (as opposed to an overall
change in the transcription of all RNA species) and that these observed
changes are meaningful. These considerations have across the board applicability to all assays with any quantitative component.

In order to at least partially address these concerns, one might consider the assay of poly(A)$^+$ material only. Although the shortcomings of this approach should be clear from the arguments presented under "The Poly(A) Caveat" (Chapter 6), loading equal amounts of poly(A)$^+$ RNA (both hnRNA and mRNA) may be considered at least one acceptable attempt to normalize samples. Another method for normalization, best suited for when the availability of very small quantities of RNA, cells, or tissue precludes poly(A)$^+$ selection, involves sythesis of a labeled poly(T) probe (Fornace and Mitchell, 1986; Hollander and Fornace, 1990; Farrell and Greene, 1992). The probe, prepared essentially by performing the synthesis of first-strand cDNA, can be used to estimate the poly(A) content of various samples; an adaptation of this high-precision technique is presented here. With this normalization approach, a minimum amount of cellular RNA (500 ng or less) is sacrificed for slot blot or dot blot analysis. The observed hybridization of the poly(T) probe is a barometer of the poly(A) content of the sample (Figure 1). When compared with the hybridization signal intensity observed in other samples under investigation, this information may be used to normalize samples of total RNA such that equivalent quantities of poly(A)$^+$ material are loaded into each lane of gel or used for any other analysis requiring precise normalization. This technique accomplishes the equivalent of UV spectrophotometric

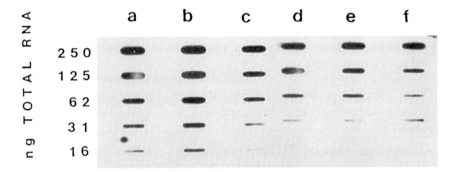

FIGURE 1

Slot blot normalization of poly(A)$^+$ content in samples of total cytoplasmic RNA. Serial twofold dilutions of 500 ng (A_{260}) were made in phosphate buffer and 100 μl was applied to each slot. Samples were then hybridized to a ^{32}P-labeled poly(T) probe to demonstrate that the samples of total RNA normalized by mass contained equal mass quantities of poly(A)$^+$ mRNA in the absence of oligo(dT) affinity selection. Autoradiography, coupled with scanning densitometry (not shown), revealed that although an equal mass of total RNA was applied to each slot, the poly(A)$^+$ component of these samples varies from two- to fourfold.

normalization of samples of purified poly(A)$^+$ mRNA; failure to normalize with respect to poly(A)$^+$ content may result in observed variations that result directly from loading different quantities of poly(A)$^+$ RNA into each lane.

The most important point to remember with respect to the normalization of several samples is that RNA electrophoresis and Northern analysis, while historically venerable techniques, are very close to the bottom of the sensitivity index (Chapter 20) with respect to the assay of RNA as a parameter of gene expression. Northern analysis, in which RNA electrophoresis is an initial step, should be used to roughly gauge the behavior of transcripts in the cell; the need for fine-tuned analysis of gene expression should then lead the investigator to supersensitive RNA PCR-based assays.

Protocol

Sample Preparation

1. Remove 500 ng (A_{260}) of RNA from each sample to be normalized and dissolve each in 200 μl ice-cold 50 mM phosphate buffer, pH 6.8 (50 mM Na$_x$H$_x$PO$_4$ prepared from the monobasic and dibasic salts; refer to Table 1).

2. Using a slot blot or dot blot manifold (Schleicher & Schuell), apply half of each sample (100 μl = 250 ng) to a nylon filter that was pre-wet first in H$_2$O and then in phosphate buffer.

Note: In this case the use of the slot configuration is preferred to the dot configuration because the smaller area into which the sample is concentrated results in more quantitative data.

3. To the remaining 100 μl of RNA sample, add a fresh 100-μl aliquot of phosphate buffer, effecting a twofold dilution. Usually five twofold dilutions blotted onto a filter membrane are sufficient to accurately normalize the sample.

4. Wash each well twice with 300 μl phosphate buffer.

5. Positive controls may be generated with the dilution of commercially available pure poly(A), which, at greater dilutions, can serve as a standard for estimating the absolute poly(A)$^+$ content of the sample. Wells into which tRNA or non-poly(A+T)-rich plasmid DNA is added serve as excellent negative controls.

6. Immobilize the RNA on the filter paper as described by the manufacturer, ideally by UV crosslinking. Begin prehybridization or store filter in a cool, dry place until ready to use.

Prehybridization—Option 1

7a. Church blots (Church and Gilbert, 1984): Equilibrate filters for a minimum of 20 min in prehybridization solution (1% bovine serum albumin; 1 mM EDTA; 0.5 M Na$_x$H$_x$PO$_4$, pH 7.2; 7% SDS) at 65°C.

Prehybridization—Option 2

7b. Soak filter(s) in 5× SSPE for at least 15 min and then prehybridize in a volume of at least 100 µl/cm^2 prehybridization buffer (5× SSPE, 5× Denhardt's solution, 0.1% SDS, 100 µg/ml denatured salmon sperm DNA, 50% formamide). Allow prehybridization to proceed for 2–3 hours at 42°C.

Note: The inclusion of SSPE rather than SSC in the prehybridization buffer tends to result in improved background because the phosphate in the SSPE buffer mimics the phosphodiester backbone of nucleic acids and can actually help to block the filter. In addition, gentle rocking of the filter, as with an orbital shaker incubator (Lab-Line), will yield superior data.

Synthesis of Poly (T) Probe

8. The poly(T) probe necessary for this experiment can be synthesized essentially as first-strand cDNA by the reverse transcription of a poly(A) template (polyadenylic acid). A typical reaction (100 µl volume) consists of 250 ng/µl poly(A); 2 mM DTT; 50 mM Tris–Cl, pH 8.3; 40 mM KCl; 10 mM MgCl$_2$; 0.5U/µl RNasin (Promega); 1 mM TTP; 0.05 µg/µl oligo(dT)$_{12\text{-}18}$; 100 µCi [α-^{32}P] TTP (3000 Ci/mmol)[4]; 5 U/µl AMV reverse transcriptase. Any first-strand cDNA synthesis kit is satisfactory. The probe length should be about 200–300 bases. Alternatively, a nonisotopic poly(T) probe can be generated using any of the methods described in Chapter 12.

9. Incubate at 42°C for 1.5 h.

10. Terminate the reaction with the addition of 100 mM NaOH and 10 mM EDTA. Incubate at 65°C for 30 min to hydrolyze the poly(A) template.

Historical note: This is analogous to the removal of the RNA template following first-strand synthesis during classical cDNA synthesis.

11. Neutralize the alkaline reaction mixture with 100 mM Tris–Cl, pH 7.5, for 15 min at room temperature.

---

[4]Alternatively, 250 µCi of [^{35}S]thymidine 5′-(α-thio)triphosphate can be used to label the probe.

12. Radiolabeled probe must be separated from unincorporated radionucleotides and other reaction components. This is easily accomplished by spun-column chromatography (Sambrook *et al.,* 1989) through Sephadex G-25 (Pharmacia Biotech) or by using another commercially available concentrating/purification device. Unused probe (and previously used hybridization buffer containing probe) can be stored for up to several weeks at –20°C for future use.

Hybridization

13. Replace prehybridization buffer with a fresh aliquot of the same prehybridization buffer, using no more than 100 μl/cm² filter. Add the poly(T) probe at a concentration of 100 ng/ml **or** 1×10^6 cpm/ml.

14. Hybridize blots with gentle agitation for 3–4 h at 44°C.

Posthybridization Washes

15. At the conclusion of the hybridization period remove the probe solution. The probe can be saved (frozen) for subsequent hybridization analyses. If filters were prehybridized/hybridized in Church blot buffer (Option 1), then wash blots twice for 10 min each in 200 mM Na⁺ at room temperature, once for 30 min at 44°C, and finally once in 75 mM NaCl at room temperature for 5 min. If filters were hybridized in SSPE-containing buffer (Option 2), then wash the filters twice in 2× SSPE, 0.1% SDS for 10 min at room temperature and then wash twice in 0.1× SSPE, 0.1% SDS for 10 min at 37°C. Use a volume of at least 100 ml of buffer for each wash.

16. Wrap the filters, while still damp, in plastic wrap and then set up autoradiography.[5] With this type of hybridization a single overnight exposure at –70°C with an intensifying screen is quite sufficient.

Note: After an autoradiogram has been generated, it may also be helpful to measure the extent of hybridization by Cerenkov counting, in which individual dilutions (i.e., the dots or slots) are cut from the membrane and dropped into scintillation cocktail.

17. The poly(A) content of each sample, as determined by the degree of poly(T) probe hybridization, permits samples of total cellular or total cytoplasmic RNA to be normalized to that of a reference sample. In many cases this may simply be a control or untreated sample. All samples are then prepared for electrophoresis.

---

[5]Detection mechanics will be different for chemiluminescence. See Chapter 14.

Formaldehyde

Formaldehyde and formamide are very commonly used denaturants of RNA. Both are carcinogens and teratogens; these should be handled in a chemical fume hood and avoided altogether by pregnant women. To ensure that the RNA migrates only with respect to molecular weight, samples of RNA are denatured with both formaldehyde and formamide prior to electrophoresis, with formaldehyde being added to the gel to maintain the denatured state during electrophoresis. Formaldehyde is routinely supplied as a 37% stock solution (12.3 M), containing 10–15% methanol as a preservative. Supplied in a dark brown bottle, formaldehyde should be stored at room temperature and out of direct sunlight. Formaldehyde oxidizes when it comes into contact with the air. As such, the pH of formaldehyde should be checked prior to each use, and must be at pH greater than 4.0 for RNA work. At pH below 4.0 RNA is likely to experience degradation. If necessary, deionize formaldehyde in the same manner as formamide (Appendix E). In many laboratory settings, formaldehyde gels furnish greater detection sensitivity than the glyoxal denaturing system, due in part to difficulties associated with the complete removal of glyoxal from the RNA prior to hybridization.

The running buffer of choice for formaldehyde gels is 1× MOPS[6] buffer. This buffer is conveniently prepared as a 10× stock (0.2 M MOPS, pH 7.0; 50 mM sodium acetate; 10 mM EDTA, pH 8.0) and, if made properly, assumes a characteristic straw color when autoclaved. Alternatively, MOPS buffer can be twice filtered through a nitrocellulose membrane to eliminate nuclease activity and to extend its shelf life.

Gels containing formaldehyde are less rigid than other agarose gels and are considerably more slippery. It is always wise to support such gels from beneath, as with a spatula, when moving them from place to place. If you are running an RNA gel for the first time, the information at the end of this chapter may be very useful.

Protocol

1. **Wear gloves!**
2. To prepare 200 ml of molten agarose solution (more than enough for two 12 × 14-cm gels), melt 2.4 g of agarose in 170 ml sterile H_2O in an Erlenmeyer flask on a hot plate or in a microwave oven. Be certain that the

---

[6]MOPS, 3-[N-morpholino]propanesulfonic acid; 10× MOPS buffer = 0.2 M MOPS, pH 7.0; 50 mM NaOAc; 10 mM EDTA.

agarose has melted *completely* by gently swirling the flask and looking for undissolved beads of agarose "fish-eyes." When the gel has cooled to 55–60°C, add a mixture of 20 ml prewarmed 10× MOPS buffer and 10 ml prewarmed 37% formaldehyde. This will produce a solution of 1.2% agarose in 1× mops buffer and 0.6 *M* formaldehyde. Keep this molten agarose solution covered and in a 55–60°C water bath until ready to cast the gel.

> *Note: Certain older protocols recommend a final formaldehyde concentration of 2.2 M, which is much too high. Most investigators use a maximum final concentration of 0.6 M, which, if desired, may be further reduced to 0.22 M.*

3. Cast a 0.5- to 0.75-cm-thick gel using the solution prepared in Step 2. Be sure to cast the gel in a fume hood to minimize formaldehyde fumes in the room. Remember that the thinner the gel, the better the resolution of the RNA.

4. Prepare each sample for electrophoresis by mixing the following in a sterile microfuge tube:

4.7 μl RNA (up to 20 μg)

2.0 μl 5× MOPS buffer

3.3 μl formaldehyde

10.0 μl formamide

Be certain to mix thoroughly by pipetting and then pulse centrifuge to collect the reaction components at the bottom of the microfuge tube.

> *Note 1: Do not exceed 20 μg RNA per lane, beyond which one runs the risk of overloading the lanes of the gel and losing resolution. If rare or very low abundance mRNAs are of interest it may be worthwhile to load as much as 4–5 μg of poly(A)$^+$ RNA per lane instead.*

> *Note 2: Use deionized formaldehyde and formamide.*

5. Prepare 4–5 μg of RNA molecular weight standard for electrophoresis, as described in Step 4.

> *Note: These standards, prepared by* in vitro *transcription, are single stranded and need to be denatured, too.*

6. Denature all samples and the molecular weight marker by heating to 55°C for 15 min. Alternatively, heat the sample to 65°C for 10 min.

7. During the incubation of the RNA and once the gel has solidified, immerse the gel in running buffer (1× MOPS).

Note: Be sure to use the smallest amount of buffer necessary to completely cover the gel. Carefully remove the comb by pulling it upward slowly and evenly. Covering the gel with buffer prior to removal of the comb will lubricate the wells and reduce the vacuum effect when the comb is pulled. If the gel is not immersed in buffer, the vacuum may be so great as to pull out the bottom of the wells that are formed by the comb.

8. Following denaturation, add 2 µl of 10× loading buffer (50% glycerol; 1 mM EDTA, pH 8.0; 0.25% bromophenol blue; optional: 0.25% xylene cyanol) to the samples and mix thoroughly by micropipetting. Load samples onto the gel immediately.

Note: Because some condensation may form on the inside of the tube during the denaturation incubation period be sure to briefly microcentrifuge the tube just prior to the addition of the loading buffer.

9. Electrophorese the samples at a maximum of 5 V/cm distance between the electrodes. Continue electrophoresis until the bromophenol dye front has migrated to within 1–2 cm from the distal edge of the gel. It is especially important to stop the electrophoresis at this point when one is evaluating RNA species of unknown size. In subsequent experiments, RNA gels may be electrophoresed for longer periods once it has been determined empirically that the RNA species of interest will not migrate off the edge of the gel.

10. At the conclusion of the electrophoresis, the gel may be stained in a solution of 0.5 µg/ml ethidium bromide (EtBr; in sterile water or in a fresh aliquot of 1× MOPS buffer) or 1× SYBR Green II (in 1× TBE or 1× TAE buffer) in order to visualize the RNA within the gel. Staining techniques and options are found later in this chapter.

Note: One phenomenon commonly associated with the EtBr staining of formaldehyde gels is the quenching of the fluorescence of the ethidium bromide in gels that contain excessive amounts of formaldehyde. Gels so treated frequently appear to have taken on a strong orange/pink hue to the extent that any nucleic acid in the gel is almost completely obscured. Such gels usually require extensive destaining in sterile (DEPC-treated) H$_2$O or MOPS buffer for 1 h or longer.

11. Regardless of whether the gel was stained/destained, it should be soaked in 1× MOPS buffer or sterile H$_2$O to remove formaldehyde from the gel. Typically, a 30-min soaking in several changes of buffer or sterile

H_2O is adequate for formaldehyde removal from the gel. Alternatively, the gel may be stored in 300 ml sterile H_2O overnight at 4°C.

Glyoxal/Dimethyl Sulfoxide

Another common method of denaturing RNA in preparation for electrophoresis is to treat the sample with a mixture of 1 M glyoxal and 50% (v/v) DMSO at 50°C for 1 h (Bantle *et al.*, 1976; McMaster and Carmichael, 1977). Glyoxalation (Salomaa, 1956) introduces an additional ring into guanosine residues and, in so doing, sterically interferes with G:::C base pairing (Hutton and Wetmur, 1973). At high concentrations, glyoxal has been shown to react with all bases of both RNA and DNA (Nakaya *et al.*, 1968), the guanosine–glyoxal adduct being the most stable (Broude and Budowsky, 1971; Shapiro *et al.*, 1970). The preparation of RNA in this fashion precludes some of the potential health hazards associated with the use of formaldehyde (and certainly methylmercuric hydroxide). RNA samples are denatured with glyoxal prior to electrophoresis although, unlike the formaldehyde system, glyoxal is not added to the gel.

Glyoxal oxidizes very rapidly; therefore, glyoxal stock solutions (40% glyoxal = 6 M) must be deionized to neutral pH before use (see Appendix E). If not deionized, the oxidation product, glyoxylic acid, will cause fragmentation of the RNA sample. For the convenience, small aliquots of deionized glyoxal can be stored at −20°C in tightly capped tubes. Thawed aliquots of deionized glyoxal should be used only once, and the unused portion discarded.

Glyoxalated DNA and RNA molecules are electrophoretically equivalent (McMaster and Carmichael, 1977); therefore, small DNA molecules of known molecular weights can be used reliably along side RNAs of unknown size. For example, triple digestion of the 4361-bp plasmid pBR322 with the *Pst*I, *Pvu*II, and *Hind*III restriction endonucleases yields three double-stranded fragments that are 779, 1545, and 2037 bp in length. In the absence of suitable RNA size standards, glyoxalation of these fragments may be of great utility. Moreover, if the hybridization probe is cloned into pBR322 or any vector that shares sequence homology with pBR322, then the markers will act as target sequences, hybridize with the probe, and produce a permanent marker image, autoradiographic or otherwise, during detection. pBR322 size markers generated in this fashion are also extremely well suited for Southern analysis.

Gels resolving glyoxalated RNA should be run more slowly than formaldehyde gels to prevent the rapid formation of a pH gradient; glyoxalated nucleic acids also intrinsically exhibit decreased electrophoretic mobility than an identical sample denatured with formaldehyde. In return, the banding of the glyoxalated RNA tends to be sharper than that in formaldehyde denaturing systems. The major drawback of the typical glyoxal gel is that the buffering capacity of the standard running buffer, 10 mM sodium phosphate, pH 7.0, is quite poor. This buffer has been used in the past because low ionic strength electrophoresis buffers at pH 7.0 maintain the maximum glyoxal denaturation effect and preclude the need for adding a denaturant to the gel itself. The inability of the sodium phosphate buffer to resist radical changes in pH, however, necessitates recirculation of the running buffer throughout the electrophoresis period to maintain pH 7.0. This is imperative because glyoxal will dissociate from RNA (and DNA) at pH > 8.0 (Thomas, 1980), not to mention the fact that the RNA sample itself is susceptible to alkaline hydrolysis. If recirculation is not possible, the power supply should be disconnected and the buffer changed every 30 min during the run, though this approach is usually quite unsatisfactory. Alternatively, glyoxal gels may be electrophoresed in 1× MOPS buffer, rather than phosphate buffer, thereby eliminating the need to recirculate the buffer. If you are running an RNA gel for the first time, the information at the end of this chapter may be very useful.

Protocol

In advance: Prepare sodium phosphate buffer by combining stock solutions of Na_2HPO_4 (dibasic) and NaH_2PO_4 (monobasic) in the appropriate ratio to yield a solution of the desired pH (Table 1). The resulting sodium phosphate buffer is often described chemically as $Na_xH_xPO_4$, referencing the inclusion of both the monobasic and dibasic salts in the formulation of this buffer.

Prepare 0.2 M stock solutions of Na_2HPO_4 and NaH_2PO_4, and then autoclave or filter sterilize. Next, combine Na_2HPO_4 and NaH_2PO_4 in the proportions indicated in Table 1 to produce the desired pH of the final mixture. Then add 90% of the remaining water required to dilute the mixture. Check the pH of the diluted phosphate buffer. It is almost always necessary to make minor pH adjustments: add NaH_2PO_4 to lower the pH or Na_2HPO_4 to raise the pH. Important: fine-tuning the pH must be performed with phosphate buffer that has been diluted to the desired final concentration, rather than using the 0.2 M stock solutions. Finally, add sterile H_2O to produce the final volume.

TABLE 1		

Preparation of Na$_x$H$_x$PO$_4$

NaH$_2$PO$_4$	Na$_2$HPO$_4$	Resulting pH
93.5	6.5	5.7
92.0	8.0	5.8
90.0	10.0	5.9
87.7	12.3	6.0
85.0	15.0	6.1
81.5	18.5	6.2
77.5	22.5	6.3
73.5	26.5	6.4
68.5	31.5	6.5
62.5	37.5	6.6
56.5	43.5	6.7
51.0	49.0	6.8
45.0	55.0	6.9
39.0	61.0	7.0
33.0	67.0	7.1
28.0	72.0	7.2
23.0	77.0	7.3
19.0	81.0	7.4
16.0	84.0	7.5
13.0	87.0	7.6
10.5	90.5	7.7
8.5	91.5	7.8
7.0	93.0	7.9
5.3	94.7	8.0

Note. These are the theoretical proportions of 200 m*M* stock solutions of NaH$_2$PO$_4$ and Na$_2$HPO$_4$ required to prepare pH-defined sodium phosphate buffer. Be certain to check pH after mixing and diluting NaH$_2$PO$_4$ and Na$_2$HPO$_4$. Minor adjustments in pH are frequently required. See text for details.

Gel and Sample Preparation

1. **Wear gloves!**

2. Melt agarose (final concentration of 1.2%) in an appropriate volume of 10 m*M* Na$_x$H$_x$PO$_4$, pH 7.0.

3. When molten agarose has cooled to 55–60°C, cast a 12 × 14-cm gel and allow to solidify.

Note: Do not add ethidium bromide to glyoxal gels, as this dye will react with the glyoxal. The inclusion of SYBR Green should also be avoided.

4. Denature the RNA in an autoclaved microfuge tube by adding:

$\quad$ 3.7 μl RNA (up to 10 μg per lane)

$\quad$ 2.7 μl 6 M glyoxal (freshly deionized)

$\quad$ 8.0 μl DMSO

$\quad$ 1.6 μl 100 mM Na$_x$H$_x$PO$_4$, pH 7.0

These proportions will render final concentrations of 1 M glyoxal, 50% DMSO, and 10 mM Na$_x$H$_x$PO$_4$.

5. Close the tube tightly and incubate the mixture at 50°C for 1 hr.

6. After the gel has solidified, immerse it in running buffer (10 mM Na$_x$H$_x$PO$_4$, pH 7.0). This should be done while the RNA is being glyoxalated.

Note: Be sure to use only the smallest amount of buffer necessary to completely cover the gel. Carefully remove the comb by pulling it upward slowly and evenly. Covering the gel with buffer prior to removal of the comb will loosen the comb and minimize the vacuum created when the comb is pulled. If the gel is not immersed in buffer, the vacuum may damage the wells that are formed by the comb.

7. At the conclusion of the incubation period, cool the mixture to 20°C and add 2 μl of loading buffer (50% glycerol; 10 mM Na$_x$H$_x$PO$_4$, pH 7.0; 0.25% bromophenol blue). If necessary, pulse centrifuge the tubes for 2–3 s to collect the entire sample at the bottom of the tube.

8. Immediately load the sample onto the gel.

9. Electrophorese at 3–4 V/cm distance between the electrodes, with constant buffer recirculation. Electrophoresis should continue until the bromophenol blue tracking dye has migrated about 80% along the length of the gel.

Note: It is especially important to stop the electrophoresis at this point when one is evaluating RNA species of unknown size. In subsequent experiments, RNA gels may be electrophoresed for longer periods once it has been empirically determined that the RNA species of interest will not migrate off the edge of the gel.

10. At the conclusion of the electrophoresis, the gel can be used for Northern transfer without prior staining. If desired, the gel can be stained in a solution of 0.5 μg/ml ethidium bromide made up in 100 mM ammonium acetate or in a solution of 2× SYBR Green II made up in 1× TBE buffer.

Note 1: The investigator should be aware of potential interaction be-tween glyoxal and ethidium bromide, which may change its spectral properties (Johnson, 1975; McMaster and Carmichael, 1977) and re-duce transfer efficiency.

Note 2: As an alternative to EtBr and SYBR Green, the gel may be stained in a 30 μg/ml solution of acridine orange (made up in DEPC-treated 10 mM sodium phosphate, pH 7.0). This is not recommended, however, because acridine orange results in notoriously high levels of background in the gel, requiring considerable destaining. The destain-ing is ideally performed for about 1 h at room temperature in an enam-eled metal pan (McMaster and Carmichael, 1977) because enamel ad-sorbs acridine orange. To remove acridine orange from the enamel, rinse with running hot tap water for 10–15 min. Polyacrylamide gels are destained in the dark for 2 h at room temperature or at 4°C overnight.

11. Glyoxalated RNA may be transferred from the agarose gel to a ny-lon or nitrocellulose membrane without any pretreatment to remove gly-oxal. Glyoxal is removed from the RNA during a posttransfer, prehy-bridization wash of the filter paper (described in Chapter 11).

Note: Soaking the gel in alkali to remove the glyoxal is not recom-mended because such treatment can reduce transfer efficiency and, if not monitored, hydrolyze the RNA to a point at which it is no longer hybridization-competent.

Molecular Weight Standards

The accurate determination of the size of RNA species is just as im-portant as deduction of the molecular weight of DNA, proteins, and sug-ars. Because of the intrinsic chemical differences between DNA and RNA, although both are nucleic acids, the length of RNA molecules in-creases their rate of migration faster than that of double-stranded DNA size standards of comparable length.[7] This is especially true of agarose gels that contain formaldehyde (Wicks, 1986). Clearly, it is more desir-able to electrophorese RNA molecular weight standards in an RNA gel, as opposed to the use of the popular bacteriophage λ (*Hind*III digest) and φX174 (*Hae*III digest) genomic DNAs. DNA and RNA are not elec-

---

[7]Some investigators do rely on double-stranded DNA standards for the sizing of small RNAs, typically fewer than 400 bases.

trophoretically equivalent unless both have been glyoxalated (McMaster and Carmichael, 1977); even then DNAs are useful as molecular weight standards (also known as size standards or markers) only if they fall within the range of the majority of cellular RNAs.

Useful markers for RNA electrophoresis are not at all difficult to come by even in laboratories with minimal funding. There are two basic approaches for the size determination of RNAs of interest:

1. **External molecular weight standards** are nucleic acid molecules of known size that are electrophoresed on the same gel as experimental samples in an unoccupied lane. These standards are commercially available from most molecular biology vendors. For RNA gels, molecular weight standards typically consist of a mixture of 5–10 single-stranded RNAs that were produced by *in vitro* transcription. Because these markers are, in fact, single stranded, they must be denatured by the same method used to denature the samples. Care must be taken to protect RNA molecular weight markers from degradation; these are just as susceptible to RNase as the experimental RNA.

Another possibility, for glyoxal systems, is to glyoxalate a triple digest (*Pst*I, *Pvu*II, *Hind*III) of the 4361-bp pBR322 plasmid for use as a marker (discussed earlier in this chapter). This digestion will yield three fragments which are 779, 1545, and 2037 bp in length.

2. **Internal molecular weight standards** are RNA species of known size that are actually part of the RNA experimental sample, most often the large and small ribosomal RNAs. As these markers are part of the sample, they will be denatured during preparation of the sample for electrophoresis.

The molecular weights of the ribosomal RNAs are known for many organisms, and their relative positions on a gel are almost always indicated to provide the reader with at least an approximate frame of reference. Whereas more precise measurements of molecular weight are possible using a mixture of RNA size standards, the use of the ribosomal 28S and 18S RNAs[8] alone or in conjunction with other size standards is nearly universal. In addition to rRNA, Northern blots are often subjected to hybridization with a probe for a "housekeeping" gene, that is, a gene whose expression is not expected to change as a function of an experimental stimulus or manipulation. Popular housekeeping genes include β-actin, fibronectin, histone, GAPDH, and transferrin receptor mRNAs; in addi-

---

[8]In prokaryotic cells, the 23S and 16S rRNAs are the counterparts of the eukaryotic 28S and 18S rRNAs.

tion one may use the total mass of tRNA as a method by which to verify normalization. The reader is cautioned, however, that there is no single transcript that is optimal for normalization because different cells respond differently to various experimental stimuli: Subtle transcriptional and/or posttranscriptional variations may accompany experimentally induced permutations in the cellular biochemistry. See "Normalization of Nucleic Acids," this chapter, for details and suggestions for normalizing samples with respect to poly(A)$^+$ content.

Proper Use of Size Standards

Either internal or external size standards can be used with confidence to discern the sizes of RNAs under investigation. Relying on one type or the other, or both, has several advantages as well as drawbacks. With respect to external standards, the key decisions that must be made are:

1. The amount of the standard to use;
2. The positioning of the standard in the gel;
3. The size range of the standard;
4. The method of visualizing the markers (staining vs. end-labeling vs. detection by hybridization).

One characteristic of several brands of premade RNA size standards is that the DNA template upon which the transcripts are synthesized is not always completely removed from the *in vitro* transcription reaction. Further, many of these templates exhibit limited sequence homology with plasmid vectors commonly used for molecular cloning. A possible consequence of this type of construction is manifested when a nucleic acid probe is labeled and used for hybridization without cutting the probe itself out of the vector, or a probe that is removed from a vector is still flanked by limited sequences from the vector. Although one would certainly expect (hope) that the probe sequence would hybridize with great discrimination, vector sequences may well hybridize to residual template in the marker lane. Because this hybridization is probably an exact match (nucleotide for nucleotide), even subnanogram quantities of template are more than sufficient to generate a strong signal upon hybridization detection that may actually bleed over and obscure signals from hybridization events that may have occurred in adjoining lanes. If there is reason to believe that this might occur, it is judicious to leave a blank lane in the gel between the size standards and the experimental samples, and to use nanogram quantities of the RNA marker.

On the other hand, it will be necessary to load about 3–4 μg/lane in order to have sufficient mass to photograph the gel after staining. The cross-hybridization phenomenon will not occur if vector sequences are not present, or if completely unrelated vector sequences are involved. If the pBR322 triple-digest format is used, these fragments produce strong hybridization signals when pBR322-associated vectors are involved.

If no vector sequences are involved or if it is known that the probe was cloned into a vector that does not hybridize to components of the marker, then one may confidently load 3–4 μg/lane, a more than sufficient mass for photodocumentation. If vector sequences are involved, as little as 15 ng of marker will hybridize very efficiently to complementary vector sequences and render sharp bands at the detection stage.

Just as the mass of marker is an important parameter, so is the placement of the marker. If nothing else, and especially if running gels for the first time, always load the gel asymmetrically. Very little is more frustrating than obtaining a significant banding pattern and not knowing which lane is which. The confusion can be further compounded by the fact that some investigators invert the gel during the transfer, and then fail to maintain the orientation of the filter paper after it is removed from the surface of the gel following the transfer. Were this not bad enough, inversion of the exposed X-ray film during the developing steps adds yet another level of confusion. When the markers are loaded in the center of the gel, the distinction between left and right becomes further obscured. Consistently loading the gel in an asymmetrical fashion will eliminate at least one source of ambiguity.

The range of RNA size standards differs from one source to the next. New systems are best evaluated using a set of markers that spans the range of the RNAs in the sample (500–10,000 bases is a good place to start). After some definition has been given to "typical results" from a model system, one may opt to select a preparation of very large or very small standards, for extremely precise size determination of hybridized RNA species.

For internal standards, the highly abundant rRNAs can be very useful as:

1. Molecular weight markers;
2. An indication of the integrity of the sample by visual inspection;
3. Evidence that an equivalent mass of total RNA has been loaded into each experimental lane. This can be performed by densitometric analysis or through the use of image analysis software. For an historical perspective, see Bonini and Hofmann (1991) and Correa-Rotter *et al.* (1992).

Eukaryotic ribosomes consist of two main components, namely the large and small subunits. The larger, or 60S subunit, and the smaller, or 40S subunit, consist of rRNA and more than 50 different proteins. In the course of an RNA extraction, the various ribosomal proteins are stripped away from the rRNA; these, in turn, copurify with the other cellular RNA species. Removal of the proteins from the 60S and 40S subunits yields 28S and 18S rRNA, respectively. A 5S and 5.8S rRNA, also integral components of eukaryotic ribosomes, are too small to be useful as molecular weight markers and they generally migrate along with the tRNA at the leading edge of the gel in the 300–500 base range. Because rRNA is by far the most prevalent product of transcription (80–85% of cellular RNA), an aliquot of total cellular or total cytoplasmic RNA should electrophoretically resolve only two very distinct bands, the 28S and the 18S rRNAs. The appearance of these bands is convincing evidence of the integrity of the sample, that is, that the sample has not been degraded. In mammals, the intensity of the 28S rRNA is about twice that of the 18S rRNA; it is interesting to note that this is not the case for all eukaryotes. The ratio approaches 1 as one moves down the evolutionary ladder (Figure 2).

Other than the 28S and 18S rRNAs, an intact RNA sample manifests its mRNA component as a significantly lighter smearing slightly above, between, and below the rRNAs. This is the normal appearance of cellular mRNA because of its extremely heterogeneous nature and because, collectively, the mRNA is usually less than 4% of the total mass of the sample. Often, in order to achieve maximum resolution of larger molecular weight RNAs, electrophoresis is allowed to continue to the extent that the small 5S and 5.8S rRNA and tRNA species actually run off the edge of the gel and into the running buffer. Although this approach is certainly not recommended when evaluating a new system, after a while it will become clear to the investigator whether resolution can be improved by extended electrophoresis.

Samples of oligo(dT)-purified poly(A)$^+$ RNA almost always manifest trace amounts of rRNA, the banding intensity of which may or may not support size calibration. The very fact that the rRNAs are visible at all is evidence supporting the integrity of the sample. One common mistake made during the selection of poly(A)$^+$ mRNA is the discarding of the poly(A)$^-$ material, which consists primarily of rRNA and tRNA. If this fraction were reprecipitated and electrophoresed along with experimental samples, the poly(A)$^-$ rRNA would provide suitable size standards and also act as an excellent negative control: The absence of poly(A)$^+$ material should allow the investigator to demonstrate that hybridization signals observed in other lanes are not artifactual or indicative of nonspecific hy-

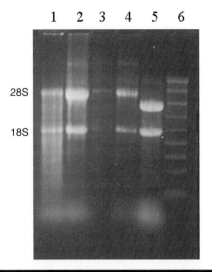

FIGURE 2

Comparison of eukaryotic RNA profiles from different species. Samples of RNA purified from CHO-K1 cells (lanes 1–4) and yeast (*S. uvarum*) on a denaturing agarose gel. (Lane 1) Total cytoplasmic RNA, purified by nonionic lysis. (Lane 2) Total cellular RNA purified using guanidinium-acid-phenol. (Lane 3) Poly(A)$^+$ RNA, showing traces of carryover rRNA. (Lane 4) Poly(A)$^-$ RNA molecular weight standard from Promega. Notice the shift (greater mobility) of the 28S rRNA in the yeast sample in lane 5, compared to the CHO-K1 samples. Notice also the presence of hnRNA and the precursor 45S rRNA in lanes 2 and 4.

bridization. This is especially important when the RNA species of interest are located very close to either the 28S or 18S rRNA. An excellent gel-loading strategy is to run, in order, total RNA, poly(A)$^+$, and poly(A)$^-$ RNA, as is frequently performed in this laboratory.

Samples in which there has been almost any degree of degradation usually fail to manifest the characteristic formation of the 28S and 18S rRNAs and light smearing of the mRNA. Obliterated samples appear as heavy smears that are localized almost entirely below the level at which the 18S rRNA bands in intact samples (Chapter 7, Figure 1). Smears that appear along the length of the lane into which a particular sample was loaded and that are heavier than normal are usually indicative of a limited degree of degradation; less frequently this aspect may suggest that the sample was incompletely denatured.

Another bit of information that can be conveyed by examining an aliquot of RNA comes from the appearance of fluorescence within the well into which the sample was loaded. Such an appearance suggests that genomic DNA is present in the sample. Enormous fragments of DNA derived from the nucleus and generated by shearing forces are frequently unable to enter the gel during the course of the electrophoresis. Further, because UV spectrophotometry does not distinguish between pure samples of RNA and DNA-tainted samples, normalization based on A_{260} may be compromised due to the contribution to the total mass of the sample made by genomic DNA. When other than optimal conditions are suggested by the appearance of a sample on a gel, it is wise to determine the full extent of the problem as well as the quality of the remaining RNA before going on to any subsequent experiments involving a dubious preparation. At this point it should be clear to the investigator, and to the reader, that running a gel, even a minigel, is the single best diagnostic available for assessing the integrity and probable utility of a sample.

There is no reason why commercially available size standards cannot be used with internal rRNA size standards. Having knowledge of the sizes of the 28S and 18S rRNAs, one need only to measure the distance that each of these species has migrated—recall that only two points are needed to draw a straight line—if external standards are unavailable. This is most often performed by image analysis software that can instantly make these measurements from a Polaroid photograph of a gel or by video capture of a stained wet gel. In all cases it is essential to have photographic documentation of even the most preliminary data. If image analysis software is not available, it is quite useful to place a millimeter ruler next to the object being photographed to facilitate measurements; one must ensure that the placement of the ruler is consistent from one gel to the next. In instances when UV light is used to generate an image (e.g., SYBR Green, ethidium bromide), a fluorescent millimeter ruler (see Chapter 5, Figure 4) placed beside or on top of the gel is especially helpful.

The utility of the ribosomal species as markers is completely dependent on knowledge of the sizes of these RNAs either from the literature or from prior empirical determination. It is unwise to proceed in this manner unless the molecular weight of the ribosomal RNAs is known for the species in question. There is considerable variation in the sizes of the rRNAs even among the mammals and, in particular, with respect to the 28S rRNA. There is much less variability observed among 18S rRNAs. In general, 18S rRNAs range from 1.7–1.9 kb, while the range of 28S rRNAs is about 4.6–5.2 kb.[9] As a supplementary tool, the location of the tracking dye(s), most commonly bromophenol blue and xylene cyanol, may be used as an indicator of the progress of the migration of the samples. While performing electrophoresis, bromophenol blue generally runs

[9]In humans the sizes of the 28S and 18S rRNAs are 5025 and 1868 bases, respectively.

with the leading edge of the migrating species and xylene cyanol runs just behind the 18S ribosomal RNA subunit in 1% agarose gels. When characterizing a sample by Northern analysis for the first time it is judicious to continue electrophoresis just until the bromophenol blue has traveled three-fourths along the length of the gel. This will produce very reasonable separation of RNA in medium size gels (12 × 14 cm), which will simplify accurate size determination of hybridized species while ensuring that small RNAs of possible interest are not lost over the distal edge of the gel into the running buffer.

The reliability of data is only enhanced when measurements are derived from both internal and external standards, especially if the RNAs of interest are very large or very small. This is true for two fundamental reasons. First, the logarithmic separation of molecules in a gel is not linear over the entire length of the gel. Very large or very small RNAs that end up outside the linear range of the gel and whose sizes are derived from markers that are within the linear portion of the gel may be sized inaccurately. When external markers are used, it is likely that at least one of the multiple species will electrophorese closer to the RNA of interest; if that RNA is outside the linear range of the gel, then part of the marker itself may run outside the linear range as well. This translates into fewer errors in size determination. Second, one should never rule out the possibility of uneven transfer of sample material during Northern transfer. For example, an air bubble trapped between the gel and the filter membrane or between the gel and the wicking material will drastically reduce the transfer of sample in the vicinity of the bubble. In one unfortunate scenario, several samples purportedly exhibit differential modulation of gene X following posthybridization detection. In reality, signal modulation among the samples may be the result of uneven transfer: Potential target sequence molecules may have been left behind in the gel! In order to assess whether efficient and complete transfer has occurred, one or more of the following may be useful:

1. At the conclusion of the blotting period, the gel can be stained again (EtBr; SYBR Green) and inspected for residual RNA (there should be none).

2. The membrane itself can be reversibly stained with methylene blue to identify directly the positioning of the ribosomal 18S and 28S rRNAs and, in so doing, serve to monitor transfer efficiency.

3. If the gel was stained prior to transfer and small amounts of EtBr or SYBR Green remained in the gel, then visualization of the rRNAs directly on the filter membrane may be possible if the filter paper is briefly placed on a transilluminator or observed by overhead irradiation with a low-intensity, handheld UV light source.

Ultimately, a size calibration curve can be constructed by plotting the $\log_{10}$ of the sizes of the RNA standards against distance migrated from the origin (distance from the well into which the sample was loaded) on semi-log paper (or with a computer). A curve generated in this fashion can then be used to ascertain the sizes of RNA species observed directly in the gel or that may subsequently be detected by nucleic acid hybridization.

Gel Staining Techniques

Ethidium Bromide

Ethidium bromide is currently the most widely used dye for the visualization of nucleic acids in agarose gels. It is a phenanthridinium intercalator, structurally similar to propidium iodide. EtBr binds DNA with no apparent sequence preference once every 4–5 base pairs [*J. Mol. Biol.* **13,** 269 (1965)]. In view of the emergence of SYBR Green as a realistic alternative method for staining, EtBr is really no longer the best choice. In addition, the issue of *when* to apply the dye is a source of considerable intra-laboratory debate. Before electing to use EtBr as a staining tool, it is incumbent upon the investigator to weigh three key disadvantages associated with the use of EtBr into this decision:

1. Ethidium bromide is a powerful mutagen. Extreme care must be exercised when preparing and manipulating ethidium bromide stock solutions and dilutions thereof. Thus, contaminated buffers must be treated as toxic waste and equipment likewise treated as such; EtBr must be inactivated and disposed of properly (Appendix B; Lunn and Sansone, 1987, 1990).

2. The presence of ethidium bromide in a gel or added to samples just prior to loading the gel will retard the electrophoretic migration rate of nucleic acids by about 15%. Although this may not seriously delay the progress of an investigation, it should by no means be used as an excuse to increase the applied voltage to compensate for this phenomenon. Remember that slower running gels demonstrate enhanced resolution.

3. The presence of ethidium bromide in a gel will reduce its transfer efficiency. If a traditional method of transfer is utilized, such as passive capillary diffusion (Chapter 11), the transfer period may need to be extended for unacceptably long periods. Destaining the gel may facilitate transfer, although even minute amounts of ethidium bromide remaining in the gel can result in poor transfer.

The only advantage of running samples in the presence of ethidium bromide is that the progress of electrophoresis can be monitored by disconnecting the power and briefly irradiating the casting tray upon which the gel is resting by placing it directly on a transilluminator or irradating it from above with a low intensity UV light source. (Remember eye and skin protection!) If greater separation is required, the gel can be placed back in the gel box and the power supply reconnected for a suitable interval.

If the decision is made to use or continue to use EtBr then the following may be helpful:

1. **Always** wear gloves when handling ethidium bromide-contaminated solutions, pipet tips, and so forth. In this laboratory, EtBr-contaminated micropipet tips are placed in conical 15-ml tubes before being placed in the autoclave bag.

2. EtBr is most conveniently prepared as a 10 mg/ml stock solution in water, is foil-wrapped, and should be stored at 4°C. Tubes containing stock EtBr should be placed in small storage boxes and also maintained at 4°C.

3. Do not exceed a final EtBr concentration of 0.5 µg/ml in the gel, the running buffer, or the staining buffer (in instances when the dye is to be used after electrophoresis).

4. Consider staining only the lane(s) containing the size markers and do so after electrophoresis has been completed. To do this, cut the marker lane(s) directly from the gel with a razor blade. The drawback is that one will not be able to assess the integrity of the experimental material unless the membrane is reversibly stained.

5. If adding EtBr directly to molten agarose, the agarose temperature must first cool to 60°C because it is heat labile. In this instance EtBr should be added to the running buffer; otherwise, the EtBr will electrophorese out of the gel. Do not forget to dispose of the residual running buffer at the end of the electrophoresis period.

6. Do not add EtBr to glyoxal gels, as this dye will react with the glyoxal.

7. Adding EtBr to the samples alone, rather than to molten agarose, will certainly reduce the total mass of the dye required; however, EtBr will still be a transfer obstacle unless it is drawn out of the gel by destaining.

8. Adding EtBr to gels containing formaldehyde can be especially troublesome because formaldehyde gels exhibit very high background fluorescence when stained with EtBr. Such gels often require extensive destaining before even the abundant rRNA species can be observed.

9. Destain gels in the same buffer that was used to stain the gel such as running buffer or sterile H_2O. Alternatively, destain the gel in the buffer that will be used for Northern transfer (5× SSPE, for example).

10. Make certain that the staining and destaining buffers were at least autoclaved to remove residual RNase; otherwise, the sample may become partially degraded.

SYBR Green

The novel nucleic acid cyanine-binding dyes SYBR Green I and SYBR Green II (Molecular Probes) have become popular stains for in-gel visualization of DNA and RNA, respectively. These dyes are usually purchased as a 10,000× stock solution in DMSO. Just prior to use, SYBR Green is diluted to 1× in a Tris-containing buffer, usually 1× TBE or 1× TAE, although TE buffer or sterile H_2O are also acceptable. Unlike EtBr, the presence of formaldehyde in a gel does not have a negative influence on the background: SYBR Green will fluoresce only when bound to the sample.

Although there are many significant advantages conferred through the use of SYBR Green (Table 2), it should not be added to the gel. Samples electrophoresed in the presence of SYBR Green, as opposed to staining the gel afterward, routinely show very wavy, irregular banding in all lanes and at most molecular weights (Figure 3). This makes mass and molecular weight determinations extremely difficult, if not impossible. In this laboratory, various concentrations of SYBR Green, both above and below the recommended working concentration, have been tested by inclusion in the gel. None have provided satisfactory results. While SYBR Green[10] is used here extensively, with remarkable success, gels are always stained at the conclusion of the electrophoresis. SYBR Green is also compatible with a variety of in-gel enzymatic manipulations; if desired, it can be removed by standard ethanol precipitation.

Silver Staining

Silver staining is a very sensitive method for detecting small amounts of proteins and low-molecular-weight nucleic acids. In general, the method is about 100 times more sensitive than Coomassie blue for many proteins and 2–3 times as sensitive as EtBr for polynucleotides. It has the advantages of allowing visualization of the sample without any special-

---

[10]Related stains for proteins: SYPRO orange and SYPRO red are displacing traditional Coomassie brilliant blue and silver staining techniques.

Comparison of Ethidium Bromide and SYBR Green

	Ethidium bromide	SYBR Green I & II
Stock solution	10 mg/ml in H_2O	10,000× in DMSO
Working concentration	0.5–1.0 μg/ml in H_2O	1–2× in Tris buffer (e.g., 1× TAE, 1× TBE, or TE buffer)
Binding affinity for nucleic acids	Moderate to high	Moderate to high
Add directly to gel?	Optional	Should be avoided; always stain gel after electrophoresis
Gel background	High	Low to none
Excitation	UV transilluminator	UV transilluminator
Color	Pink/Orange	Green
Excitation-induced bleaching of the dye	Yes	No
Filtration	No. 15 deep yellow	No. 15 deep yellow
Sensitivity[a]	1.5 ng RNA	0.5 ng RNA
Mutagenicity[b]	High	Less than EtBr; must be handled with the same degree of caution.
Disposal[c]	Filtration through activated charcoal	Filtration through activated charcoal
Cost	Low to moderate	Moderate to above average
Staining time	5–10 min	25–30 min
Destaining helpful?	Yes	Limited benefit

[a]Limit of sensitivity is even greater with epi-illumination (Haugland, 1996).
[b]Limited toxicity data available for SYBR Green, though preliminary characterization suggests a reduced toxicity compared to EtBr [*Proc. Natl. Acad. Sci. USA* **70**, 2281 (1973); *Mutation Res.* **113**, 173 (1983)].
[c]The Extractor (Schleicher & Schuell; Cat. No. 448031) is an excellent, economical device for the decontamination of large volumes of ethidium bromide (and SYBR Green). Filtrates are poured down the drain, and the filter itself is autoclaved and disposed. See Appendix B for details.

ized equipment and rendering gel backgrounds that are virtually colorless. Silver staining of agarose gels is not recommended mainly because of the drastically reduced sensitivity in this matrix. Briefly, after acid fixation, polyacrylamide gels are impregnated with soluble silver ions (Ag^+), followed by reduction to metallic silver. The initial deposit of the insoluble metallic silver initiates the autocatalytic deposit of more silver,

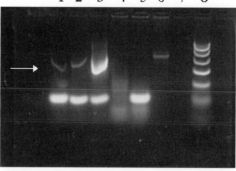

FIGURE 3

Effect of adding SYBR Green to agarose before casting the gel. DNA and RNA at nearly all molecular weights experience profound band distortion, confounding mass and molecular weight determinations. In contrast, staining gels with SYBR Green *after* electrophoresis results in superior staining.

manifested ultimately by visualization of the sample. All of the reagents needed for silver staining are readily available in kits from most major suppliers.

Acridine Orange

An alternative technique for visualizing nucleic acids molecules in a gel is to stain with the cationic dye acridine orange (McMaster and Carmichael, 1977; Carmichael and McMaster, 1980), historically used in conjunction with glyoxalated RNA and also as a cytochemical stain (Kasten, 1967). Acridine orange does not exhibit qualities that inhibit nucleic acid transfer from gels to the extent that is observed with ethidium bromide.

Acridine orange will either (1) bind electrostatically to the phosphate groups of single-stranded molecules and fluoresce red at about 650 nm (Bradley and Wolf, 1959) or (2) intercalate into double-helical molecules and fluoresce green at about 525 nm (Lerman, 1961, 1963). Thus, in the same gel, single-stranded nucleic acid molecules appear red/orange, while double-stranded molecules such as DNA and non-denatured RNA appear green. Given the versatility of this dye, the investigator may also detect incomplete denaturation or partial renaturation of an RNA sample before moving on to subsequent experiments. The main reason that acridine orange has not found widespread use is because of the immenseness of the background in gels so stained and the requirement for unrealistic efforts at destaining.

The staining of agarose gels is accomplished by soaking them in a solution of 30 μg/ml acridine orange in 10 mM sodium phosphate buffer, pH 7.0, for 30 min. Polyacrylamide gels may also be stained, and in as few as 15 min. To reduce the very high background produced by staining with acridine orange, gels are destained by soaking them in 10 mM sodium phosphate buffer for 1–2 h. It should be easy to detect as little as 0.05 μg of double-stranded nucleic acid and 0.1 μg single-stranded nucleic acid upon UV illumination. Other applications for glyoxal have been described elsewhere (Broude and Budowsky, 1971; Carmichael and McMaster, 1980; Sambrook *et al.,* 1989).

Methylene Blue

As another alternative to the commonly used ethidium bromide and SYBR Green methods for staining of nucleic acids in gels prior to blotting, Herrin and Schmidt (1988) described a technique by which RNA transferred to nylon or polyvinylidene difluoride (PVDF) filter membranes can be reversibly stained with a solution of methylene blue prior to hybridization. This approach has several advantages:

1. The RNA can be visualized directly on the filter membrane and the locations of the ribosomal RNA species and other size standards identified and marked appropriately.
2. The gel can be stained with ethidium bromide after the transfer has been completed, rather than before, in order to verify that efficient transfer has occurred.
3. The absence of ethidium bromide in the gel prior to blotting will facilitate much more efficient transfer of RNA from the gel.
4. Methylene blue does not pose the potential health hazards and disposal difficulties associated with ethidium bromide, SYBR Green, and acridine orange, though it does not offer nearly the same level of sensitivity of these other dyes, either.

RNA that has been transferred to nylon should be crosslinked, with a calibrated UV light source, to the filter paper prior to staining or baked onto the filter paper after staining. RNA may be observed directly on a filter paper within 3 min when placed in a solution of 0.02% methylene blue and 0.3 M sodium acetate, pH 5.5. If RNA was crosslinked to the filter, the stain is easily removed by washing the filter paper in 1× SSPE for 15 min or in a solution of 20% ethanol/80% H_2O for 3–5 min. If the filter was stained before baking the RNA onto it, the stain may be removed from the RNA immediately prior to hybridization by washing the filter in a solution of 0.2× SSPE, 0.1% SDS for 15 min at room temperature, with gentle shaking.

Running Agarose Gels for the First Time: A Few Tips

Essential Vocabulary

Casting tray: The portion of the gel box into which the gel is poured. Gel casting trays are either stationary or removable, depending on the model. The ends of the tray must be sealed, either by taping or through the use of rubber gaskets, until the gel has solidified completely.

Gel box: The apparatus in which the electrodes are mounted and in which the actual electrophoretic separation is carried out.

Loading buffer: A dense reagent that is added to a sample prior to electrophoresis in order to increase the density of the sample. This will allow the investigator to load the wells of the gel after the gel has been completely submerged in running buffer. Loading buffers routinely contain a colored dye, such as bromophenol blue and/or xylene cyanol, the position of which can be used to track the progress of the electrophoresis. Colloquialisms for loading buffer include sample buffer and blue juice.

Power supply: The electronic device that will generate constant voltage to conduct the electrophoretic separation. The power supply is connected to the gel box after all of the samples have been loaded and the safety cover closed; only then is the power supply plugged into an electrical outlet, after which it is switched on. Keep in mind that power supplies are potentially very dangerous; investigators using such an instrument for the first time should seek assistance from qualified personnel.

Running buffer: The electrolyte solution used to conduct the electric current and in which the gel is submerged in order to dissipate heat. A running buffer is useful only to the extent that it resists pH changes and ion depletion during the electrophoresis.

Points to Keep in Mind

1. There is a potential for contaminating the electrophoresis apparatus, especially the teeth of the comb, with RNase, while preparing the gel.

While nucleases may be inactive in the presence of some denaturants, they can be transferred to and interfere with latter components of an experiment. If nothing else, the comb should be soaked in a solution of 3% H_2O_2, or in a solution of 0.5% SDS, 50 mM EDTA, and then rinsed with copious amounts of autoclaved H_2O prior to use.

2. A variety of gel boxes are commercially available. Many models are outfitted with rubber gaskets on both sides of the casting tray; this precludes the necessity of taping the tray when pouring the gel (see Appendix K, Figure K-1). In either case, the ends of the tray should be securely sealed so that molten agarose does not leak out.

3. Watch out for agarose boil over, due to superheating, especially when melting agarose in a microwave. Boiling agarose can cause serious burns.

4. Molten agarose is clear. Once molten agarose has solidified in the casting tray of the gel box, the gel will take on an opaque appearance. A casting tray into which molten agarose has been poured should not be disturbed until it is clear that the gel has solidified completely.

5. Gels should be cast to a thickness of 0.5–0.75 cm. To accommodate samples in large volumes, use combs with wider teeth to increase the capacity of the wells. Standard tooth sizes are 1, 1.5, and 2 mm. Gels that are cast more than 0.75 cm thick will impede transfer efficiency and require exceptionally long blotting periods to recover all of the sample from the gel.

6. Make certain that the casting tray is resting on a level surface before pouring the gel.

7. The best way to cast a gel is to pour the molten agarose solution slowly and evenly into the center of the casting tray before putting the comb in place. Stop pouring as soon as the entire surface of the casting tray has been covered. This will usually generate a gel with an appropriate thickness. Then insert the comb without delay. Following these steps in this order will yield a very homogeneous gel with no hidden air bubbles. If air bubbles do form, they can be popped or pushed to the side of the tray using a sterile micropipet tip.

8. Agarose solutions containing denaturants should be prepared and cast in a fume hood to minimize exposure and possible health hazards.

9. Be certain to allow enough time for the gel to solidify, and then completely cover it with running buffer *before* attempting to remove the comb. Agarose gels do not have nearly the physical strength of polyacrylamide gels. The vacuum forces created when the comb is pulled upward are minimized if the gel is submerged in the running buffer. Otherwise, one increases the chances of damaging the bottom of the well. If this is

not detected, the RNA sample will be lost as soon as it is applied to the well.

10. If the bottom of the wells of the gel are repeatedly damaged upon removal of the comb, let the gel "soak" in the running buffer for several minutes to lubricate and loosen the comb.

11. Submerge the gel in only the minimum volume necessary to *completely* cover the gel. This is because the current will travel the path of least resistance, which is around the gel, rather than through it. This will greatly reduce the electrophoresis rate. However, be sure to look at the edges of the gel that were formed against the sides of the casting tray. Agarose has a tendency to "smile" by clinging to the sides of the tray. Failure to completely submerge these portions of the gel can result in high temperature build-up and melting of the agarose because there is no running buffer covering it to dissipate the heat. In addition, avoid loading samples into the outside lanes that, because of their nonuniform thickness, will cause nonconforming sample migration in those lanes (edge effects).

12. When possible, use small sample volumes for electrophoresis. In the context of this chapter, a small volume will cause the RNA to enter the gel in the narrowest cross section, thereby maximizing resolution.

13. All gel loading buffers contain a dense material such as glycerol or sucrose. Addition of the loading buffer to the denatured RNA sample imparts an added density to the sample and, in so doing, allows it to fall through the running buffer and into well. Therefore, all that is required to successfully load the well is that the investigator merely break the surface of the running buffer above the well and carefully extrude the sample from the micropipet tip (Figure 4). If there is difficulty seeing the well because of reflections from the surface of the buffer, it is useful to place a piece of dark tape or a piece of dark paper under the gel box just beneath the wells.

14. Be certain to eliminate any air space from the end of the pipet tip containing a sample, before attempting to load it into the well of a gel. Injecting air into a well ahead of a sample may result in the loss of the sample.

15. When all samples have been loaded into the gel, make sure that the cover of the gel box is secured, to prevent accidental electrocution. Connect the leads to the power supply, making use of gel boxes and power supplies that are appropriately matched. Only then should the power be turned on. Electrophorese at a maximum of 5 V per centimeter of gel length. Because the gel can rest in the gel box in either orientation in most models, make certain that the negatively charged RNA loaded into the wells runs from negative to positive.

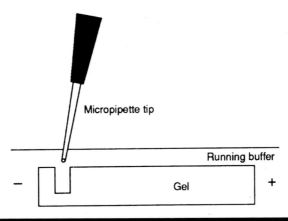

Micropipette tip

Running buffer

−   Gel   +

FIGURE   4

Proper positioning of micropipet tip for loading samples into the wells of an agarose gel.

16. When inspecting the gel or terminating electrophoresis, be sure to turn the power off before handling either the gel box or the power supply. Further, disconnect power leads one at a time and never handle both at once.

17. Never, ever, attempt to load a gel while the power supply is on, even at minimal voltages. The author is painfully reminded of one scientist who admitted that "loading the gel at low voltage" improved her data. This is a very dangerous technique that is to be frowned upon and discouraged in every instance.

References

Alwine, J. C., Kemp, D. J., Parker, B. A., Reiser, J., Renart, J., Stark, G. R., and Wahl, G. M. (1979). Detection of specific RNAs or specific fragments of DNA by fractionation in gels and transfer to diazobenzyloxymethyl paper. *Methods Enzymol.* **68,** 220.

Alwine, J. C., Kemp, D. J., and Stark, G. R. (1977). Method for detection of specific RNAs in agarose gels by transfer to diazobenzyloxymethyl paper and hybridization with DNA probes. *Proc. Natl. Acad. Sci. USA* **74,** 5350.

Bailey, J. M., and Davidson, N. (1976). Methylmercury as a reversible denaturing agent for agarose gel electrophoresis. *Anal. Biochem.* **70,** 75.

Bantle, J. A., Maxwell, I. H., and Hahn, W. E. (1976). Specificity of oligo(dT)-cellulose chromatography in the isolation of polyadenylated RNA. *Anal. Biochem.* **72,** 413.

Berger, S. L., and Kimmel, A. R. (1987). "Guide to Molecular Cloning Techniques." Academic Press, San Diego, CA.

Berk, A. J., and Sharp, P. A. (1977). Sizing and mapping of early adenovirus mRNAs by gel electrophoresis of S1 endonuclease-digested hybrids. *Cell* **12,** 721.

Boedtker, H. (1971). Conformation independent molecular weight determinations of RNA by gel electrophoresis. *Biochim. Biophys. Acta* **240**, 448.

Bonini, J. A., and Hofmann, C. (1991). A rapid, accurate, nonradioactive method for quantitating RNA on agarose gels. *BioTechniques* **11**(6), 708.

Bradley, D. F., and Wolf, M. K. (1959). Aggregation of dyes bound to polyanions. *Proc. Natl. Acad. Sci. USA* **45**, 944.

Broude, N. E., and Budowsky, E. I. (1971). The reaction of glyoxal with nucleic acid components. III. Kinetics of the reaction with monomers. *Biochim. Biophys. Acta* **254**, 380.

Carmichael, G. G., and McMaster, G. K. (1980). The analysis of nucleic acids in gels using glyoxal and acridine orange. *Methods Enzymol.* **65**, 380.

Church, G. M., and Gilbert, W. (1984). Genomic sequencing. *Proc. Natl. Acad. Sci. USA* **81**, 1991.

Correa-Rotter, R., Mariash, C. N., and Rosenberg, M. E. (1992). Loading and transfer control for northern hybridization. *BioTechniques* **12**(2), 154.

Farrell, R. E., Jr., and Greene, J. J. (1992). Regulation of c-*myc* and c-*Ha-ras* oncogene expression by cell shape. *J. Cell. Physiol.* **153**, 429.

Favaloro, J. R., Triesman, R., and Kamen, R. (1980). Transcriptional maps of polyoma virus-specific RNA: Analysis by two dimensional nuclease S1 gel mapping. *Methods Enzymol.* **65**, 718.

Fornace, A. J., and Mitchell, J. (1986). Induction of B2 RNA polymerase III transcription by heat shock: Enrichment for heat shock induced sequences in rodent cells by hybridization subtraction. *Nucleic Acids Res.* **14**, 5793.

Gruenwedel, D. W., and Davidson, N. (1966). Complexing and denaturation of DNA by methylmercuric hydroxide. I. Spectrophotometric studies. *J. Mol. Biol.* **21**, 129.

Haugland, R. P. (1996). "Handbook of Fluorescent Probes and Research Chemicals," 6th ed. Molecular Probes, Inc., Eugene, OR.

Herrin, D. L., and Schmidt, G. W. (1988). Rapid, reversible staining of Northern blots prior to hybridization. *BioTechniques* **6**(3), 196.

Hollander, M. C., and Fornace, A. J., Jr. (1990). Estimation of relative mRNA content by filter hybridization to a polythymidylate probe. *BioTechniques* **9**, 174.

Hutton, J. R., and Wetmur, J. G. (1973). Effect of chemical modification on the rate of renaturation of deoxyribonucleic acid. Deaminated and glyoxalated deoxyribonucleic acid. *Biochemistry* **12**, 558.

Johnson, D. (1975). A new method of DNA denaturation mapping. *Nucleic Acids Res.* **2**, 2049.

Junghans, R. P. (1983). A review of the toxicity of methyl mercury compounds with application to occupational exposures associated with laboratory uses. *Environ. Res.* **31**, 1.

Kasten, F. H. (1967). Cytochemical studies with acridine orange and the influence of dye contaminants in the staining of nucleic acids. *Int. Rev. Cytol.* **21**, 141.

Landers, T. (1990). Electrophoresis apparatus maintenance. *Focus* **12**(2), 54.

Lehrach, H., Diamond, D., Wozney, J. M., and Boedtker, H. (1977). RNA molecular weight determinations by gel electrophoresis under denaturing conditions: A critical reexamination. *Biochemistry* **16**, 4743.

Lerman, L. S. (1961). Structural considerations in the interaction of DNA and acridines. *J. Mol. Biol.* **3**, 18.

Lerman, L. S. (1963). The structure of the DNA-acridine complex. *Proc. Natl. Acad. Sci. USA* **49**, 94.

Lewin, B. (1997). "Genes VI." Oxford University Press, New York.

Lunn, G., and Sansone, E. B. (1987). Ethidium bromide: Destruction and decontamination of solutions. *Anal. Biochen.* **162,** 453.

Lunn, G., and Sansone, E. B. (1990). Degradation of ethidium bromide in alcohols. *BioTechniques* **8(4),** 372.

Maniatis, T., Jeffrey, A., and Kleid, D. G. (1975). Nucleotide sequence of the rightward operator of the phage lambda. *Proc. Natl. Acad. Sci. USA* **72,** 1184.

McMaster, G. K., and Carmichael, G. C. (1977). Analysis of single- and double-stranded nucleic acids on polyacrylamide and agarose gels by using glyoxal and acridine orange. *Proc. Natl. Acad. Sci. USA* **74,** 4835.

Melton, D. A., Krieg, P. A., Rebagliati, M. R., Maniatis, T., Zinn, K., and Green, M. R. (1984). Efficient *in vitro* synthesis of biologically active RNA and RNA hybridization probes from plasmids containing a bacteriophage SP6 promoter. *Nucleic Acids Res.* **7,** 1175.

Nakaya, K., Takenaka, O., Horinishi, H., and Shibata, K. (1968). Reactions of glyoxal with nucleic acids, nucleotides and their component bases. *Biochim. Biophys. Acta* **161,** 23.

Rave, N., Crkvenjakov, R., and Boedtker, H. (1979). Identification of procollagen mRNAs transferred to diazobenzyloxymethyl paper from formaldehyde gels. *Nucleic Acids Res.* **6,** 3559.

Reijnders, L., Sloof, P., Siral, J., and Borst, P. (1973). Gel electrophoresis of RNA under denaturing conditions. *Biochim. Biophys. Acta* **324,** 320.

Salomaa, P. (1956). Two volumetric methods for the determination of glyoxal. *Acta Chem. Scand.* **10,** 306.

Sambrook, J., Fritsch, E. F., and Maniatis, T. (Eds.). (1989). "Molecular Cloning: A Laboratory Manual," 2nd ed. Cold Spring Harbor Laboratory Press, Cold Spring Harbor, NY.

Shapiro, R., Cohen, B. I., and Clagett, D. C. (1970). Specific acylation of the guanine residues of ribonucleic acid. *J. Biol. Chem.* **245,** 2633.

Sharp, P. A., Berk, A. J., and Berget, S. M. (1980). Transcriptional maps of adenovirus. *Methods Enzymol.* **65,** 750.

Southern, E. M. (1975). Detection of specific sequences among DNA fragments separated by gel electrophoresis. *J. Mol. Biol.* **98,** 503.

Spohr, G., Mirault, M. E., Imaizumi, T., and Scherrer, K. (1976). Molecular-weight determination of animal-cell RNA by electrophoresis in formamide under fully denaturing conditions on exponential polyacrylamide gels. *Eur. J. Biochem.* **62,** 313.

Thomas, P. S. (1980). Hybridization of denatured RNA and small DNA fragments transferred to nitrocellulose. *Proc. Natl. Acad. Sci. USA* **77,** 5201.

Wicks, R. J. (1986). RNA molecular weight determination by agarose gel electrophoresis using formaldehyde as denaturant: Comparison of RNA and DNA molecular weight markers. *Int. J. Biochem.* **18,** 277.

Zinn, K., DeMaio, D., and Maniatis, T. (1983). Identification of two distinct regulatory regions adjacent to the human β-interferon gene. *Cell* **34,** 865.

CHAPTER 10

Photodocumentation and Image Analysis

Rationale

The absolute necessity for maintaining accurate records of experiments performed and the resulting data is indisputable. Electrophoresis gels contain information critical for accurate interpretation of data derived from the latter components of an experimental design. In both analytical and preparative settings, the information obtained by electrophoresis per se may be invaluable. Gels deteriorate in a relatively short time, however, and the dyes used to stain nucleic acids in gels deteriorate even more rapidly. This mandates a photographic record of gels (an electrophoretogram) as well as a photographic record of any associated X-ray films generated upon hybridization detection (by chemiluminescence or autoradiography). This is the only realistic option for providing a stable, long-term record of results for future analysis, for in-house presentations, and for publication.

A variety of systems have gained widespread acceptance and are currently in use to record and analyze the various types of images commonly encountered in a molecular biology setting. These systems fall into two broad categories: (1) the traditional photograph generated on Polaroid film or, less commonly, on thermal paper; and (2) digital imaging, in which the image of the gel or X-ray film is analyzed by and stored in a computer and/or on a disk. Both approaches have numerous advantages and disadvantages, as described below.

Photodocumentation

In the traditional sense, photodocumentation refers to the investigator physically taking a picture of a gel soon after electrophoresis and staining. This is most commonly performed using Polaroid instant imaging products. To this end, perhaps the two most widely used systems are the stationary Polaroid MP-4+ Multipurpose System and the handheld Polaroid DS-34 camera (Figure 1), the latter of which has proven to be an extremely versatile, low-cost instrument. In either case the investigator takes a picture of a gel, tapes it into a laboratory notebook, and then annotates the picture appropriately. In the context of some of the newer digital technology, described below, Polaroid photographs continue to play and important role: They can be scanned, digitized, recalled, and archived as needed. There will always be a place for this approach. Computers are convenient, but they also crash from time to time, making a hard copy of an electrophoretogram an absolute necessity. Moreover, from a time-

Polaroid DS-34 handheld camera. Snap-on hoods allow photodocumentation of a variety of gel sizes and they also position the camera for sharp focus. A built-in lens ensures proper magnification and focal length.

saving perspective, it is much easier to open a lab notebook and look at a Polaroid photograph than to

- wait for the computer to become available;
- close the currently running program;
- open a gel analysis program;
- find the directory in which the image was stored; and
- print out a copy of the stored image, if needed.

The take-home lesson: Image analysis software is, for the most part, a highly efficient and time-saving tool, but having a photographic hard copy of a gel, which can be digitized later, is just as important.

In contrast to the Polaroid instant imaging approach, newer instrumentation has been developed and is firmly established in many laboratories. Using these systems, the image is digitized and a hard copy generated on heat-sensitive thermal paper, much as is generated on fax paper. Generally speaking, these printouts lack the definition and longevity of traditional photographs. These systems also require a significant up-front investment. It is often argued that these digital systems offer superior speed and efficiency in the manipulation of data; although this is true in many cases, one can also just as easily digitize a Polaroid photograph and proceed from there.

Sample Visualization

Typically, either instant photographs or instant slides are prepared either in black and white or in color, with or without a negative. In the case of nucleic acids, a gel is stained with SYBR Green, ethidium bromide, or other dye, UV irradiated from below by placing the gel on the filter plate of a standard laboratory transilluminator, and photographed (Figure 2). The distribution of the sample in the gel is then characterized by digital analysis of the resultant images. This and related Polaroid systems accept a variety of instant film formats, and feature snap-on hoods that accommodate gels of several dimensions. These produce sharp electrophore-

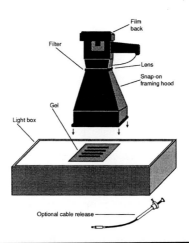

FIGURE 2

Irradiation of a gel from below permits direct examination of nucleic acids in the gel. This camera further accommodates photodocumentation while the gel is fully supported and evenly irradiated. Courtesy of Polaroid.

tograms by maintaining the camera at the correct focal distance from the gel.

While focusing is not required, it is necessary to set the amount of time that the shutter in the lens will be open (the shutter speed) as well as the aperture setting itself, that is, the *f*/stop (how wide the lens will open). If uncertain as to the correct parameters, try photographing ethidium bromide-stained gels for 0.5 s at an aperture of *f*/8, making corrections as needed. For SYBR Green-stained gels, one should begin with 0.5 s at *f*/5.6. Of course, the exact settings will be empirically determined for each gel and are directly related to the mass of nucleic acid in the gel, the staining time, the destaining time (if any), and the magnitude of background fluorescence.

Filtration

Unless the bands stand out in a photograph of a gel, the images will not convey a useful level of detail. Since most electrophoresis photography involves black and white film, sufficiently high contrast is essential. It should be noted, however, that color photographs of SYBR Green-stained gels are stunning. Regardless of the camera or film used, color filters enhance detail and make electrophoresis bands appear more sharply defined.

A filter absorbs its own complementary colors and transmits its own color, allowing the photographer to manipulate the relative brightness of colors that would otherwise appear similar in black and white. For example, a red filter makes green and blue appear darker, and red appears relatively lighter. A useful mnemonic by which to remember pairs of complementary colors and, thereby, the correct filter for use, is

"Red Cadillac by General Motors"

that is

Red Cadillac: red and cyan

by: blue and yellow

General Motors: green and magenta

Table 1 lists some of the common electrophoretic stains[1] for nucleic acid and protein gels, the corresponding filters that offer the best results, and approximate aperture setting and shutter speeds. Because filters ab-

---

[1] For acridine orange staining and color photography, use Polaroid Type 108 color film and a yellow filter; for black and white photography use Polaroid Type 105 or Type 107 film with a red filter.

Common Electrophoretic Stains and the Corresponding Color Filters

Light source (wavelength)	Stain	Filter	Aperture	Shutter speed[a] (s)
UV (302 nm)	Ethidium bromide	No. 22 Deep orange or No. 15 Deep yellow	$f/16$	1/4
UV (302 nm)	SYBR Green	No. 15 Deep yellow	$f/5.6$	1
Epi-illumination (254 nm)	SYBR Green	No. 15 Deep yellow	$f/5.6$	1
White (400–700 nm)	Coomassie Blue	No. 8 Medium yellow or No. 9 Deep yellow	$f/16$	1/30
White (400–700 nm)	ELISA	No. 8 Medium yellow	$f/16$	1/30
White (400–700 nm)	Silver	No. 58 Green	$f/16$	1/30

[a]Based on Polaroid Type 667 and 57 film (ISO 3000). Longer exposure times or wider apertures are necessary with lower speed films.

sorb light, it is necessary to ensure that the gel is properly illuminated; this means uniform irradiation from beneath. Burned out UV bulbs in transilluminators should be replaced by qualified personnel as soon as possible. To an extent, one may compensate for the effect of a filter by increasing the exposure time, using a wider aperture, or both. Moreover, because UV light sources[2] have a low light output, high-speed films are particularly well suited for this purpose. In each application, the precise parameters are best determined empirically.

Safety First

UV transilluminators and handheld UV light sources are both sources of intense, dangerous ultraviolet radiation. Exposure to UV light from these standard laboratory instruments, which commonly emit UV radiation at 302 or 312 nm, and less commonly at 254 nm, and other sources as well, can cause serious damage to the cornea, retina, and other structures of the eye, not to mention the damage to exposed skin. Because the

[2]**Caution:** Although the actual light output from UV sources may be low, the energy associated with short- and medium-wave emission can rapidly cause permanent damage to unprotected eyes and skin. Always wear UV-rated eye protection and avoid exposure of skin to all UV light sources.

damage is cumulative, minute amounts of exposure can be catastrophic in both the short and the long term. Be sure to wear gloves and protective eye wear/face shield at all times, even when using transilluminators with a UV protective cover, such as the type that covers the gel while the UV light is on. Although these covers usually block most of the UV light, there may still be a small amount of transmission. Moreover, cracks, improper placement, and open areas permitting direct exposure to UV light are all serious potential health and safety risks.

When photographing gels resting on the filter plate of a UV transilluminator, the gel is usually covered entirely by the hood of handheld cameras. The UV light should be switched on immediately prior to physically taking the picture and should be turned off immediately thereafter, even before removing the exposed film from the camera; because it is so easy to forget that the UV light remains on, personnel in the room could be exposed to dangerous UV light. This occurs frequently when the room lights are on while pictures of the gel are being taken. Exposed sections of the transilluminator, no matter how small, pose a great danger to the operator; in addition the transilluminator surface becomes hot very quickly. One should also be aware that not all safety glasses are UV rated; it is critically important to have access to the correct safety equipment. Finally, it is important to avoid contact with the dyes with which the gels are impregnated, as all DNA and RNA binding dyes carry significant mutagenic potential. The dangers associated with photodocumentation should not be misjudged.

Tips for Optimizing Electrophoretograms

1. *Maximize the size of the image.* When using the DS-34 camera, be sure to select the hood that most closely approximates the size of the gel. Minigels (e.g., 7×8 cm) photographed with a large hood (15×15 cm) will cause the gel to look like a postage stamp in the middle of large black background. An image in which the gel more or less fills the photograph will always offer the greatest level of detail, in terms of both sensitivity and resolution.

2. *Use higher speed films with UV light.* Whenever possible, use films with ISO 3000, such as Polaroid Type 667 and Type 57 films, both of which are particularly well suited for most routine applications. High-speed films preserve depth of field and image sharpness. Not only will a shorter exposure time minimize exposure of the investigator to UV light, but it will also allow for multiple photographs of the same gel, important because ethidium bromide, in particular, is rapidly bleached by UV light

to the extent that it will no longer fluoresce. Shorter exposure times will also minimize nicking of the phosphodiester backbone of nucleic acids due to overirradiation of the gel. While minimal exposure to UV randomly nicks the backbone of RNA (and DNA), thereby enabling efficient transfer out of the gel, excessive nicking will render the sample incapable of nucleic acid hybridization in subsequent steps.

3. *Keep camera rollers clean.* To prevent spotting and other processing defects, clean camera rollers periodically with warm water and a cotton swab or Kimwipe. The need to clean the rollers becomes obvious when three or four dots, evenly spaced, appear on the film (Figure 3). These dots are the result of debris on the rollers, which places uneven pressure on the film as it is pulled from the camera.

4. *Follow film development instructions.* Pay close attention to the development time and temperature recommended on the package of instant film. Underdeveloping or overdeveloping film, a common error, can adversely affect the characteristics of the image. In general, the most commonly used black and white films require a 30-s development after the film is pulled out of the camera. Overdevelopment will enhance contrast, and this can be good or bad, depending on the extent of variation in mass among bands on the gel. By increasing the contrast, bands with the great-

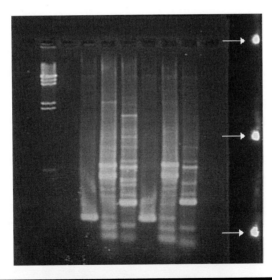

FIGURE 3

Effect of dirty rollers in a Polaroid camera. Notice the three heavy white dots, evenly spaced, along the extreme right-hand edge of the image.

est mass will stand out even more (Figure 4). However, bands in which the mass of sample is minimal may actually begin to blend into the background of the image, leading the viewer to believe that no such bands exist. For the sake of consistency, pay careful attention to the aperture setting, shutter speed, and developing time.

5. *Avoid temperature extremes.* The ideal development temperature for Polaroid films is room temperature, generally 21–24°C (70–75°F). If the lab is much warmer or cooler, or if the film was stored refrigerated until just prior to use, an adjustment in development time or exposure time may be necessary. Refer to the film package insert for guidance.

6. *Use small apertures.* The smaller the aperture (the higher the f/stop), the greater the depth of field and the more sharply defined the resulting images. Ideally, one should begin with f/16 and optimize by incrementally decreasing the aperture, as much as practical. In this laboratory, ethidium bromide-stained gels are routinely photographed for $\frac{1}{4}$ s at f/8. SYBR Green gels are photographed at f/5.6 for $\frac{1}{4}$ or $\frac{1}{2}$ s, adjusted as needed.

7. *Minimize camera movement.* Gels photographed with slower films require long exposure times when stained with ethidium bromide and, in particular, SYBR Green. To minimize camera movement in this lab, a cable release has replaced the pistol-grip shutter release, which easily unscrews from the handheld DS-34 camera. It is also far more practical to use faster films (see Item 2).

8. *Eliminate ambient light.* If using a handheld camera, be sure to rest the camera hood directly on the light box; this will prevent ambient light from reaching the film. It is usually not necessary to turn off the lights in the room with this type of camera. Be certain, though, to turn off the transilluminator immediately after making the exposure.

9. *Use filters.* Consider filtration a standard component of photodocumentation. Color filters can improve the contrast of electrophoretograms, as described. Ideally, one should use screw-in glass filters whenever possible. Acetate filters can sometimes distort the optical path and, for this reason, they should be avoided.

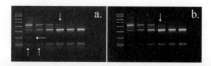

FIGURE 4

Effect of overdeveloping a Polaroid electrophoretogram. (a) Correct developing time. (b) Overdevelopment of film causes loss of low-intensity bands.

10. *Photograph on a clean background.* Use a light source with a clean surface to illuminate gels. The surface of the transilluminator (the filter plate) should be wiped clean and dried just prior to use.

11. *Avoid fogging the lens.* Pay attention to the amount of time that the transilluminator has been turned on. Because the surface of the transilluminator becomes hot in a short time, some of the liquid from the gel will vaporize and then condense on the lens located on the inside of hood. This effectively fogs the camera lens and compromises greatly the quality of the photograph. This is easily remedied by wiping the lens dry with a Kimwipe.

12. *Recover the film properly.* Film that has been exposed is to be pulled straight out of the camera, slowly and evenly, and perpendicular to the user. The film should not be pulled upward or downward from the camera. It has been suggested that the user intone "Pol-ar-oid," the cadence of which is used as a vocal guide to be followed as the film is pulled. When the recommended developing time has elapsed, turn the film/receiver sheet complex over, take hold of the corner of the *film* where the number is printed, and peel it away from chemistries impregnated on the receiver sheet. Type 667 film does not require any type of coating; the film may be slightly arched at first, but will flatten out within 4–5 min. Do not touch the negative itself or the sheet to which it is attached: Both contain an alkali that should not come into contact with skin.

Finally, for the non-photographically minded, the following should be taken to heart:

1. The shutter speed scale found on the lens is labeled "B, 1, 2, 4, 8, 16, 32, and 64." The "B" setting means that the lens will be open as long as the shutter release is depressed by the user. The numbers that follow are the reciprocal of the actual shutter speed (i.e., "2" is actually 1/2 (0.5 s); "4" is actually 1/4 (0.25 s), and so forth).

2. The terms "aperture setting" and "*f*/stop" are used interchangeably and refer to how wide the lens shutter will open. Obviously, this setting influences the amount of light that will enter the camera and expose the film. The *f*/stop scale is also "anti-intuitive," meaning that the smaller the number, the larger the opening of the shutter.

Inherent Limitations of Photographic Film

In photographic systems, the method of image recording is an energy conversion process whereby the sensitivity of silver halide crystals is used to record the level of photons emitted by or reflected from a source or object. The photons are absorbed by the silver crystal and when the level of energy is sufficient, the crystal forms a deformation known as a latent image site. At this site, the silver and halogen bond of the crystal is broken, resulting in free halogen and metallic silver. If the crystal is exposed to too much energy in too short a time, it will be unable to react fast enough to record the event "correctly"; likewise, if a low level of energy strikes the crystal for a prolonged time, then the crystal will react differently than it would under normal conditions.

At some point, the crystal will be exposed to sufficient energy to fully saturate it with metallic silver. Beyond this point it can no longer record further information, becoming uniform in response at whatever level of response that system can produce. In the case of Polaroid film, a system referred to as direct positive response, this level of saturation relates to the lowest density, known as D-Min. When the crystal reaches such a level of low absorption of photon energy that it no longer is capable of producing a latent image, it acts as if no energy has been absorbed. This produces a uniform response of maximum density, known as D-Max. Between the D-Max and D-Min are regions where the silver crystal is recording the highlight and shadow details, in a nonlinear manner, as the crystal begins to approach the linear portion of its ability to record the energy absorption. The region of useful recording between the highlight and shadow region of the film is known as the dynamic range of the film. Although ideally the film and the photon source would have the same range, thus allowing the system to record in a true-to-life manner, it is often true that the photon source exceeds the range of the film. In this case, the user of the imaging system must choose which areas are of greatest importance. By adjusting the exposure of the system, the investigator defines the range of the image that will be recorded, while losing highlights and shadows outside the dynamic range of the film, thus defined. In the molecular biology laboratory, where both strong and weak bands are often observed in the same image, it is laudable to make two or three exposures of the same image, using different aperture settings, in order to derive as much useful data from a single image as possible.

Digital Image Analysis

Many of the traditional methods for the interpretation of photographs of gels (electrophoretograms), including visual inspection and densitometry, have been abandoned in favor of the plethora of computer software for rapid, reproducible image analysis. Useful measurements that are now automated include determination of molecular weight, concentration, relative abundance, and integrated optical density. The computer also provides a means of cataloging stored images. Meaningful image analysis requires an understanding of the basic terminology and the benefits and deficiencies associated with this approach for analyzing gels and archiving data; thus, a short overview of this emerging technology and practical concerns are presented here.

An image is a visual representation of an object, and image processing manipulates information within an image to make it more useful. Digital image processing is a specific type of image processing performed with a computer. Succinctly, image analysis may be defined as the presentation and quantification of images.

The digitization process divides an image into a horizontal grid, or array, of very small regions, each of which is called a "picture element" or pixel (Figure 5). In the computer, the image is represented by this digital grid, or bitmap. A pixel usually represents a very small region within an image, often 1/300th of a square inch, or less. Each pixel in the bitmap is identified by its position in the grid, referenced by its row (x coordinate) and column (y coordinate) number. By convention, pixels are referenced from the upper-left position of the bitmap, which is considered position 0,0 (row 0, column 0).

When an image, such as a photograph of a gel, is digitized, it is examined in grid fashion. This means that each pixel in the image is individually sampled, and its brightness is measured and quantified. This measurement results in a value for the pixel, usually an integer, which represents the brightness or darkness of the image at that point. This value is stored in the corresponding pixel of the computer's image bitmap. When the image is digitized, the width and height of the array is chosen and fixed. Together, the bitmap's pixel width and height are known as its spatial resolution.

Depending on the capability of the measuring hardware and the complexity of the image, anywhere from 1 to 32 bits might be used to store each pixel value. Pixel values for simple images, such as line drawings, containing only black and white information, are represented by a single bit: 0 = black, 1 = white. In contrast, photographic-like images contain much more information; 24 bits are needed to represent all the possible colors that might occur in a true color image. The number of bits used to present the pixel values in an image is referred to as its pixel depth, or bits per pixel (BPP). The number of bits per pixel used to represent each pixel value determines the image's class.

While the bit depth (BPP) indicates how many unique colors an image

pixel 0,0 ——▶

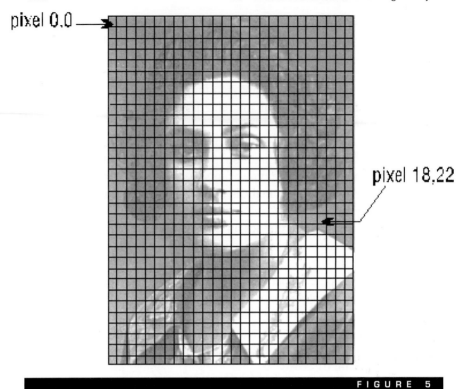

pixel 18,22

Pixel bitmap. Courtesy of Media Cybernetics.

can possess, it does not indicate which colors are actually contained within the image. Color interpretation is determined by bit depth or image class, which include

- gray scale 8;
- gray scale 12; and
- gray scale 16

Gray scale pixel values represent a level of grayness or brightness, ranging from completely black to completely white. This class is sometimes referred to as "monochrome." In an 8-bit gray scale image, a pixel with a value of 0 is completely black, and a pixel with a value of 255 is completely white (Figure 6). A value of 127 represents a gray color exactly halfway between black and white (medium-gray), and a pixel value of 64 has a gray color halfway between medium-gray and black. Although gray scale images with bit depths of 2, 4, 6, 12, 16, and 32 exits, 8 BPP gray

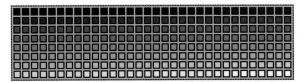

FIGURE 6

256 shades of gray.

scale images are the most common because (1) its 1 byte per pixel size makes it easy to manipulate with a computer, and (2) it can faithfully represent any gray scale image because it provides 256 distinct levels of gray.[3]

Often it becomes necessary to enhance an image of a gel in order to maximize the amount of useful information that can be derived from it. Whereas there are numerous methods for enhancing images, the three most commonly used in digital image analysis are brightness, contrast, and gamma correction.

Brightness is a term used to describe the overall amount of light in an image. When brightness is increased, the value of every pixel in the image is increased, moving each pixel closer to 255, or white. When brightness is decreased, the value of each pixel is likewise decreased, moving it closer to 0, or black.

Contrast is a term used to denote the degree of difference between the brightest and darkest components in an image. An image has poor contrast if it contains only harsh black and white transitions, or contains pixel values within a narrow range. For example, an image with values ranging from 100 to 140 would have poor contrast. An image has good contrast if it is composed of a wide range of brightness values from black to white. The amount of the intensity scale used by an image is called its "dynamic range." An image with good contrast will have a broad dynamic range.

Gamma correction is a specialized form of contrast enhancement that is designed to enhance contrast in the very dark or very light areas of an image. It does this by changing midtone values, particularly those at the low end, without affecting the highlight (255) and shadow (0) points. Gamma correction can be used to improve the appearance of an image, or to compensate for differences in the way different input and output de-

---

[3]The human eye can distinguish fewer than 200 gray levels.

vices respond to an image. Gamma correction involves applying standard, nonlinear γ curves to the intensity scale (Figure 7). A γ value of 1 is equivalent to the identity curve, which has no effect on the image. An increase in the γ value (γ > 1) will generally lighten an image and increase the contrast in its dark areas. A decrease in the γ value (γ < 1) will generally darken the image and emphasize contrast in the lighter areas.

Another broad category of operations frequently used to collect data from gel images is known as density analysis. The key term for understanding the all important quantitative aspect of digital image analysis is **optical density** (OD). Optical density analysis is a common image processing application used to determine the amount of matter in a material by measuring the amount of light it transmits (lets pass through it). Because OD analysis measures the amount of light passing through a material, OD measurements are meaningful only in the analysis of images that have been captured with the light source radiating from *behind* them; gels UV irradiated from beneath constitute a premier example. Optical density measurements are not useful in the analysis of images captured under reflected light. In general, OD of a band or other element of a digitized image assumes an exponential decay of light inside the transmitting material and is calculated as

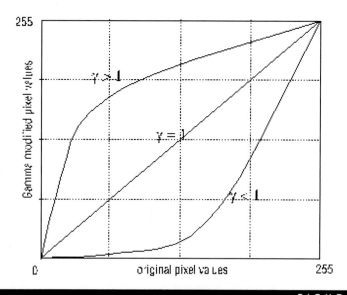

FIGURE 7

Gamma correction curve. Effect of γ curves on pixel values 0 through 255.

$$\text{Optical density } (x,y) = \frac{-\log(\text{intensity } (x,y) - \text{black})}{\text{incident} - \text{black}},$$

where

intensity (x,y) is the intensity at pixel (x,y)

black is the intensity generated when no light goes through the material

incident is the intensity of the incident light.

A material through which no light is transmitted has an infinite optical density, and one through which all light is transmitted has zero optical density. Thus the OD scale is inversely related to the intensity scale, that is, on an optical density scale, dark pixels produce high values, and light pixels produce low values.

Integrated optical density (IOD) is the sum of pixel values minus a background value for each pixel within a user-defined area of interest. In a given area so defined, the increased size of a band or the increased brightness of a spot relates to higher integrated optical density measurements. IODs are very useful when comparing band intensities on images in which mass standards have not been included.

Image Formats

Digitized images can be stored on a computer hard drive or on a floppy disk and can always be recalled for further analysis. Because digital images occupy a fairly large amount of disk space, it is usually best to crop the image as much as possible, though obviously not to the detriment of the interpretation of the data. When images are stored to a disk, they are stored in a particular format. These formats vary mostly in the types of image data that they support and the types of compression that they make available to reduce the storage size of the image. Digitized images are most commonly stored as "tiff" files (*.tif), an acronym for "tagged image file format" (also, "tagged information field format").[4] Once stored, these images can be retrieved as any other file would be and even transmitted electronically over the Internet.

Practical Considerations

The primary concern of most scientists considering the acquisition of image analysis software and hardware is "How much?" Not surprisingly,

[4]Other image storage formats include BMP, PCX, GIF, and JPEG.

there is enormous variation in the format of equipment, the variety of functions that can be performed, the system requirements, the amount of supporting hardware required for the system to run and, of course, the cost. Many systems are completely self-contained, are elegant to behold, and have a great many functions preprogrammed. Of these, many are very expensive and often beyond the budget of many laboratories: At best these might be purchased at the department level and used by everyone. Other digital image analysis products are designed to be loaded directly into a computer already in the laboratory; beyond the purchase price of the software, the only investment required is the purchase of some type of image acquisition device, such as a scanner for Polaroid photographs, a charge-coupled device camera, or a frame grabber. Of the products that fall into this category, Gel Pro Analyzer[5] is of superior quality. This software uses the point and click approach to derive the desired information, without demanding computer literacy of the user, and many of the functions including identifying lanes, finding bands, and performing user-defined macros are automated. While limited input is required from the user (Figure 8), the capacity for complete override of all automated functions allows maximum flexibility. One point that is abundantly clear is that, at all levels of sophistication, the added benefits of image analysis software easily justify the costs (Table 2).

A key advantage of using image analysis software is the automation of tedious and time-consuming determinations of the location and mass distributions of molecules in a gel. Calculations commonly associated with the antiquated method involve manually generating a size calibration curve and concomitant densitometry for mass determinations. Digital image analysis reduces intralaboratory error and obviously favors maximum throughput in laboratories generating large numbers of images. Further, better quality image analysis software is fully compatible with images of gels as well as images on X-ray films and chromogenic detection methods, and is fully able to analyze bands as well as signals in dot blot format.

One area of gross misunderstanding about image analysis software pertains to the ability of such software to improve upon the image itself. One must realize first and foremost that if the investigator scans a poor image of a gel, then the image analysis software has a poor image with which to work. If a gel was not electrophoresed for a sufficiently long period, the software cannot further resolve the bands. A good deal of suc-

[5]Media Cybernetics (Silver Spring, MD). http://www.mediacy.com.

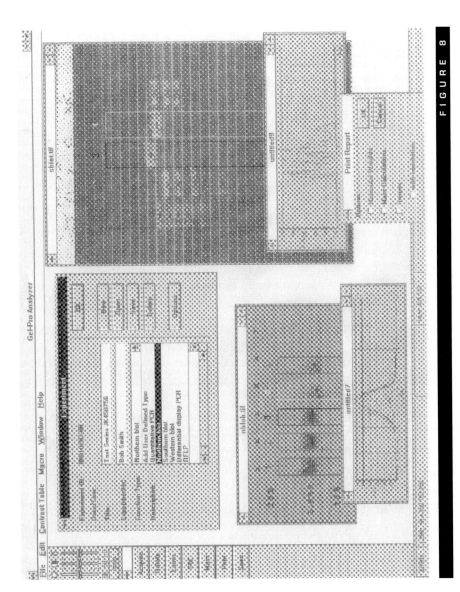

Northern analysis under investigation using Gel Pro Analyzer. Courtesy of Media Cybernetics.

T A B L E   2

Advantages of Image Analysis Software

Perform standard and customized measurements
Automate and unify image analysis in the lab
Save time and increase productivity
Data analysis is reproducible[a]
Perform intensity and background calibrations
Compensate for nonuniform background
Calculate integrated optical density of each band, as opposed to the error-prone method
   of density readings of small areas with old-style densitometers

[a]This aspect of image analysis is especially important because most people tend to overestimate the
integrated optical density of bands on gels and autoradiograms.

cessful image analysis depends on the quality of the image provided by the investigator.

Background correction is one area of acute controversy, particularly when high precision quantitation of band and lane mass are required. The importance of and difficulties associated with this aspect of image analysis are further compounded by the fact that the background is never uniform in a gel or on an image; rather, a gradient of background exists from left to right and top to bottom. As such, there are three basic approaches:

1. Examine numerous small areas in the image, to deal with "local background" only (a so-called "area of interest").
2. Position the cursor at some neutral location on the gel, sample the background at that point, and then perform a global background subtraction.
3. Position numerous background sampling symbols at various points in the image, and subtract the measured background from the nearest lane in a gel or series of spots.

Of the three, the third method offers the greatest degree of accuracy and versatility for the user.

Because of the inherent limitations of photographic films in general, as described above, one must also be aware of the potential for obscuring the true quantitative differences between bands under investigation on the same gel. Recall that 255 is the largest possible pixel value, and areas within an image containing these values are said to be saturated. While pixel values below 255 will continue to increase with longer exposures, these will also become saturated eventually. Since longer exposure times will not/cannot increase pixel values beyond 255, calculations on bands containing saturated pixels are likely to generate data that are lower than expected, compared to non-pixel-saturated bands. Quantitative measurements are, therefore, compromised when saturated pixels are involved.

Finally, the better software will prevent, or at least make difficult, changing the actual appearance of the image. It is one thing to mathematically add, delete, and straighten lanes and bands, it is quite another to physically modify the image.

References

Baxes, G. A. (1994). "Digital Image Processing—Principles and Applications." Wiley, New York.

Baxes, G. A. (1988). "Digital Image Processing—A Practical Primer." Cascade Press, Denver, CO.

Castleman, K. R. (1996). "Digital Image Processing." Prentice-Hall, Englewood Cliffs, NJ.

Chellappa, R. (1992). "Digital Image Processing." IEEE Computer Society Press, Los Alamitos, CA.

Gonzalez, R. C., and Wirtz, P. (1987). "Digital Image Processing." Addison-Wesley, Reading, PA.

Media Cybernetics. (1996). *In* "Gel Pro Analyzer Reference Guide," 2nd ed. Media Cybernetics, Silver Spring, MD.

Pratt, W. K. (1991). "Digital Image Processing." 2nd ed. Wiley, New York.

Russ, J. C. (1995). "The Image Processing Handbook." CRC Press, Boca Raton, FL.

The Northern Analysis

Rationale

An enormous number of variations on the theme of Northern analysis have emerged since the original descriptions of the technique (Alwine *et al.*, 1977, 1979). At the heart of the Northern analysis is the transfer, or blotting, of electrophoretically separated RNAs from a gel to a filter membrane for subsequent fixation and hybridization to a specific probe or probes. Because the RNA is transferred in exactly the same configuration as it was separated in the gel, hybridization signals generated by Northern analysis provide a qualitative and semiquantitative profile of the sample under investigation. This method, which is compatible with tissue samples and cells grown in culture, continues to be a commonly used technique to assess messenger RNA length.

Currently, there are various methods for isolating RNA, preparing and running denaturing agarose gels, and transferring the nucleic acid from the gel. In addition, selections must be made regarding the type of filter membrane, method of immobilization, type of probe, method of probe labeling, hybridization recipe, and method of detection. Obviously, quite a few decisions must be made. When optimized in one's own laboratory, many of the technical difficulties and issues pertaining to reproducibility resolve themselves.

What then constitutes Northern analysis? Generically, it is the electrophoresis of RNA, transfer and immobilization on a filter, hybridization with a labeled DNA or RNA probe, and finally analysis of hybridization events. Previous chapters have described methodologies for RNA isolation, quantification, and electrophoresis. This chapter provides options with respect to filter membranes, transfer techniques, and methods for nucleic acid fixation on the filter. Subsequent chapters describe hybridization and detection methods.

Choice of Filter Membrane

One important decision associated with the study of RNA by Northern analysis is the selection of the solid support matrix to which electrophoretically separated RNA is transferred prior to hybridization. This selection must be influenced by the knowledge that different types of matrices exhibit wide variations in nucleic acid binding capacity and support different transfer and immobilization methodologies. Moreover, because the RNA sample on a filter paper may well represent an enormous, labor-intensive effort, many investigators wish to do multiple probings with a battery of probe sequences. The hybridized probe may be removed, following autoradiography, by a very high-stringency wash, as described by the manufacturer of the filter paper and some general guidelines that are described here. The filter may then be rehybridized with a completely unrelated probe. Thus, knowledge of whether one type of filter or another is able to withstand this regimen is invaluable.

Nitrocellulose

The original RNA blotting technique described the use of diazobenzyloxymethyl paper for nucleic acid immobilization (Alwine *et al.*, 1977). This technique evolved fairly rapidly to accommodate the use of nitrocellulose, the so-called "classical membrane" of molecular biology. The underlying chemical basis by which nucleic acids become fixed onto nitrocellulose is poorly understood, though it is believed to be hydrophobic in nature. Perhaps the most critical parameter is the need for very high ionic strength transfer buffer, for the binding of single-stranded molecules (RNA or denatured DNA). The most common pore size for nitrocellulose used in blotting is 0.45 μm, although 0.22-μm membranes can also be used, particularly in the study of smaller nucleic acid molecules. Nitrocellulose is still used today in many laboratories for nucleic acid analysis, although there are intrinsic qualities that make the use of other membranes far more desirable:

1. Nitrocellulose has a relatively low binding capacity for nucleic acids (about 80 μg/cm^2).
2. Nitrocellulose has a very poor binding capacity for smaller nucleic acid molecules (typically less than 1000 bases).
3. Nitrocellulose has a slight net negative charge that mandates the use of a relatively high ionic strength transfer buffer, usually 20× SSC or 20× SSPE.
4. To immobilize nucleic acids onto nitrocellulose, baking in a vacuum oven at 80°C is necessary. Baking nitrocellulose in a vacuum oven is a nonnegotiable requirement because humidity will interfere with the immobilization process.
5. Nitrocellulose becomes very brittle after baking, which is required for nucleic acid immobilization. This makes subsequent manipulations very difficult and requires handling of the membrane with extreme care.
6. The frailty of this membrane following the baking step makes repeated probing impractical.

Nylon

Nylon filters were developed to circumvent some of the difficulties inherent in nitrocellulose. Perhaps the most significant advantage associated with their use is that these membranes exhibit great tensile strength. This enables more assertive handling of the filter and repeated probing of the same filter. Nylon filters are widely used in the molecular biology laboratory and are sold under a variety of trade names. They fall into two categories: those with a neutral surface charge, an ancient name for which is nylon 66, and those with a net positive charge imparted by surface amines. The latter are frequently referred to as charge-modified nylon filters and denoted as nylon(+). As with nitrocellulose, many of the nylon filters used for blotting have a pore size of 0.45 μm, although 0.22-μm filters are available as well. Neutral nylon filters in particular have also proven to be very compatible with chemiluminescence-related technologies. Nylon membranes exhibit an enhanced nucleic acid binding capacity compared to nitrocellulose, as great as $400-500\mu g/cm^2$, depending on the manufacturer and the exact buffers used by the investigator. Nylon membranes also show a particular affinity for smaller nucleic acid molecules (500 bases). Compared to nitrocellulose, the transfer buffers used in conjunction with nylon filters are routinely of lesser ionic strength, to as low as 5× SSC or 5× SSPE. While the positive charge associated with nylon(+) membranes certainly increases the electrostatic attraction between target RNA and membrane, it also tends to increase the degree of background observed following the posthybridization stringency washes due to nonspecific interaction between the positively charged membrane and the negatively charged phosphodiester backbone of the probe.

The investigator is afforded much greater latitude when blotting onto nylon membranes. Immobilization of nucleic acids can be accomplished either by baking or by irradiation with UV light. Nylon filters, if desired, may be baked instead, requiring only 1–1.5 h at 65–70°C, and do not require a vacuum oven; any instrument capable of addressing these parameters will suffice. Most notably, nylon filters do not become brittle when baked, and they withstand repeated probing much better than nitrocellulose. Last, it has been reported that torn nylon filters can be repaired by spot-welding them with a hot metal implement such as a bacterial inoculating loop or soldering pen (Pitas, 1989).

Polyvinylidene Difluoride

Polyvinylidene Difluoride (PVDF) is a fluorocarbon polymer that has been used extensively for Western blotting, with only recent application for the blotting of nucleic acids. The durability of this material and chemical stability support the utility of this type of filter in all common nitrocellulose and nylon applications pertaining to nucleic acids. These membranes possess an average pore size of 0.45 μm. Nucleic acids can be permanently immobilized by crosslinking with UV light or by baking (vacuum oven not required). PVDF membranes are gaining acceptance in many nonisotopic applications as well.

Handling and Filter Preparation

No matter what type of filter membrane is selected, it should be handled only with gloves and cut with sharp scissors or a scalpel. This is necessary in order to prevent finger greases from compromising the nucleic acid binding capacity of the membrane, not to mention the accidental introduction of RNase into the system! Most filters are supplied sandwiched between two sheets of protective wrapping paper. When cutting the membrane to the appropriate size, which will usually be the size of the gel, the protective wrapping should be left in place and used to handle the filter.

Filter membranes generally require pre-wetting in a clean pan or tray of sterile water for 5 min, followed by brief equilibration in the high salt transfer buffer selected for the particular application at hand. The proper way to wet a filter is to float it on the surface of the water (preferably autoclaved or DEPC-treated). An immediate change in the appearance of the membrane can be observed: a color change from white to off-white/gray as the filter comes into contact with the water. This technique is strongly recommended because it will allow immediate observation of any area of the filter that has become mechanically damaged or touched with bare fingers; such areas appear as smudges or an obvious uneven wetting of the filter. Once it is clear that the filter has not been damaged, it can be completely submerged by rocking the tray. Allow the filter to pre-wet for at least 5 min (longer periods are fine, too).

After wetting, transfer the filter to a tray containing the same transfer buffer that will be used for the blot. A brief equilibration of about 30 s is all that is required. Then the filter can be used for the transfer itself.

Northern Transfer Techniques

The actual transfer of electrophoretically separated RNA species from the gel to a solid support can be accomplished using any of a variety of blotting formats that differ primarily in the method by which the RNA is drawn out of the gel matrix. Remember that no matter which transfer technique is chosen, intact RNA must have been electrophoresed, in a clean gel box, through a gel favoring maximum resolution. Otherwise, the resultant data are likely to be suboptimal. In general, the transfer techniques described here are applicable to both Northern and Southern analysis.

Capillary Transfer

The method requiring the fewest accessories is the traditional passive capillary transfer method, which is also used in the classical Southern analysis (Southern, 1975). In short, a stack of absorbent material, such as paper towels or blotting paper, is used to draw (wick) the transfer buffer from a reservoir, through the gel, and finally into the dry stack of paper towels. In so doing, RNA is essentially eluted from the gel and is trapped on a juxtaposed piece of membrane filter paper, as the buffer continues to move through the membrane and into the stack of paper towels. Traditionally, a typical setup of this nature involved placing the blotting paper on the top of the stack, as shown in Figure 1. Two common complaints associated with transfer in this fashion, however, are (1) poor transfer efficiency, especially of larger RNA molecules, and (2) transfer takes too long.

Regarding the former, inefficient transfer is commonly associated with the following:

1. The gel was cast too thick.
2. The gel consisted of more than 1.2% agarose.
3. EtBr was used to stain the gel, either during electrophoresis or afterward, without proper destaining. Recall that there are good alternatives to staining with EtBr.
4. The transfer period was insufficient; most capillary transfers require 16 h or longer, that is, overnight.
5. The gel was not UV irradiated to nick the backbone and promote expedient transfer.
6. The RNA was degraded to begin with, so no hybridization signal could possibly be generated.
7. Any combination of the above.

In instances where incomplete transfer simply cannot be remedied and, regarding the latter complaint of time-consuming transfer, alternative transfer techniques that accelerate and improve transfer efficiency

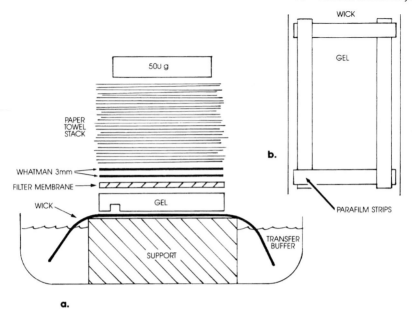

F I G U R E   1

A typical setup for capillary transfer of RNA from denaturing agarose gels to a filter membrane. (a) Side view. Invert the gel to reduce the blotting time and achieve greater resolution, since the RNA does not have to travel as far through the gel matrix to reach the membrane. (b) Top view. Place overlapping strips of parafilm directly on the wick and up against the gel (not on top of the gel). Set parafilm in place around the sides of the gel first, and then overlay filter membrane and other blotting materials.

have been developed and widely adapted. These alternative approaches include the following:

TurboBlotter

TurboBlotter (Schleicher & Schuell, Keene, NH) downward transfer systems take advantage of gravity and, in so doing, offer enhanced transfer efficiency in a much shorter time than is afforded by the older upward capillary transfer method (Figure 2). This method likewise eliminates the need for heavy weights on top of the capillary stack, thereby minimizing compression and concomitant collapsing of the pores of the gel. The technique does not require the use of any expensive equipment, such as the vacuum blot devices or electroblotting apparatuses described below.

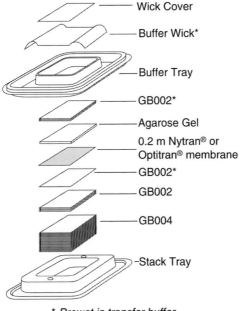

Wick Cover

Buffer Wick*

Buffer Tray

GB002*

Agarose Gel

0.2 m Nytran® or
Optitran® membrane

GB002*

GB002

GB004

Stack Tray

Prewet in transfer buffer

FIGURE 2

TurboBlotter transfer method. The principle is essentially that of an inverted capillary transfer. By taking advantage of gravity, rapid downward movement of the high salt transfer buffer accelerates the transfer and dramatically improves transfer efficiency. The absorbant blotting paper is at the bottom of the stack and the buffer reservoir on top, precisely the opposite of the classical capillary blot setup shown in Figure 1. Courtesy of Schleicher & Schuell.

Vacuum Blotting

A vacuum blotting system is one in which negative pressure is applied to accelerate the transfer process. In this technique, also known as vacuum-assisted capillary transfer, the transfer buffer is essentially sucked through the gel along with the nucleic acid sample. There are, of course, a number of different instruments for vacuum blotting, available from different sources. The advantage is that the transfer occurs far more rapidly than traditional capillary transfers (30–60 min vs. overnight). To an extent, the accelerated process reduces molecular diffusion of nucleic acid molecules on the way out of the gel, minimizing the outward (lateral) spreading of discrete bands, as observed in the gel. One must use great care when vacuum blotting, however, not to exceed the hydrostatic pressure of the gel (vacuums of 30–50 cm H_2O are optimal), because gel collapse will result in immediate ruination of the experiment. Vacuum blotting can be used for the transfer of nucleic acids from agarose gels (Medveczky *et al.*, 1987; Olszewska and Jones, 1988), but not for transfer from polyacrylamide gels.

Positive Pressure

In contrast, to vacuum blotting, positive pressure can be employed to "push" nucleic acids from the gel. The PosiBlot (Stratagene, La Jolla, CA) uses positive pressure to drive buffer through the gel and complete transfer in less time than is required for traditional capillary transfer. As with vacuum blotting, the accelerated transfer process reduces lateral molecular diffusion of nucleic acid molecules, thereby increasing the sharpness of bands. Positive pressure does not result in gel collapse as sometimes occurs when an excessive vacuum is applied.

Electroblotting

A completely different approach known as electroblotting is yet another transfer method. Electroblotting is widely used in conjunction with polyacrylamide gels, and infrequently utilized for the transfer of nucleic acids from agarose gels. In one form of this technique, also known as electrophoretic transfer, the gel is placed next to the membrane in a special cassette that, in turn, is placed in a tank of electrolyte buffer. Upon application of a voltage gradient perpendicular to the gel, the sample migrates out of the gel and onto the filter paper, much like the transfer methodology of the Western blot. In another form of this technique, known as semi-dry electroblotting, only minimal buffer is needed, essentially to saturate the membrane and blotting papers that provide electrical contact with the gel and apparatus itself. The advantage here is that low voltage and current are needed for transfer, precluding the requirement for a high current power supply. A wide selection of electroblotting apparatuses are commercially available (e.g., Hoefer Pharmacia Biotech).

The extent to which electroblotting accelerates the transfer procedure is directly related to the current; the investigator should follow the instructions of the manufacturer and exercise due caution when working with power supplies and other high voltage equipment.

Nitrocellulose is *not* a suitable filter membrane for electroblotting for two principal reasons. First, the binding of nucleic acids to nitrocellulose requires high ionic strength buffers; thus high electrical currents are produced, overheating the system and melting the gel (remember that $I = E/R$

and $P = I^2R$; see Appendix K for review). Second, nylon has a much higher affinity for smaller nucleic acid molecules, a likely situation when polyacrylamide gels are involved. For an historical perspective see Arnheim and Southern (1977) and Bittner *et al.* (1980).

Alkaline Blotting

Alkaline blotting of nucleic acids is essentially a variation on the theme of capillary transfer. The key difference between these two approaches, however, is the method of denaturation and nucleic acid transfer. Traditional approaches to capillary blotting include the use of high salt buffer (15–20× SSC or SSPE) to transfer RNA or previously alkalidenatured DNA from the gel to the filter membrane; positively charged nylon is recommended almost exclusively. In alkaline blotting of DNA, the denaturation and transfer steps are accomplished simultaneously by the use of 0.4 N NaOH, with subsequent fixation onto nylon membranes. In the case of Northern transfer, *limited* alkaline hydrolysis can facilitate the transfer of RNA from the gel to a solid support. Depending on the exact parameters used, alkaline blotting onto certain membranes can preclude the requirement for posttransfer immobilization. Moreover, the exact composition of transfer buffer and transfer time greatly influence the signal-to-noise ratio, with regard to the nonspecific binding of nucleic acid probes to rRNAs, particularly when the alkaline transfer blotting method is used. For details, see Reed and Mann (1985); Li *et al.* (1987), and Löw and Rausch (1994).

Protocol: RNA Transfer by Passive Capillary Diffusion

Note: Refer to Figure 1 as needed.

1. **Wear gloves** throughout the procedure. It is imperative to prevent nuclease contamination at every level. Remember that RNAs are not safe from nuclease degradation until they have been immobilized, by baking or crosslinking, on a filter membrane (nitrocellulose, nylon, etc).

2. Gels that have been stained or that contain formaldehyde must be destained prior to setting up the transfer. Failure to do so will result in highly inefficient transfer and loss of sample.

Note: A spatula is most convenient for transferring gels from one solution to another. Formaldehyde gels are especially slippery and should be handled with care.

2a. Formaldehyde gels: Soak the gel in an excess of 1× MOPS buffer, sterile H_2O, or transfer buffer to remove the formaldehyde. Formaldehyde is typically drawn out of unstained gels by soaking them three times for 10 min each. If the gel also contains ethidium bromide, it will likewise be drawn out of the gel. This is required to eliminate the high background fluorescence associated with EtBr-stained formaldehyde gels so they can be photographed. Thorough destaining usually requires 2 or more hours. In this laboratory, formadehyde gels are occasionally destained overnight in an excess of buffer at 4°C.

Optional, though not recommended: Soak the gel in 50 mM NaOH, 10 mM NaCl for 30 min; in 100 mM Tris, pH 7.5, for 30 min, and then in 10× SSC for 30 min. This will result in partial alkaline hydrolysis along the length of the RNA backbone; although this may facilitate transfer of very large RNA molecules out of the gel it is usually not necessary. Moreover, overexposure to NaOH can destroy the ability of the RNA to hybridize to any probe. Instead, random nicking of the RNA backbone and improved transfer efficiency may be accomplished by ir-radiating the gel on a UV transilluminator or, for greater precision, by using a calibrated UV light source.

2b. Glyoxal gels: Transfer is set up immediately after electrophoresis has been completed. Glyoxal will be removed from the RNA in a post-transfer, prehybridization wash.

3. Trim the left and right edges of the gel (but not the top or bottom) such that there is a 1-cm margin bordering the lanes of interest.

4. Cut a piece of filter membrane to the size of the gel and prepare it for transfer according to the instructions of the manufacturer, typically by pre-wetting in H_2O for 5 min and then in transfer buffer for 30 s (use 5× SSC for nylon). Never handle a filter without gloves. Finger greases will compromise the binding capacity of most filters. To cut the filter mem-brane and other sheets of blotting paper to the size of the gel, place a clean, dry, gel casting tray directly on the wrapping paper that protects the membrane itself, and trace the outline of the casting tray with a pen-cil. The gel casting tray acts as a template so that all materials (filter, Whatman paper, paper towels) can be cut without a ruler, while the gel is destaining or soaking in buffer. For laboratories with enhanced budgets, all required blotting materials may be purchased pre-cut to the proper di-mensions.

5. Typically, nylon and nitrocellulose membranes are first floated on the surface of RNase-free water and then submerged for a minimum of 5 min. Old and/or soiled filters do not wet evenly, showing irregular patch-

es of wetting when floated on the surface of water, and should be discarded. Follow the recommendations of the manufacturer for optimal filter membrane utility.

6. After wetting the filter paper for 5 min, the filter is equilibrated in transfer buffer (5× SSC or 5× SSPE for nylon). The filter may remain in transfer buffer until it is used.

7. Cut one sheet of Whatman 3MM paper (or the equivalent) approximately 1 in. wider than the gel and at least 6 in. longer than the gel. This is the wick and it will act as the contact between the transfer buffer reservoir and the gel itself; a wick that proves too long can be trimmed later.

8. Saturate the wick by soaking it in transfer buffer for 10 s. Next, drape the wick over a plexiglass sheet supported by a baking dish or similar implement.

9. Fill the baking dish with enough transfer buffer such that a minimum of 1–2 in. of each end of the wick is submerged in the transfer buffer.

10. Any air bubbles that are trapped under the wick are easily removed by gentle rolling with a sterile pipet. Failure to remove all air bubbles will compromise transfer efficiency at that location.

11. Place the gel, wells facing down if possible, in the middle of the wick, and again remove any air bubbles trapped between the gel and the wick. Mask the area surrounding the gel with strips of Parafilm to prevent short-circuiting[1] the system during transfer (Figure 1b).

12. Carefully place the filter membrane on top of the gel and remove any air bubbles trapped beneath it.

13. Cut two or three sheets of Whatman 3MM paper (or the equivalent) to the same size as the gel. Pre-wet at least one of these sheets in the transfer buffer.

Note: Pre-wet at least one of these sheets in transfer buffer just prior to use. In this lab, pre-wetting the sheet that will rest directly on the filter membrane appears to prime the wicking action efficiently.

14. Place the wet sheets of Whatman 3MM paper on top of the filter membrane and eliminate air bubbles. Then place the dry piece on top.

---

[1]Occasionally, the paper towels that draw the buffer upward from the reservoir and through the gel will sag down below the gel as they become wet. This may occur when the paper towels are even slightly larger than the gel. If the paper towels touch the wick along the side of the gel, they will draw the transfer buffer up *around* the gel rather than through it. Less buffer moving through the gel causes slowing of the transfer process. Parafilm along the periphery of the gel will prevent direct contact between wick and paper towels.

15. Cut a stack of paper towels (2–3 in. when compressed) to the size of the gel. This can be done well ahead of time. Position the paper towels on top of the sheets of Whatman 3MM paper. Finally, cover the entire stack with a 500-g weight or an obscure textbook to compress the entire setup.

Note: Avoid paper towels with creases such as the C-fold type as ineffi-cient transfer from the gel over the area of the crease is almost guaran-teed. If no other paper towels are available, alternate the paper towels in the stack so that the creases are not aligned over any particular part of the gel. Uniformity of capillary action is an absolute requirement. As an alternative to paper towels, use pre-cut blotting paper, such as GB002 and GB004 (Schleicher & Schuell).

16. Allow the transfer to proceed for several hours or overnight, if possible.

17. At the conclusion of the transfer period, carefully peel the filter membrane away from the gel with forceps or a gloved finger. Clip one corner of the filter to mark it asymmetrically and indicate orientation. This is essential for accurate posthybridization interpretation of data.

18. Follow the posttransfer washes suggested by the manufacturer. Typically, wash the membrane for 30 s in 5× SSC or 5× SSPE, depending on the buffer that was used for the transfer.

Note: The brief, posttransfer washing of the filter removes any pieces of agarose that may be clinging to the filter as well as any other resid-ual materials that could interfere with immobilization of the sample and/or hybridization specificity.

19. Nylon membranes should be UV irradiated while still damp. Ni-trocellulose membranes should be air-dried and then baked as soon as possible.

Protocol: TurboBlotter Downward Transfer of RNA

Note: Refer to Figure 2 as needed.

1. **Wear gloves** throughout the procedure. It is imperative to prevent nuclease contamination at every level. Remember that RNAs are not safe from nuclease degradation until they have been immobilized, by baking or crosslinking, on a filter membrane (nitrocellulose, nylon, etc).

2. Gels that have been stained or that contain formaldehyde must be destained prior to setting up the transfer. Failure to do so will result in highly inefficient transfer and loss of sample.

Note: A spatula is most convenient for transferring gels from one solution to another. Formaldehyde gels are especially slippery and should be handled with care.

2a. Formaldehyde gels: Soak the gel in an excess of 1× MOPS buffer, sterile H_2O, or transfer buffer to remove the formaldehyde. Formaldehyde is typically drawn out of unstained gels by soaking them three times for 10 min each. If the gel also contains ethidium bromide, it will likewise be drawn out of the gel. This is required to eliminate the high background fluorescence associated with EtBr-stained formaldehyde gels so they can be photographed. Thorough destaining usually requires 2 or more hours. In this laboratory, formadehyde gels are occasionally destained overnight in an excess of buffer at 4°C.

Optional, though not recommended: Soak the gel in 50 mM NaOH, 10 mM NaCl for 30 min; in 100 mM Tris, pH 7.5, for 30 min; and then in 10× SSC for 30 min. This will result in partial alkaline hydrolysis along the length of the RNA backbone; although this may facilitate transfer of very large RNA molecules out of the gel it is usually not necessary. Moreover, overexposure to NaOH can destroy the ability of the RNA to hybridize to any probe. Instead, random nicking of the RNA backbone and improved transfer efficiency may be accomplished by irradiating the gel on a UV transilluminator or, for greater precision, by using a calibrated UV light source.

2b. Glyoxal gels: Transfer is set up immediately after electrophoresis has been completed. Glyoxal will be removed from the RNA in a post-transfer, prehybridization wash.

3. Select pre-cut blotting papers that are as close to the size of the gel as possible. If the gel is slightly larger, the excess will be trimmed away during the assembly of the TurboBlotter. If the dimensions of the gel are less than those of the blotting paper, the exposed portion of the blotting paper will be insulated later with Parafilm.

4. Cut a piece of nylon filter to the size of the gel and prepare it for transfer according to the instructions of the manufacturer, typically by pre-wetting in H_2O for 5 min and then in transfer buffer for 30 s (use 5× SSC for nylon). Never handle a filter without gloves. Finger greases will compromise the binding capacity of most filters.

5. Typically, nylon (and nitrocellulose) membranes are first floated on the surface of RNase-free water and then submerged for a minimum of 5 min. Old and/or soiled filters do not wet evenly, showing irregular patches of wetting when floated on the surface of water, and should be

discarded. Follow the recommendations of the manufacturer for optimal filter membrane utility.

6. After wetting the filter paper for 5 min, the filter is equilibrated in transfer buffer (5× SSC or 5× SSPE for nylon). The filter may remain in transfer buffer until it is used.

7. Place the stack tray of the transfer device on the bench, making sure that it is level.

8. Place 20 sheets of dry GB004 blotting paper in the stack tray.

9. Place 4 sheets of dry GB002 blotting paper on top of the GB004 stack.

10. Place 1 sheet of GB002 blotting paper, prewet in transfer buffer, onto the stack. Use a sterile 5-ml pipet to gently smooth the membrane, ensuring that no air bubbles are trapped beneath. Do not apply too much pressure with the pipet.

11. Place the equilibrated filter membrane on the stack.

12. Carefully place the agarose gel on top of the filter membrane and then trim the gel to the size of the membrane by cutting away any over-hanging, unused lanes. *Do not* move the gel once it has made contact with the filter. If any portion of the stack is exposed because the gel was small-er, mask the area surrounding the gel with strips of Parafilm to prevent short-circuiting the system during transfer.

13. Use a sterile 5-ml pipet to gently smooth the gel, ensuring that no air bubbles are trapped between the gel and the filter paper.

14. Wet 3 sheets of GB002 blotting paper in transfer buffer, and care-fully place these on top of the stack.

15. Use a sterile 5-ml pipet to gently smooth the stack, ensuring that no air bubbles are trapped between the gel and the sheets of blotting paper.

16. Attach the TurboBlotter buffer tray to the stack tray, guided by the circular alignment buttons to ensure proper placement.

17. Fill the buffer tray with transfer buffer. Do not pour any transfer buffer into the center of the stack tray where the blotting papers and gel are located.

18. Wet the wick in transfer buffer. Initiate the transfer by connecting the gel stack with the buffer tray: Place the wick across the stack so that the shorter dimension of the wick competely covers the blotting stack and both ends of the long dimension extend into the buffer tray.

19. Cover the stack with the wick cover (provided) or simply with Parafilm. This is strongly suggested to prevent evaporation.

20. Allow transfer to proceed until the buffer has been drawn out of the buffer tray. While this usually requires 3 h, it has been determined in

this laboratory that for most routine blots more than 50% of the sample in the gel transfers within the first hour (data not shown).

21. At the conclusion of the transfer period, carefully peel the filter membrane away from the gel with forceps or a gloved finger. Clip one corner of the filter in order to mark it asymmetrically to indicate orientation. This is essential for accurate posthybridization interpretation of data.

22. Follow the posttransfer washes suggested by the manufacturer. Typically, wash the membrane for 30 s in 5× SSC or 5× SSPE, depending on the buffer that was used for the transfer.

Note: The brief, posttransfer washing of the filter removes any pieces of agarose that may be clinging to the filter as well as any other residual materials that could interfere with immobilization of the sample and/or hybridization specificity.

23. Nylon membranes should be UV irradiated while still damp. Nitrocellulose membranes should be air-dried and then baked as soon as possible.

Posttransfer Handling of Filters

Formaldehyde Denaturing Systems

Although filters can be stored confidently before hybridization, it is necessary to immobilize the sample on the filter as soon as possible after transfer. Only then is the RNA sample stabilized and rendered resistant to degradation by RNase. Once immobilization is complete, filters can be used immediately for hybridization or stored for months. Place air-dried filters between sheets of Whatman 3MM paper and store dry and in the dark until further use. Formaldehyde will have already been removed from the system during gel destaining in preparation for gel blotting.

Glyoxal Denaturing Systems

The glyoxal that denatured the RNA prior to electrophoresis must now be removed from the filter.

Option 1: Air-dry and then bake the filter membrane according to the recommendations of the manufacturer. This will destroy some of the glyoxal and immobilize the RNA at the same time. The remaining glyoxal can be removed by immersing the filter in

200 ml of 20 mM Tris, pH 8.0, preheated to 90°C, and then immediately cooling to room temperature.

Option 2: After immobilization, remove glyoxal from RNA by washing the filter for 15 min at 65°C in 20 mM Tris, pH 8.0.

Immobilization Techniques

Thorough immobilization of target RNA and DNA on a filter membrane, after the transfer, is just as important as ensuring that the transfer process itself is complete. It is most important point to recognize that nucleic acids, especially RNA molecules, are not safe from nuclease or natural biological degradation until the sample has been immobilized onto a solid support. Failure to completely immobilize the sample will invariably result in the loss of the sample during the prehybridization, hybridization, and/or posthybridization washing and detection steps. A variety of methods for nucleic acid immobilization have been described, each of which has defined advantages, limitations, and applicability. In this laboratory, nylon filters are used almost exclusively because of the great experimental flexibility offered by this technique. In every instance, it is strongly advisable to follow the recommendations of the manufacturer of a particular filter membrane. A few general guidelines are presented here.

Baking

The classical technique for nucleic acid immobilization involves baking the sample onto the surface of nitrocellulose. Although the exact nature of interaction between the matrix and nucleic acids is not clear, it is believed to be hydrophobic. This method, in spite of its numerous shortcomings, persists today. The investigator allows the nitrocellulose filter to air-dry completely, after the posttransfer wash, and then bakes the membrane for 2 h at 80°C *in vacuo*. When working with nitrocellulose, the use of a vacuum oven is a nonnegotiable requirement because humidity interferes with the immobilization process. Whereas baking will certainly do the job, this method is not a completely permanent process: After a few reprobings of the filter, some of the sample is lost from the filter and all subsequent assays are no longer quantitative.

With nylon membranes, nucleic acid samples can be immobilized by baking or by UV irradiation. The baking process usually requires 1–1.5 h, with temperatures only as high as 65°C. Moreover, baking of nylon does not require a vacuum oven. Some protocols suggest that nylon filters may be baked while still damp, although in this case, the application of a vacuum may accelerate the drying process and improve retention on the filter.

Crosslinking by UV Irradiation

A more reliable method for immobilization is the use of a calibrated UV light source that can effectively crosslink nucleic acids[2] to nylon membranes (Church and Gilbert, 1984, Khandjian, 1987); the practice of UV irradiation of nitrocellulose is to be avoided because it does not result in the comparable retention of nucleic acids achievable by crosslinking to nylon and, more importantly, over-irradiation of nitrocellulose poses a serious risk of fire. Don't do it.

Crosslinking of nucleic acids to filters with UV light has been a viable option since it was described as part of the Church and Gilbert (1984) protocol for genomic sequencing. Short-wave and medium-wave UV light (254 and 302 nm, respectively) activate thymine and uracil, and other bases to a lesser degree, which become highly reactive and form covalent bonds with the surface amines that characterize many nylon matrices (Saito *et al.*, 1981). UV crosslinking is usually more efficient with damp nylon membranes, which generally require a total exposure of $1.2–1.6 \, kJ/m^2$. Alternatively, air-dried membranes require approximately 160 J/m^2.

Increased stability of nucleic acids on the matrix, compared to baking procedures, is an important dividend associated with this method of fixation. Many laboratories have reported an increase in sensitivity with UV-crosslinked filters compared to baked filters. Moreover, fixation by crosslinking appears to be a far more permanent immobilization technique than baking. This is of critical importance if the investigator plans to screen a filter more than once; repeated high-stringency removal of hybridized probe usually results in the loss of target sequence from filters that were baked.

A variety of methods have been derived to accomplish fixation. The most precise energy output is, of course, delivered by a calibrated UV light source (e.g., Stratalinker, Stratagene, La Jolla, CA). Such instruments have an auto-crosslink feature preset to deliver a UV dose of about

---

[2]In addition to a crosslinking function, irradiation of agarose gels is an alternative method of randomly nicking the phosphodiester backbone of RNA and DNA. This takes the place of limited hydrolysis by immersion in a dilute solution of NaOH, and is more reproducible. These random nicks will facilitate blotting from the gel onto the membrane.

120 mJ/cm^2, which is ideal for damp membranes; this approach works very efficiently with damp filters. Alternatively, a standard laboratory transilluminator may suffice and is likely to involve considerable trial and error in the absence of a calibration instrument. Most transilluminators in the molecular biology laboratory emit at one of two wavelengths, either 302 nm (medium-wave energy) over 254 nm (short-wave energy). Crosslinking with a 302-nm emitter is strongly recommended over 254 nm-emitting instruments because overirradiation with the shorter wave energy will reduce sensitivity due to gross damage to the blotted nucleic acid molecules. Damp filters are placed directly on the surface of the transilluminator and irradiated.

Protocol

1. Wash filters post-Northern transfer as described above and blot excess liquid by placing the RNA-containing filter membrane on top of one or two sheets of Whatman 3MM paper. Do not allow filters to dry out completely; it is best to irradiate filters that are still damp.

2. Place filter(s) face down directly on the surface of the transilluminator, the surface of which has been wiped clean just prior to use.

CAUTION   Be sure to wear proper eye protection, as serious permanent damage can occur if eyes (and skin) are not properly shielded.

3. The appropriate exposure time is a direct function of the wavelength and age of the transilluminator. With 302-nm-emitting instruments, a total of 2 min is usually more than adequate. For 254-nm-emitting instruments, crosslinking is complete in less than 1 min.

Note 1: These are only general guidelines; the exact parameters must be empirically determined for each instrument.

Note 2: In this laboratory, filters are irradiated for one-half of the estimated time required, rotated 90°, and then irradiated again for the remaining interval. This is suggested because of the parallel orientation of the bulbs within the transilluminator, some of which may be burned out in older instruments. Depending on the exact specifications of a particular instrument, this may or may not be necessary.

4. Store filters sandwiched between Whatman 3MM paper in a dry location out of direct light until further use.

In most applications, UV crosslinking, which generally requires less than 1 min to perform, has superseded baking in a vacuum oven for 2 h. In addition to the crosslinking application, calibrated UV light sources can be used to nick EtBr-stained RNA or DNA gels to facilitate complete blotting.

Postfixation Handling of Filters

The time frame for Northern analysis changes dramatically after transfer and immobilization have been accomplished. It should be abundantly clear by now that up to this point, and beginning with cell lysis, all manipulations involving labile RNA should revolve around the actual date of RNA isolation. This pertains to poly(A)$^+$ purification (if necessary), determination of concentration and purity, electrophoresis, blotting, nuclease protection analysis (Chapter 18), conversion into complementary DNA (Chapter 15), and so forth. Once a sample is immobilized on a solid support (filter membrane), it can be stored for months until hybridization (or rehybridization) is convenient.

The important thing to remember is that filters that have yet to be subjected to hybridization should be stored dry or in a vacuum oven to prevent moisture and opportunist growth from encroaching. Filters that have been subjected to hybridization, which will be rescreened, should be washed at very high stringency to remove the old probe as soon after completing hybridization detection as possible. Do not allow filters to dry out until the previously hybridized probe has been removed. Until then, keep the filter damp and wrapped in plastic. It is extremely difficult to completely remove hybridized probe from filters that have been allowed to dry.

References

Alwine, J. C., Kemp, D. J., Parker, B. A., Reiser, J., Renart, J., Stark, G. R., and Wahl, G. M. (1979). Detection of specific RNAs or specific fragments of DNA by fractionation in gels and transfer to diazobenzyloxymethyl paper. *Methods Enzymol.* **68,** 220.

Alwine, J. C., Kemp, D. J., and Stark, G. R. (1977). Method for detection of specific RNAs in agarose gels by transfer to diazobenzyloxymethyl-paper and hybridization with DNA probes. *Proc. Natl. Acad. Sci. USA* **74,** 5350.

Arnheim, N., and Southern, E. M. (1977). Heterogeneity of ribosomal genes in mice and men. *Cell* **11,** 363.

Bittner, M., Kupferer, P. and Morris, C. F. (1980). Electrophoretic transfer of proteins and nucleic acids from slab gels to diazobenyloxymethyl cellulose or nitrocellulose sheets. *Anal. Biochem.* **102,** 459.

Church, G. M., and Gilbert, W. (1984). Genomic sequencing. *Proc. Natl. Acad. Sci. USA* **81,** 1991.

Khandjian, E. W. (1987). Optimized hybridization of DNA blotted and fixed to nitrocellulose and nylon membranes. *Biotechnology* **5,** 165.

Li, J. K. K., Parker, B., and Kowalik, T. (1987). Rapid alkaline blot-transfer of viral dsR-NAs. *Anal. Biochem.* **163,** 210.

Löw, R., and Rausch, T. (1994). Sensitive, nonradioactive Northern blots using alkaline transfer of total RNA and PCR-amplified biotinylated probes. *BioTechniques* **17**(6), 1026–1029.

Medveczky, P., Chang, C. W., Oste, C., and Mulder, C. (1987). Rapid vacuum drive transfer of DNA and RNA from gels to solid supports. *BioTechniques* **5**(3), 242.

Olszewska, E. and Jones, K. (1988). Vacuum blotting enhances nucleic acid transfers. *Trends Genet.* **4,** 92.

Pitas, J. W. (1989). A simple technique for repair of nylon blotting membranes. *BioTechniques* **7**(10), 1084.

Reed, K. C., and Mann, D. A. (1985). Rapid transfer of DNA from agarose gels to nylon membranes. *Nucleic Acids Res.* **13,** 7207.

Saito, I., Sugiyama, H., Furukawa, N., and Matsuura, T. (1981). Photochemical ring opening of thymidine and thymine in the presence of primary amines. *Tetrahedron Lett.* **22,** 3265.

Schleicher & Schuell (1987). "Transfer and Immobilization of Nucleic Acids to S & S Solid Supports." Schleicher & Schuell, Keene, NH.

Southern, E.M. (1975). Detection of specific sequences among DNA fragments separated by gel electrophoresis. *J. Mol. Biol.* **98,** 503.

Nucleic Acid Probe Technology

Rationale

In an investigation involving hybridization between complementary molecules, the selection of the type of nucleic acid probe best suited for a particular application is just as important as the methodology by which hybridization events will be localized and quantified. The role of the probe is to hybridize to every complementary sequence present in a hybridization reaction.

Nucleic acid probes have numerous applications. They can be used to detect:

- Quantitative or qualitative changes in gene expression
- Gene amplifications and deletions
- Gene rearrangements
- Chromosomal translocations
- Point mutations
- Presence of new genetic sequences in cells (pathogens)

Moreover, oligonucleotides used as primers to support the polymerase chain reaction (Chapter 15) can usually be used interchangeably as nucleic acid probes in the classical sense.

A nucleic acid probe is little more than a polynucleotide that carries some type of label or tag, allowing the investigator to follow it throughout the experiment, the goal being the capacity for at least semiquantitative detection. Historically, probe synthesis (i.e., the incorporation of label) and detection were entirely dependent on the efficient incorporation of radiolabel accompanied by sensitive autoradiographic techniques. The major decision to be made was the method by which deoxynucleoside triphosphate (dNTP)-radiolabeled precursor(s) were to be incorporated. Now, however, the pressing question in each new study is whether to use isotopes at all.

Perhaps an even more pressing question, the answer to which is not as straightforward as the molecular biology novice might like, is "When can nonisotopic labeling and detection by chemiluminescence be used in place of radiolabeling and autoradiography?" The not-so-direct answer is "Often, but much depends on the precise application." For the most part, nonisotopic procedures are quite compatible with nearly all molecular biology applications involving sample fixation on filter membranes (not nitrocellulose). These techniques include, but are not limited to, Northern analysis, Southern analysis, Western analysis, dot and slot blot analyses, colony hybridization, and plaque screening. The rule of thumb that this author frequently proffers is "If you can see bands by autoradiography after the third day under film,[1] then it is likely that you can switch to chemiluminescence without loss of sensitivity or resolution. Autoradiographic analysis requiring more than 3 days may or may not be fully compatible with nonisotopic methods."

Some of the techniques that offer enhanced sensitivity compared to the filter-based systems, including nuclease protection assays and the nuclear runoff assay, are best performed with radiolabel, though nonisotopic alternatives are under development in this laboratory. Several nonisotopic alternatives of sufficient sensitivity have been developed to offer a nonisotopic alternative for *in situ* hybridization.

Probe Classification

Generally speaking, nucleic acid probes can be either homogeneous or heterogeneous in nature. Homogeneous and heterogeneous probes may be either DNA or RNA molecules. In a hybridization solution, a probe preparation is said to be homogeneous if all probe molecules are the same. For example, every probe molecule is a human *fos* oncogene complementary DNA (cDNA); the goal of the experiment is assessment of the prevalence of *fos* transcripts in the sample(s). Heterogeneous probes consist of mixtures of two or more sequences that may be closely related in nucleotide sequence or may be completely dissimilar. For example, in the preparation of a subtraction library, the goal is to identify messenger RNA (mRNA) sequences that are unique to one population of cells by subtracting or removing all sequences expressed in common with a reference (control, untreated, undifferentiated) population of cells. Thus, in a subtraction hybridization the probe might be made of thousands of different types of members. Another example is an instance where several oligonucleotides are present in a hybridization reaction because of sequence ambiguities pertaining to the target molecules of interest. When planning to work with heterogeneous probes, however, one must ensure that no repetitive sequences are present (e.g., Alu) so as not to generate gross and unsatisfactory hybridization signals.

Selection of Labeling System

The key considerations in the selection of a probe labeling system are:

- Required sensitivity and resolution
- Method of label incorporation

[1] Autoradiography performed at $-70°C$ with an intensifying screen.

- Probe stability after labeling
- Type of hybridization
- Desired method of detection

There are two basic types of labeling of nucleic acids: (1) radiolabeling and (2) nonisotopic methods, including those that support detection by chemiluminescence, chemifluorescence, chromogenic techniques, or traditional fluorescence tagging. These techniques are compatible with the labeling of DNA, RNA, and/or oligonucleotides for most applications. The methods, merits, and drawbacks of the permutations are described here. Further, numerous systems are available commercially for synthesis of nucleic acid probes, and each has a recommended detection methodology for optimal sensitivity and resolution.

One may further subclassify nucleic acid probes by the distribution of label, either isotopic or nonisotopic, in the probe, a classification that is clearly a function of the type of labeling technique performed. Probes may be classified as being either end-labeled or continuously labeled, and multiple varieties of each are given below. End-labeled probes are those to which the label has been added at the 5′ end (for example, by the kinasing reaction) or at the 3′ end (for example, using terminal transferase). Continuously labeled probes are those to which the label has been added along the length of the backbone of the probe, at fairly regular intervals. Examples of continuously labeled probes include, but are not limited to, those generated by random priming, polymerase chain reaction (PCR), *in vitro* transcription, or crosslinkage to psoralen–biotin. The choice between end labeling and continuously labeling techniques depends on

1. Whether the probe is DNA or RNA
2. Whether the probe is single stranded or double stranded
3. Whether the probe, if double stranded, has recessed or protruding 5′ and 3′ ends
4. The required level of label incorporation (relates directly to the required level of sensitivity)
5. The length of the probe to be labeled
6. Whether the probe, if double stranded DNA, is linear or a covalently closed circle (a plasmid)

These permutations are addressed in detail below.

Isotopic Labeling

CAUTION   Caution and common sense should always be used when working with or storing radioactive materials. A lab coat, disposable gloves, and eye protection are essential at all times, regardless of the isotope or its specific activity. Be certain to adhere closely to departmental or institutional regulations concerning shielding and laboratory safety when handling radiochemicals or any other hazardous material.

In many applications, isotopes offer excellent sensitivity and compatability with many labeling techniques. Some of the inherent disadvantages of isotope labeling include the half-life[2] of the isotope, potential health hazards, containment of radioactivity, purchase cost, disposal cost, probe stability and usefulness after labeling, and the need for relatively long detection periods.

The most frequently used isotopes for nucleic acid probe synthesis are ^{32}P, ^{3}H, and ^{35}S and, less frequently, ^{33}P. Nucleotide triphosphate precursors are labeled in the appropriate position (α or γ), thereby supporting enzyme-mediated transfer of the portion of the nucleotide containing the isotope to the molecules being labeled. The intended application and the precise method of detection dictate the selection of isotope. For example, the long path length associated with β-emission from ^{32}P-labeled probes makes them useful for recording hybridization events on X-ray film; for precisely the same reason, anatomical resolution is inadequate for RNA target localization *in situ*. This necessitates the use of an alternative labeling and detection system.

The ready availability of ^{33}P has assuaged some of the dangers inherent in and fears associated with the use of ^{32}P. ^{33}P nucleotides are an alternative for use in general molecular biology applications, especially DNA sequencing (Zagursky *et al.*, 1991). The half-life of ^{33}P[3] is 25.4 days with a β E_{max} of 0.25 MeV, which is about five fold less than that of ^{32}P (E_{max} = 1.6 MeV). NEN Life Science Products (formerly DuPont-NEN) has indicated that ^{33}P can be handled on the bench-top using routine safety practices but without elaborate shielding.

Minimizing Decomposition Problems[4]

The shelf-life of a nucleotide depends on two factors: its radioactive half-life and its biological half-life. Whereas the radioactive half-life is a constant, the rate of biological degradation varies depending on the handling of the material and the temperature. Storing vials of radioactive nucleotides in a refrigerator or worse, at room temperature, can greatly in-

---

[2]For the convenience of the reader, half-life data for commonly used isotopes are presented in Tables 1–4 (data courtesy of NEN Life Science Products).

[3] ^{33}P data courtesy of NEN Life Science Products.

[4]Adapted in part from NEN Life Science Products' "Guide to Storing and Handling NEN® Radiochemicals."

^{32}P Decay Data: Physical Half-Life = 14.29 Days

	Days									
	0.0	0.5	1.0	1.5	2.0	2.5	3.0	3.5	4.0	4.5
0	1.000	0.976	0.953	0.930	0.908	0.886	0.865	0.844	0.824	0.804
5	0.785	0.766	0.748	0.730	0.712	0.695	0.678	0.662	0.646	0.631
10	0.616	0.601	0.587	0.573	0.559	0.545	0.532	0.520	0.507	0.495
15	0.483	0.472	0.460	0.449	0.438	0.428	0.418	0.408	0.398	0.388
20	0.379	0.370	0.361	0.353	0.344	0.336	0.328	0.320	0.312	0.305
25	0.297	0.290	0.283	0.277	0.270	0.264	0.257	0.251	0.245	0.239
30	0.233	0.228	0.222	0.217	0.212	0.207	0.202	0.197	0.192	0.188
35	0.183	0.179	0.174	0.170	0.166	0.162	0.158	0.155	0.151	0.147
40	0.144	0.140	0.137	0.134	0.130	0.127	0.124	0.121	0.118	0.116
45	0.113	0.110	0.107	0.105	0.102	0.100	0.098	0.095	0.093	0.091
50	0.088	0.086	0.084	0.082	0.080	0.078	0.077	0.075	0.073	0.071
55	0.069	0.068	0.066	0.065	0.063	0.062	0.060	0.059	0.057	0.056
60	0.054	0.053	0.052	0.051	0.049	0.048	0.047	0.046	0.045	0.044

^{33}P Decay Data: Physical Half-Life = 25.4 days

	Days									
	0	1	2	3	4	5	6	7	8	9
0	1.000	0.973	0.947	0.921	0.897	0.872	0.849	0.826	0.804	0.782
10	0.761	0.741	0.721	0.701	0.683	0.664	0.646	0.629	0.612	0.595
20	0.579	0.564	0.549	0.534	0.520	0.506	0.492	0.479	0.466	0.453
30	0.441	0.429	0.418	0.406	0.395	0.385	0.374	0.364	0.355	0.345
40	0.336	0.327	0.318	0.309	0.301	0.293	0.285	0.277	0.270	0.263
50	0.256	0.249	0.242	0.236	0.229	0.223	0.217	0.211	0.205	0.200
60	0.195	0.189	0.184	0.179	0.174	0.170	0.165	0.161	0.156	0.152
70	0.148	0.144	0.140	0.136	0.133	0.129	0.126	0.122	0.119	0.116
80	0.113	0.110	0.107	0.104	0.101	0.098	0.096	0.093	0.091	0.088
90	0.086	0.084	0.081	0.079	0.077	0.075	0.073	0.071	0.069	0.067
100	0.065	0.064	0.062	0.060	0.059	0.057	0.055	0.054	0.053	0.051
110	0.050	0.048	0.047	0.046	0.045	0.043	0.042	0.041	0.040	0.039
120	0.038	0.037	0.036	0.035	0.034	0.033	0.032	0.031	0.030	0.030

^{35}S Decay Data: Physical Half-Life = 87.4 Days

					Days				
	0	3	6	9	12	15	18	21	24	27
0	1.000	0.976	0.954	0.931	0.909	0.888	0.867	0.847	0.827	0.807
30	0.788	0.770	0.752	0.734	0.717	0.700	0.683	0.667	0.652	0.636
60	0.621	0.607	0.592	0.579	0.565	0.552	0.539	0.526	0.514	0.502
90	0.490	0.478	0.467	0.456	0.445	0.435	0.425	0.415	0.405	0.395
120	0.386	0.377	0.368	0.359	0.351	0.343	0.335	0.327	0.319	0.312
150	0.304	0.297	0.290	0.283	0.277	0.270	0.264	0.258	0.252	0.246
180	0.240	0.234	0.229	0.223	0.218	0.213	0.208	0.203	0.198	0.194
210	0.189	0.185	0.180	0.176	0.172	0.168	0.164	0.160	0.156	0.153
240	0.149	0.146	0.142	0.139	0.136	0.132	0.129	0.126	0.123	0.120
270	0.118	0.115	0.112	0.109	0.107	0.104	0.102	0.099	0.097	0.095
300	0.093	0.090	0.088	0.086	0.084	0.082	0.080	0.078	0.077	0.075
330	0.073	0.071	0.070	0.068	0.066	0.065	0.063	0.062	0.060	0.059
360	0.058	0.056	0.055	0.054	0.052	0.051	0.050	0.049	0.048	0.046

^{3}H Decay Data: Physical Half-Life = 12.28 Years

						Months					
	0	1	2	3	4	5	6	7	8	9	10	11
0	1.000	0.995	0.991	0.986	0.981	0.977	0.972	0.968	0.963	0.959	0.954	0.950
1	0.945	0.941	0.936	0.932	0.928	0.923	0.919	0.915	0.910	0.906	0.902	0.898
2	0.893	0.889	0.885	0.881	0.877	0.873	0.869	0.865	0.860	0.856	0.852	0.848
3	0.844	0.841	0.837	0.833	0.829	0.825	0.821	0.817	0.813	0.810	0.806	0.802
4	0.798	0.794	0.791	0.787	0.783	0.780	0.776	0.772	0.769	0.765	0.762	0.758
5	0.754	0.751	0.747	0.744	0.740	0.737	0.733	0.730	0.727	0.723	0.720	0.716
6	0.713	0.710	0.706	0.703	0.700	0.697	0.693	0.690	0.687	0.684	0.680	0.677
7	0.674	0.671	0.668	0.665	0.661	0.658	0.655	0.652	0.649	0.646	0.643	0.640
8	0.637	0.634	0.631	0.628	0.625	0.622	0.619	0.616	0.614	0.611	0.608	0.605
9	0.602	0.599	0.597	0.594	0.591	0.588	0.585	0.583	0.580	0.577	0.575	0.572
10	0.569	0.567	0.564	0.561	0.559	0.556	0.553	0.551	0.548	0.546	0.543	0.541
11	0.538	0.535	0.533	0.530	0.528	0.526	0.523	0.521	0.518	0.516	0.513	0.511
12	0.509	0.506	0.504	0.501	0.499	0.497	0.494	0.492	0.490	0.487	0.485	0.483

crease the rate of radiolysis and biological degradation. Decomposition products consist primarily of nucleotide monophosphates and inorganic phosphate, although nucleotide diphosphates can also be present. Nucleotides are shipped to arrive on dry ice and should be stored frozen below –20°C, or at –80°C if possible. Storage in frost-free freezers is not recommended.

The following is a partial list of recommendations from NEN Life Science Products[5] to help radiochemical users minimize decomposition problems. More complete details pertaining to nonnucleotide radiochemicals can be found in NEN's "Guide to Storing and Handling NEN[R] Radiochemicals."

1. Use the radiochemical as soon as possible after receiving it. Prolonged storage allows time for additional nuclear decay, causing increased radical generation, and therefore, greater amounts of decomposition. Such decomposition will occur even if the radiochemical container is left unopened and stored under the most ideal conditions.

2. Store the sample properly. Pay close attention to the recommended storage conditions, as outlined on the technical data sheet that accompanies each product.

3. Follow the technical data sheet instructions when opening an ampoule containing radiochemicals. Refer to these instructions for storage information.

4. Thaw, at room temperature, until melting, ^{3}H-labeled radiochemicals, stored in aqueous solvents, and ^{35}S-, ^{125}I-, and ^{32}P-labeled biochemicals, except that ^{35}S- and ^{32}P-labeled nucleotides, which should be quick-thawed. For radiochemicals stored at very low temperatures (–20°C to –80°C), slowly thawing the sample in the refrigerator (or on ice) is recommended.

5. Do not allow radiochemicals to sit at room temperature for prolonged periods of time, because this often increases the rate of radiochemical decomposition significantly.

6. Minimize the number of times the primary container is opened, and reseal the container immediately after each use. If a compound will be used several times, aliquot the required amounts into separate vials for storage. Each time a solution is handled, impurities, especially oxygen and water, may be introduced.

7. Use clean syringes and pipet tips when withdrawing aliquots from

[5]Reproduced with permission.

the primary container, since many buffers and salts used in biological systems are harmful to the radiochemical.

Specific Activity

The specific activity (SA) of ^{32}P nucleotides on any day *prior* to the calibration date can be calculated according to

$$SA\ (Ci/mmol) = \frac{SA\ cal.}{Df + [SA\ cal.\ (1\text{-}Df)/9120]},$$

where

SA cal. = specific activity on the calibration date;

Df = decay factor obtained from a decay chart; it is the fraction of current radioactivity that will remain on the calibration date;

9120 = theoretical specific activity of carrier-free ^{32}P.

The specific activity of ^{32}P-labeled nucleotides on any day *after* the calibration date can be calculated using

$$SA\ (Ci/mmol) = \frac{Df}{1/SA\ cal - (1\text{-}Df)/9120}.$$

Nonisotopic Labeling

An ever-increasing variety of nonisotopic labeling and detection methodologies have been developed in an attempt to minimize the use of radioisotopes and the inherent dangers associated with their presence in the laboratory. Truly amazing refinements have been made in many of these systems over the past 3–4 years, facilitating extraordinary levels of detection in a completely nonisotopic manner. In some, but not all, applications the sensitivity achievable is equivalent to that of ^{32}P, though much depends on the patience of the user. Many of the commercially available systems offer similar levels of sensitivity and resolution, though they differ markedly in (1) their complexity, from a user's point of view, and (2) the number of different steps that must be performed by the user, especially pertaining to the detection component of the system. The most common nonisotopic methods include biotinylation, labeling with digoxigenin (DIG), and labeling with fluorescein. Biotin, DIG, and fluorescein labeling support detection by chromogenic (colorimetric) methods or chemiluminescence. Because of the technical difficulties intrinsic in most chromogenic techniques, this type of detection chemistry never gained widespread popularity. Now, however, efficient nonisotopic labeling has become standard fare in molecular biology laboratories, utilizing chemiluminescence and some chromogenic detection technologies. Although the notion of eliminating radioactive probes is enticing, to say the least, the investigator must also be aware that the mechanics of labeling, probe purification, and detection may be drastically modified in many cases. The most significant change that most experienced ^{32}P users will notice is the amount of dedicated time that must be allocated to perform the detection, in particular. Usually, multiple washes and a posthybridization blocking are required, and once the chemiluminescence substrates are applied, the filter cannot be washed at higher stringency unless one repeats the series of posthybridization washes and blocking steps in its entirety. However, the benefits reaped by the elimination of isotopes, from a safety and economics perspective, far outweigh the additional attention that this type of detection requires.

Biotin

Biotin is a small water-soluble vitamin that can be readily conjugated to a number of biological molecules. Biotinylation can be accomplished in any of a number of ways:

a. Enzymatically, through the use of biotinylated nucleotides;
b. Photochemically, using photoactivatable biotin;
c. Photochemically, using psoralen–biotin;
d. Chemically, during the synthesis of oligonucleotides.

From a detection point of view, it really does not matter how the biotin becomes part of the probe, though 5′ biotinylation of oligonucleotides is one of the easiest and most efficient methods to accomplish labeling. Biotinylation almost never interferes with biological activity and, in the case of biotinylated nucleotides (Bio-11-dUTP[6]), linker arms between the biotin itself and the backbone of the probe effectively minimize steric interference. Biotinylated probes are quite stable at –20°C for at least 1 year after labeling.

Few molecules exhibit the high affinity observed between biotin and streptavidin ($K_d = 10^{-15}\ M^{-1}$). Streptavidin is a tetrameric protein (MW

---

[6]dUTP is universally used as a nonisotopically labeled nucleotide, functioning as a dTTP analog. The designation "Bio-11-dUTP" indicates that the nucleotide precursor is dUTP, that the dUTP is biotinylated, and that there is an 11-atom spacer arm separating the biotin itself from the nucleoside monophosphate that physically becomes part of the probe. Techniques that support this type of labeling include nick translation, random priming, and the polymerase chain reaction. RNA probes can be synthesized using Bio-11-UTP, DIG-11-UTP (digoxigenin labeled), or Fl-11-UTP (fluorescein-labeled).

60,000), isolated from the bacteria *Streptomyces avidinii,* and has four biotin binding sites. Unlike avidin (from egg), streptavidin has a neutral isoelectric point at physiological pH, with few charged groups, and contains no carbohydrate. These properties reduce nonspecific binding and background problems that would otherwise be experienced and, in so doing, enhance the sensitivity of many forms of this assay. Many biotin–streptavidin applications in nucleic acid hybridization detection and related techniques have been described (Leary *et al.,* 1983; Wilchek and Bayer, 1984; Hoffman and Finn, 1985). Thus, one of the first steps in biotin-based detection systems is the binding of biotin with a preparation of streptavidin. In most systems, streptavidin is modified, existing as a conjugate, usually to alkaline phosphatase (AP), though new systems feature a streptavidin–horseradish peroxidase (HRP) conjugate. Depending on the specific substrate to which the AP or HRP is subsequently exposed, hybridization events are then localized and quantified by light emission captured on X-ray film (chemiluminescence) or the formation of a color precipitate directly on the filter (chromogenic detection).

Digoxigenin

Digoxigenin (Boehringer Mannheim Biochemicals, Indianapolis, IN) is a derivative of the cardiac medication digitalis. It is widely used as nucleic acid label and supports both chromogenic detection and chemiluminescence. DNA probes are enzymatically labeled, usually by random priming with digoxigenin-dUTP (DIG-11-dUTP), and RNA probes are synthesized by *in vitro* transcription with digoxigenin-UTP (DIG-11-UTP). These nucleotides are linked via a spacer arm to the steroid hapten DIG. The resulting DIG-labeled molecules then function as hybridization probes in much the same manner as any other type of probe. Recently, it has become stylish to simply have oligonucleotides DIG labeled during their synthesis, thereby precluding cumbersome, inefficient labeling reactions at some later date. As with biotinylation, DIG labeling does not interfere with biological activity, and probes so labeled are stable at –20°C for at least 1 year after labeling. DIG labeling supports posthybridization detection by chemiluminescence or by formation of an insoluble color precipitate directly on the filter membrane.

Following the posthybridization stringency washes, DIG-labeled probes are detected by enzyme-linked immunoassay, using an antibody conjugate (anti-DIG-alkaline phosphatase). As with systems involving biotin, the emission of light or the formation of precipitate is mediated by AP dephosphorylation of a substrate compatible with the method of detection, either colorimetric or chemiluminescence.

Fluorescein

Fluorescein is a hapten that can be incorporated by standard probe synthesis reactions, or during oligonucleotide synthesis. Fluorescein-labeled nucleotides (Fl-11-dUTP) can be used to generate probes much as biotin- and DIG-labeled nucleotides are used. Following hybridization, high-affinity antibodies prepared against fluorescein are used to localize hybridization events. The anti-fluorescein antibodies have been modified, existing as an enzyme conjugate with either AP or HRP, chemistries that support chemiluminescence and chromogenic detection. It is also possible to make use of the intrinsic fluoresence of the hapten to monitor the incorporation of label into the probe, thereby assessing the efficiency of the labeling reaction.

DNA Probes

DNA probes can be a diverse lot indeed. Examples of DNA probes include previously cloned double-stranded cDNA or genomic DNA sequences, first-strand cDNA synthesized directly from a mixture of mRNAs in the presence of labeled dNTPs, and oligonucleotide probes:

1. Cloned cDNA and genomic DNA sequences require denaturation into their constituent single strands after labeling, and before they can be used as probes. A description of a number of labeling techniques follows. An enormous variety of DNA probes, some already labeled, are available commercially from nearly every major biotechnology supplier. One disadvantage of DNA probes is that they are less thermodynamically stable than RNA probes (see Chapters 1 and 13 for commentary on stringency), though this often has negligible impact on the outcome of an experiment.

2. First-strand cDNA can be synthesized from mRNA (or from an enriched fraction thereof) in the presence of labeled dNTP precursors. This will generate a heterogeneous probe that can be used directly for hybridization. Heterogeneous probes such as these are particularly useful for the identification of sequences that are uniquely present in one of two or more RNA populations. For the convenience of the reader, a discussion of subtraction hybridization, as this approach is generically known, is presented in Chapter 16.

3. Oligonucleotides are artificially synthesized single-stranded DNA. "Oligos" were used as standard probes for hybridization for years before PCR was developed. Thus, the primers that an investigator might now be using to support amplification of a sequence of interest by PCR could easily function as stand-alone probes for standard nucleic acid hybridiza-

tion. Oligonucleotides are classified as "long" oligomers, containing as many as 100 or more bases, or "short" oligomers, usually containing fewer than 30 bases. The advantage of working with oligonucleotides is that the investigator has complete control over the sequence. The main disadvantage of using oligonucleotides is the same as the main advantage: Because the investigator has control over the sequence of the oligonucleotide, there must be knowledge of *which* sequence to select. When working with oligonucleotide probes, the short oligonucleotides offer the greatest flexibility and experimental latitude.

Oligonucleotide sequence information can come from previously documented nucleotide sequence data, knowledge of the nucleotide sequence of a related gene in either the same or a different species, or knowledge of the amino acid sequence of the protein encoded by the gene of interest. The problem with the latter approach is the degeneracy of the genetic code: Some amino acids have multiple codons. When designing oligonucleotide probes based on peptide sequence information, one strategy is to examine the peptide for clusters of amino acids with only one or two possible codons. Single-codon amino acids include methionine and tryptophan; two-codon amino acids include glutamic acid, glutamine, aspartic acid, asparagine, phenylalanine, tyrosine, histidine, cysteine, and lysine. Moreover, the investigator may have knowledge of preferential usage of one codon over another in a particular species.

In the event that a definitive oligonucleotide probe sequence is not discernible, it is possible to use several oligonucleotides simultaneously in what might be thought of as a hybridization cocktail; alternatively, one may elect to use only one oligonucleotide, though at a lower stringency, by lowering the hybridization and washing temperatures. While this strategy reduces the concentration of the correct complementary sequence and may favor semi-nonspecific hybridization, the important point is that authentic hybridization events do not go undetected. See Chapter 13 for hybridization details for oligonucleotide probes.

DNA Probe Synthesis

A variety of methodologies for labeling DNA have been described. In short, these methods are used to generate end-labeled or continously labeled probes. Most enzyme-mediated labeling techniques are very much dependent on polymerase activity, which is responsible for incorporation of the label. It is worth noting that the substitution of some of the more traditional enzymes for *Taq* or another of the thermostable DNA polymerases used for PCR permits labeling reactions to be performed at higher temperatures, thereby reducing the incidence of enzyme-mediated point mutations during probe synthesis. Of course, one must also consider the natural error rate of the *Taq*. Proofreading enzymes and enzyme blends, described in Chapter 15, may be helpful in this regard.

A brief description of the more common techniques follows. All of these labeling reactions are commercially available in the form of kits that contain all necessary reagents, with the exception of isotopes for radiolabeling techniques. In general, most labeling systems used to generate radioactive probes can also be used to generate a variety of nonisotopic probes. When radiolabeling, be sure that the position of the isotope in the nucleotide precursor (i.e., α or γ label) supports the intended labeling method.

Polymerase Chain Reaction[7]

PCR is an excellent method for probe synthesis, requiring exquisitely small quantities of template material. In the presence of the appropriate precursor, molecules are labeled as they are being synthesized, including radiolabeling and biotinylation. Alternatively, the primers themselves may be labeled during their own synthesis, negating the requirement for the inclusion of labeled nucleotide precursors as part of the reaction mix.

PCR as a labeling method offers several advantages: the speed, versatility, and efficiency of the reaction, as well as the ready access of most labs to this technique. Further, the synthesized probe molecules, generated from minute quantities of starting material, are of uniform length. Because PCR-generated probes can be continuously labeled, they generally show a high degree of label incorporation. For a comprehensive review of reaction parameters and other PCR-related strategies, see Dieffenbach and Dveksler (1995), Innis *et al.* (1990), McPherson *et al.* (1991), and McPherson and Hames (1995).

Random Priming

Random priming is a type of primer extension in which a mixture of small oligonucleotide sequences, acting as primers, anneal to a heat-denatured double-stranded template (Feinberg and Vogelstein, 1983; 1984). The annealed primers ultimately become part of the probe itself, because the Klenow fragment of DNA polymerase I extends the primers in the 3' direction and, in so doing, incorporates the label. Random

---

[7]The polymerase chain reaction (PCR) process is covered by patents issued to Hoffman-LaRoche. Use of the PCR process requires a license under their patents.

priming works significantly better with linearized DNA molecules; attempts to label covalently closed, supercoiled DNA typically result in probe specific activities 20- to 30-fold less than those of the corresponding linear DNA. DNA molecules between 200 and 2000 bp are the best candidates for random priming, although template length is not a critical parameter in this labeling reaction. When labeling with ^{32}P, random priming typically produces probes with specific activities of about 1–3×10^9 cpm/μg when 2.5–10 ng of starting template is used. Labeling usually requires 10–30 min.

Nick Translation

Historically, nick translation (Rigby *et al.*, 1977) is one of the oldest probe labeling techniques. It involves randomly nicking the backbone of a double-stranded DNA with dilute concentrations of DNase I. At extremely low concentrations, nicking occurs in approximately four or five locations per molecule, producing a free 3′-OH primer at each nicking location. Next, the enzyme DNA polymerase I removes the native nucleotides from the probe molecules in the 5′ → 3′ direction (exonuclease activity) while replacing them with dNTP precursors by virtue of its 5′ → 3′ polymerase activity. Nick translation is efficient for both linear and covalently closed DNA molecules, and labeling requires about 1 h. When labeling with ^{32}P, nick translation produces probes with specific activities of approximately 3–5×10^8 cpm/μg.

5′ End Labeling

One alternative method to generating continuously labeled probes is to label the 5′ end of the molecule with the addition of a radiolabeled phosphate. This method of 5′ end labeling is colloquially known as the kinasing reaction; it specifically involves the transfer of the γ phosphate of ATP (not dATP) to a 5′-OH substrate group on dephosphorylated DNA molecules (forward reaction). This labeling reaction is catalyzed by the enzyme T4 polynucleotide kinase and is an excellent method for labeling short oligonucleotides. One common error in the laboratory pertaining to this labeling reaction involves the purchase of α-labeled nucleotides (needed for continuously labeled and 3′ end-labeled probes) rather than the required γ-labeled ATP.

The forward kinasing reaction is far more efficient than the exchange reaction, which involves the substitution of 5′ phosphates. While kinasing RNA is possible, it is so inefficient that it is now merely an historical footnote. DNA kinasing reactions typically require 30 min, and this type of labeling is renowned for generating extremely high specific activity probes. Finally, a variety of nonisotopic labeling kits are available from several vendors for 5′ end modification.

3′ End Labeling

Probe synthesis by 3′ end labeling involves the addition of nucleotides to the 3′ end of either DNA or RNA. DNA 3′ end labeling is most often catalyzed by the enzyme terminal deoxynucleotidyl transferase, or simply terminal transferase. Single- and double-stranded DNA molecules are labeled by the addition of dNTP to 3′-OH termini; double-stranded, blunt-ended fragments and fragments with 3′ overhang structure are labeled best. Because there is no template strand requirement, any one dNTP (or dNTP mixture) can be selected; thus terminal transferase might be thought of as a non-template-dependent polymerase. This type of labeling results in the production of a 3′ overhang, usually consisting of fewer than five to six extra nucleotides. Depending on the application, these nonblunt ends could then be polished with the Klenow fragment of DNA polymerase I to support blunt end ligations, though the loss of label may occur. The *Taq* DNA polymerase also has terminal transferase activity, as do other nonproofreading polymerases.

The disadvantage of this labeling technique can be pronounced when the labeling of short oligonucleotides is involved. The addition of nucleotides changes the length of the molecule, which may or may not modify the specificity of the oligonucleotide, as well as its thermodynamic characteristics. If an oligonucleotide is labeled by nucleotide addition to the 3′ end, it is strongly suggested that dATP be used, because of the lesser thermodynamic stability of A::T base pairing compared to G:::C base pairs.

Direct Enzyme Labeling

Each of the previously described techniques for generating DNA probes involves the substitution or addition of labeled nucleotides to the probe. In sharp contrast to these widely respected methods, a newer form of probe synthesis, one that supports detection by chemiluminescence, presents a very realistic labeling alternative. The method is known as direct enzyme labeling (ECL System; Amersham, Arlington Heights, IL) and involves crosslinking an enzyme, which will support detection by chemiluminescence, directly to the backbone of heat-denatured DNA, using glutaraldehyde as what might be thought of as a bridge or linker arm. The method is extremely efficient, requires a total of 20 min, and does not require any subsequent steps for the removal of unincorporated label. Because the probe cannot be boiled after the labeling has been completed, and because the denatured DNA will, of course, reanneal in a relatively short time, one should label only the required mass of probe, which should be added to the hybridization mix as soon as possible after labeling. Further, because the enzyme is present throughout the hybridization and subsequent washes (next chapter), stringency must be regulated by modifying ionic strength, rather than temperature.

RNA Probes

The popularity of RNA probes continues to increase because of several key advantages associated with their use. These probes are synthesized by *in vitro* transcription and can be substituted for DNA probes in nearly all applications. In the construction of the transcription template, the cDNA to be transcribed is usually flanked by two different RNA polymerase promoters from bacteriophage, which flank a multiple cloning site and are positioned in opposite orientations. The most common constructions feature the SP6, T7, and/or T3 RNA polymerase promoters (Figure 1). This construction is linearized with an appropriate restriction enzyme prior to initiating the transcription reaction, favoring the production of large amounts of efficiently labeled probe, of uniform length (Figure 2).

An integral part of the preparation of RNA probes is the transcription of antisense and sense transcripts from the same template construction.

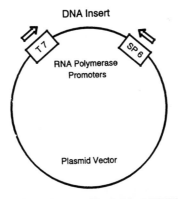

F I G U R E   1

Typical orientation of dual RNA polymerase promoters in a plasmid construction that supports *in vitro* transcription. The template for transcription, usually cDNA, is cloned between the two promoters. This plasmid construction is then linearized prior to initiation of the *in vitro* transcription reaction.

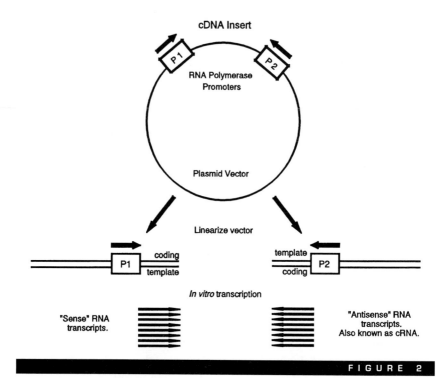

FIGURE 2

In vitro transcription templates must be linearized prior to the transcription reaction, thereby favoring the efficient production of large quantities of continuously labeled probe of uniform length. Note that the positioning of the promoters influences whether the resulting transcript will be sense or antisense RNA. If the orientation of the insert cDNA is unknown, then empirical determination will be required to discern which promoter gives which type of transcript.

Antisense RNA, also known as cRNA, is complementary to mRNA, and is therefore able to base pair to it. In addition to standard RNA transcription studies, antisense RNA can inhibit gene expression at the translational level by forming a nontranslatable double-stranded RNA strucuture. Antisense RNA has been shown to regulate gene expression *in vitro* in mammalian systems (Nishikura and Murray, 1987), in phages and bacteria (Green *et al.*, 1986), in plants (Green *et al.*, 1986), and in animals (Knecht and Loomis, 1987).

Example: An investigator wishes to characterize the histopathology of a diseased rat cerebellum by identifying which cells, if any, are synthesizing *N-myc* mRNA. To make this determination, a tissue section is prepared for *in situ* hybridization.[8] Because the interest of this study is mRNA, a complementary or antisense RNA should be used as the probe mainly because of its enhanced thermodynamic stability when hybridized; this permits very stringent assay conditions. Sense RNA probe, that is, probe with the same sequence as the mRNA, is also generated and used as a negative control to assess nonspecific hybridization. In one approach, sense and antisense probe can be synthesized in the presence of DIG-labeled UTP (Boehringer Mannheim Biochemicals, Indianapolis, IN) or fluorescein-labeled UTP (Amersham, Arlington Heights, IL), followed by chromogenic detection. The distribution of color precipitate is assessed by light microscopy and requires no isotope whatsoever. After the distribution of message *in situ* is discerned, antisense RNA can be used to obtain qualitative information pertaining to *N-myc* mRNA, by Northern analysis coupled with detection by chemiluminescence.

Characteristics of RNA Probes

1. RNA probes are single stranded and therefore do not require any type of denaturation prior to use.

2. All RNA probe molecules are available for hybridization; because of their single-stranded nature, they cannot renature as in the case of DNA, although intramolecular base pairing is always a concern.

3. RNA probes are continuously labeled as they are being transcribed, thereby generating probes with a very high degree of label incorporation.

4. RNA probes show greater thermodynamic stability than DNA probes.

5. RNA probes are synthesized by *in vitro* transcription in a template-dependent fashion; therefore, all probe molecules are of uniform length.

6. Enormous quantities of probe can be synthesized in a single *in vitro* transcription reaction.[9]

7. In the construction of the transcription template, the cDNA to be transcribed is flanked by a highly efficient bacteriophage RNA polymerase promoter. The more useful vectors contain dual RNA polymerase

---

[8]For excellent discussions of *in situ* hybridization, see Chesselet (1990); Wilkinson (1992), and Nuovo (1994).

[9]It has been reported (Krieg and Melton, 1987) that the proportion of full-length transcripts can be increased by lowering the standard reaction temperature of 37–42°C to 4°C.

promoters that flank multiple cloning sites in opposite orientations. The most common constructions feature SP6 and T7 RNA polymerase promoters. Thus, both sense and antisense RNA probes can be synthesized as needed.

8. The SP6, T7, and T3 bacteriophage RNA polymerase promoters demonstrate virtually no cross-reactivity; therefore, transcription reactions initiated from one promoter or the other are virtually free of transcripts of the opposite sense.

9. Plasmids used for RNA probe synthesis must be linearized prior to initiating *in vitro* transcription.

10. RNA transcribed *in vitro*, as with all RNA, must be treated with RNase-free reagents; failure to do so will result in rapid degradation of the probe.

11. RNA probes often produce unacceptable, high levels of background; thus, at the conclusion of the hybridization period it is a common practice to digest all probe molecules with RNase A and RNase T1. This treatment will result in degradation of all probe molecules that did not participate in duplex formation.

12. Radiolabeled RNA probes are often of such high specific activity that they may experience radiolysis if stored for extended periods.

RNA Probe Synthesis

Compared to the diverse methods for DNA probe synthesis, there for labeling RNA probes, namely *in vitro* transcription. Because of the intrinsically labile nature of RNA and the susceptibility to RNase degradation, RNA probes must be treated with the same care as any other RNA preparations. A brief description of RNA labeling techniques follows. *In vitro* transcription systems are commercially available in kit form, and contain all necessary reagents (except isotopes). These systems work well in the nonisotopic format as well.

In Vitro Transcription

In vitro transcription is the only reliable method for generating RNA probes. Large amounts of efficiently labeled probes of uniform length can be generated by transcription of a DNA sequence ligated next to an RNA promoter. One excellent strategy is to clone the DNA to be transcribed between two promoters in opposite orientations. This allows either strand of the cloned DNA sequence to be transcribed in order to generate sense and antisense RNA for hybridization studies. A number of very useful Riboprobe (Promega, Madison, WI) cloning vectors have been developed and are universally available to facilitate *in vitro* transcription of RNA probes.

RNA probes synthesized by transcription *in vitro* are assembled from NTP precursors, much as occurs *in vivo*. Transcripts are elongated by the addition of nucleotide monophosphates into the nascent backbone of the probe. Isotopic labeling with [^{32}P]UTP requires radiolabel in the α position, as with all continuously labeled probes. Since the probe is labeled as it is synthesized, the degree of label incorporation is very high.

5′ End Labeling

One alternative method to generating continuously labeled RNA probes by *in vitro* transcription is to label the 5′ end of the molecule with the addition of a radiolabeled phosphate. This method of 5′ end labeling is colloquially known as the kinasing reaction; it specifically involves the transfer of the γ phosphate of ATP to a 5′-OH substrate of RNA or DNA (forward reaction). The forward kinasing reaction is far more efficient than the exchange reaction, which involves the substitution of 5′ phosphates. This labeling reaction is catalyzed by the enzyme T4 polynucleotide kinase. The relative efficiency of this method is manyfold greater for DNA than for RNA, however, which is one reason that it is not commonly utilized to label RNA.

One common error in the laboratory pertaining to this labeling reaction involves the purchase of α-labeled nucleotides (needed for continuously labeled and 3′ end-labeled probes) rather than the required γ-labeled ATP.

3′ End Labeling

RNA can also be 3′-end labeled using the enzyme poly(A) polymerase. This enzyme, which is naturally responsible for nuclear polyadenylation of many hnRNAs, catalyzes the incorporation of AMP. Isotopic labeling requires α-labeled ATP precursors. In addition to its utility in RNA probe synthesis reactions, poly(A) polymerase can be used to polyadenylate naturally poly(A)$^-$ mRNA and other RNAs in order to support oligo(dT) primer-mediated sythesis of cDNA. Although this labeling strategy is certainly a possibility, in practical terms it amounts to little more than a last resort method for probe synthesis

Probe Purification

As with most labeling techniques, unincorporated precursor must be separated from the labeled probe. Failure to do so, especially when working with isotopes, usually results in unacceptably high levels of background. There are three basic methods for cleaning up the probe. For optimal results, follow the manufacturer's recommendations that accompany the labeling system.

In the first approach, the probe can be ethanol precipitated, in which case the unincorporated label remains in the supernatant. Whereas this is an efficient approach for purifying larger probes, attempted precipitation of oligonucleotides can be rather difficult, especially if small masses of nucleic acids are involved. Precipitation of probes is accomplished with a combination of salt and alcohol (see Chapter 5, Table 1), most often 0.1 volume NaOAc and 2.5 volumes of 95% ethanol.

The second general method for probe purification is gel filtration, or a permutation thereof. One popular approach is to make or purchase a so-called spun column (Sambrook *et al.*, 1989), which consists of little more than a 1-ml syringe packed with Sephadex (Pharmacia Biotech); Sephadex G-50 is used for larger probes, while Sephadex G-25 is reserved for the purification of labeled oligonucleotides. Spun-column chromatography is a glorified gel filtration, involving centrifugation of the column at 1000–1200 × g for approximately 4 min; the probe is rapidly eluted while the unincorporated nucleotides lag behind in the column. In this lab, more than 95% of the unincorporated label is routinely removed after the labeling reaction. Gel filtration probe purification is compatable with a wide variety of labeling methods, both isotopic and nonisotopic.

The third general method for probe purification involves the use of one of the numerous concentration or absorbent devices for probe purification. These devices include Elutip minicolumns (Schleicher & Schuell, Keene, NH), Ultrafree filtration units (Millipore, Bedford, MA), and Centricon devices (Amicon, Beverly, MA), to name but a few. All represent rapid, convenient alternatives to traditional column chromatography or alcohol precipitation, for separating radiolabeled nucleic acids from unincorporated nucleotides. The disposable nature of these devices minimizes the spread of radiolabeled waste products as well.

Probe Storage

Isotopically labeled probes are useful only as long as there is sufficient activity remaining and the probe molecules themselves remain intact. In general, radiolabeled probes should be used as soon after labeling as possible. Radiolysis of probes, especially those labeled to extremely high specific activity, becomes more problematic as the probe ages. Store probes at −20°C or −80°C until just prior to use. Avoid repeated freezing and thawing of probes.

Nonisotopic labeling systems vary with respect to the recommended postlabeling storage and handling of probes. Biotinylated, DIG-labeled, and fluorescein-labeled probes can be stored at −20°C for up to 1 year. In many labs, an entire day is invested generating nonisotopic probes for the experiments that are planned for the next several weeks or months. Probes are then available for hybridization reactions as needed. Probes should never be denatured until just prior to use.

Proper use of internal controls is discussed in Appendix M.

References

Chesselet, M. F. (1990). "*In Situ* Hybridization Histochemistry." CRC Press, Boca Raton, FL.

Dieffenbach, C., and Dveksler, G. S. (1995). "PCR Primer." Cold Spring Harbor Laboratory Press, Cold Spring Harbor, NY.

Feinberg, A. P., and Vogelstein, B. (1983). A technique for radiolabeling DNA restriction endonuclease fragments to high specific activity. *Anal. Biochem.* **132,** 6.

Feinberg, A. P., and Vogelstein, B. (1984). Addendum: A technique for radiolabeling DNA restriction endonuclease fragments to high specific activity. *Anal. Biochem.* **137,** 266.

Green, P., Pines, O., and Inouye, M. (1986). The role of antisense RNA in gene regulation. *Annu. Rev. Biochem.* **55,** 569.

Hoffman, K., and Finn, F. M. (1985). Receptor affinity chromatography based on the avidin-biotin interaction. *Ann. N.Y. Acad. Sci.* **447,** 359.

Innis, M. A., Gelfand, D. H., Sninsky, J. J., and White, T. J. (Eds.). (1990). "PCR Protocols." Academic Press, San Diego, CA.

Knecht, D. A., and Loomis, W. F. (1987). Antisense RNA inactivation of myosin heavy chain gene expression in *Dictyostelium discoideum. Science* **236,** 1081.

Leary, J. J., Brigati, D. J., and Ward, D. C. (1983). Rapid and sensitive colorimetric method for visualizing biotin-labeled DNA probes hybridized to DNA or RNA immobilized on nitrocellulose. *Proc. Natl. Acad. Sci. USA.*

McPherson, M. J., and Hames, B. D. (Eds.). (1995). "PCR 2: A Practical Approach." IRL Press/Oxford University Press, Oxford, England.

McPherson, M. J., Quirle, P., and Taylor, G. R. (Eds.). (1991). "PCR: A Practical Approach." IRL Press, Oxford, England.

Nishikura, K. and Murray, J. M. (1987). Antisense RNA of proto-oncogene c-*fos* blocks renewed growth of quiescent 3T3 cells. *Mol. Cell. Biol.* **7,** 639.

Nuovo, G. J. (1994). "PCR *in situ* Hybridization". Raven Press. New York, NY.

Rigby, P. W. J., Dieckmann, M., Rhodes, C., and Paul, P. (1977). Labeling deoxyribonucleic acid to high specific activity *in vitro* by nick translation with DNA polymerase I. J. Mol. Biol. **113,** 237.

Sambrook, J., Fritsch, E. F., and Maniatis, T. (1989). "Molecular Cloning: A Laboratory Manual," 2nd ed.. Cold Spring Harbor Laboratory Press, Cold Spring Harbor, NY.

Wilchek, M., and Bayer, E. A. (1984). The avidin-biotin complex in immunology. *Immunol. Today* **5**(2), 39.

Wilkinson, D. G. (Ed.). (1992). "*In Situ* Hybridization: A Practical Approach." IRL Press/Oxford University Press, Oxford, England.

Zagursky, R. J., Conway, P. S., and Kashdan, M. A. (1991). Use of ^{33}P for Sanger DNA sequencing. *BioTechniques* **11**, 36.

Practical Nucleic Acid Hybridization

Rationale

Perhaps the most complex, yet least understood, component of molecular biology studies pertain to the parameters that govern nucleic acid hybridization, a process first described by Marmur and Doty (1961). At the heart of molecular hybridization, two complementary polynucleotide molecules hybridize (base pair to each other) in an antiparallel fashion to form a double-stranded molecule. This phenomenon is also known as duplex formation.

Under conditions that promote nucleic acid hybridization, the strands that could potentially participate in duplex formation are known by specific names. The **probe** strand, as the name implies, usually carries some type of label that permits localization and quantitation of the probe at the conclusion of the hybridization period. The other strand, commonly known as the **target,** either may be immobilized on a solid support for filter-based analysis (e.g., Northern analysis or dot blot analysis) or may simply be suspended in buffer, free to participate in solution hybridization (e.g., S1 nuclease analysis).

Example: RNA isolated from a tissue sample contains thousands of different mRNA species. One objective of a particular study might be the assessment of *c-Ha-ras* transcription in a newly derived cell line. Addressing this question would require a suitable complementary probe, that is, one that would be able to hybridize to *c-Ha-ras*-specific transcripts. The probe itself could be a cDNA sequence, a genomic sequence, an antisense RNA sequence, or an oligonucleotide. The *c-Ha-ras*-specific probe might be labeled with ^{32}P for detection by autoradiography, or hapten labeled using any of the techniques described in Chapter 12 that support detection by chemiluminescence. Assuming that the hybridization stringency is satisfactory, *c-Ha-ras* probe molecules will only hybridize to complementary *ras* transcripts, which are the target sequences for that particular probe. Probe is always present in huge molar excess so that all complementary target strands will be base-paired. Of course, all of the RNAs present in a mixed sample are potential target sequences for other types of probes to which they are specifically complementary (or to the *c-Ha-ras* probe used in this experiment, if the hybridization stringency was too relaxed). The magnitude of hybridization is then assessed in a manner consistent with the labeling of the probe at the onset of the experiment. Further, to permanently record the location of size standards on the film (autoradiography or chemiluminescence), the investigator may wish to label a very small amount of probe that will be able to hybridize to the size markers, making detection of these nucleic acid species possible as well. This approach eliminates guesswork regarding the relative distribution of size standards compared to experimental samples.

Factors Influencing Hybridization Kinetics and Specificity

The original Southern analysis protocol (Southern, 1975) described the hybridization of a DNA probe to DNA target sequences. The resulting DNA:DNA duplex molecules are, in fact, thermodynamically less stable than the DNA:RNA or RNA:RNA duplexes that form when RNA targets, RNA probes, or both, are involved. This translates into the fact that the stringency at which the hybridization and posthybridization washes are conducted must be attuned to accommodate the chemistry of the polynucleotides involved.

Example: An investigator may wish to screen a python library using a probe derived directly from human cDNA. It may well be that a related sequence does not even exist in the python genome; alternatively a poorly conserved sequence might not be present but because of sequence divergence, the human probe might be able to hybridize under stringent hybridization conditions. To identify python sequences in a library, or by Northern or Southern analysis, it would be necessary to lower the stringency (salt, temperature, use/non-use of formamide) in order to hybridize based on partial complementarity.

Factors that influence the rate, specificity, fidelity, and probable utility of hybridization probes include (but are not limited to): temperature, ionic strength (primarily Na^+), pH, organic solvents such as formamide, guanosine and cytosine (G+C) content of probe, probe length, probe concentration, probe complexity (the total length of different probe sequences), degree of complementarity between probe and target sequences, the degree of mismatching, and viscosity of the system. It is important to realize that the influence of each of these variables is dependent on the state of the nucleic acid molecules involved: solution hybridization (S1 analysis) versus mixed-phase hybridization (Northern analysis, in which the target sequences are immobilized on a solid support).

Temperature

Perhaps the most frequently and easily manipulated variable that can either promote or prevent hybridization is the temperature of the system.

A measure of the stability of hybrids and the extent to which they are expected to form can be predicted by calculating their melting temperature (T_m). One should view the T_m as an equilibrium point; it is that temperature at which 50% of all hybrids are formed, and 50% remain dissociated into their constituent single strands. Historically, this phenomenon has been characterized by monitoring UV light absorption at 260 nm, as the temperature of a DNA solution was gradually raised. This type of assay is useful as a means of characterizing the thermodynamic stability of double-stranded DNA, because of the hyperchromic shift (absorption increase) that accompanies DNA melting or denaturation into its constituent strands (Lewin, 1997).

The maximum rate of hybridization for long probes is typically observed between 15 and 20°C below the calculated T_m in a system. In aqueous salt solutions, this corresponds to about 60–65°C. The inclusion of formamide modifies this parameter as described below. With respect to characterizing nucleic acids from one species using a probe from another, a 1°C change in melting temperature corresponds to about 1% sequence divergence (Bonner *et al.*, 1973). The inclusion of formamide, as is quite common, permits significant reduction of hybridization temperature without a relaxation of stringency.

Last, since only 50% of all possible hybrids form at the T_m of the probe, and it is unlikely that the investigator wants only 50% of all target molecules hybridized, the temperature of hybridization is usually at least 5°C below the T_m, and perhaps quite a bit more. This is especially true if there is reason the believe that the probe is not exactly matched with the target. The main concept is to get the probe to base pair; the reality of the situation is that the investigator has far more control of the stringency of the assay in the washing steps, conducted at the conclusion of the hybridization period and before initiating the detection procedures. So powerful are the posthybridization washing steps that in many cases hybridization can be performed at room temperature; this would be unlikely to have a negative impact on the outcome of the experiment as long as the filters were washed correctly prior to hybridization detection.

Ionic Strength

Stringency is dramatically influenced by increasing or decreasing the amount of salt in a hybridization buffer. In general, the rate of hybridization increases with salt concentration up to about 1.2 M NaCl, at which point the rate becomes constant; the T_m of a nucleic acid duplex changes approximately 16°C with each factor of 10 in salt concentration

(Britten and Davidson, 1985). Further, because divalent cations have a greater influence than monovalent cations at low concentrations, restrained use of chelating agents such as EGTA and EDTA is strongly recommended.

pH

One of the factors that profoundly influences the stability of double-stranded molecules is the pH of the environment. Most hybridization reactions are conducted at near neutral pH: Alkaline pH buffers promote duplex dissociation, and highly acidic pH buffers may well result in depurination of both probe and target molecules. In general, pH is among the least manipulated variables in nucleic acid hybridization.

Probe Length

The length of the probe is directly factored into the calculation of the T_m of the probe (duplex stability), according to the expression $D = 500/L$, where D is the reduction in T_m (°) and L is the number of base pairs actually participating in duplex formation (Britten and Davidson, 1985), which is usually the length of the probe itself. The shorter the probe, the more rapidly hybridization occurs and the more discriminating the probe becomes. For example, the short oligonucleotides, 20- to 25-mer sequences, can discriminate between closely related target sequences that differ by as little as 1 base. Moreover, the shorter the probe, the more influential many of the other variables become.

Probe Concentration

In preparation for nucleic acid hybridization, the investigator is afforded considerable leeway with respect to the amount of probe added to the reaction. In short, as the concentration of probe increases, the forward hybridization kinetics also increase. This means that higher probe concentrations, by accelerating hybridization, often decrease the amount of time needed to saturate all of the target sequences in the reaction. This is especially true when oligonucleotides are used as probes. The investigator must be aware, however, that the use of excessive quantities of probe can also favor nonspecific hybridization: Think about what happens in the polymerase chain reaction (PCR), for example, when excessive primer concentrations are used.

To identify the proper probe concentration, two variables must be considered: (a) the method of detection, that is, isotopic or nonisotopic, and (b) the abundance of the target, for example, single-copy genes or low abundance transcripts. The amount of probe required should be based on the guidelines shown in Table 1 and is based either on the mass of the probe or, in the case of radiolabeling, on the specific activity of the probe. In the event that excessive probe is utilized, the stringency of the assay can be easily manipulated in posthybridization washing steps.

Guanosine and Cytosine Content

The stability of a hybrid molecule is profoundly influenced by the base composition of the molecules involved. Because three hydrogen bonds occur between guanine (G) and cytosine (C), GC-rich duplexes are thermodynamically more stable than adenine/thymine (AT)-rich duplexes, which have only two hydrogen bonds between them. The importance of G+C content becomes more pronounced as the probe involved becomes shorter. For example, in the design of oligonucleotide probes, both for hybridization and for PCR, the paramount importance of the G+C content in duplex stability is underscored by the expression

$$T_m = 0.41 \ (\%GC) + 69.3,$$

where (%GC) is the percentage of the duplex consisting of guanine and cytosine residues (Marmur and Doty, 1961).

	Mass (ng/ml)	cpm/ml
TABLE 1		
Recommended Nucleic Acid Probe Concentrations		
Isotopic probes		
Single-copy genes or low abundance transcripts	20–50	10^6
Multiple-copy genes or high abundance transcripts	5–20	10^5
Oligonucleotide probes	1–5	$>10^4$
Nonisotopic probes		
Single-copy genes or low abundance transcripts	50–75	Not applicable
Multiple-copy genes or high abundance transcripts	20–50	Not applicable
Oligonucleotide probes	5–10	Not applicable

Note. Values indicated are per milliliter of hybridization solution. Probe concentration can be expressed either by actual mass added or by label incorporation. While detection sensitivity with nonisotopic probes is often comparable to that with radiolabeling, the amount of probe added per milliliter of hybridization solution is usually larger.

Mismatching

Many of the recommended hybridization conditions that accompany filter membranes and molecular biology kits assume a perfect or nearly perfect match between probe and target sequences. Under stringent conditions, the most rapid duplexes to form are those that manifest exact complementarity between probe and target. Mismatching retards the hybridization rate; at sufficiently high temperatures, for example at the T_m, mismatches may inhibit hybridization altogether. With longer probes (greater than 1 kb) the effect of mismatch may not be noticeable under standard hybridization conditions.[1] This mismatch effect is significantly more pronounced with shorter probes, especially when using oligonucleotides as probes or as primers. To promote hybridization between mismatched probes and target sequences, the stringency of the system can be relaxed, usually by decreasing the hybridization temperature. The further below the T_m of the probe, the more mismatch between probe and target will be tolerated. Below a certain sequence-dependent temperature, however, random hybridization will occur.

Probe Complexity

With respect to hybridization, the complexity of the probe is the length of different probe sequences in the hybridization buffer that could potentially base pair to complementary target molecules. In most cases, only a single type of probe would be used (e.g., a *c-myc* probe to assess the prevalence of cytoplasmic *c-myc* mRNA by Northern analysis). When using degenerate oligonucleotide probes, however, in which more than one probe sequence is present (perhaps due to target sequence ambiguity), the complexity is greater, the concentration of the correct oligonucleotide probe is reduced, and hybridization kinetics are slowed.

Viscosity

The original studies of the effect of viscosity on nucleic acid hybridization (Thrower and Peacocke, 1968; Subirana and Doty, 1966;

---

[1] Standard hybridization conditions may consist of any of a variety of multipurpose prehybridization/hybridization buffers and more than likely feature an overnight incubation. "Standard" implies that the effect of probe length is not a major variable, and abatement of the rate of hybridization due to mismatches is probably masked by the relatively long period of hybridization. As the length of the probe decreases, hybridizations become less "typical."

Chang *et al.*, 1974) showed a reduction in the rate of renaturation as viscosity (presence of sucrose, glycol) increases. One frequently used component in hybridization buffers is the anionic polymer dextran sulfate. Although dextran sulfate contributes to the viscosity of the milieu, the rate of hybridization is accelerated because the volume of the hybridization buffer occupied by dextran sulfate effectively concentrates the probe. A similar phenomenon occurs with the inclusion of Denhardt's solution[2] (Denhardt, 1966) in various hybridization recipes. It is an excellent alternative to dextran sulfate and alleviates the severe background problems frequently associated with the inclusion of dextran sulfate in hybridization recipes.

Formamide

Formamide destabilizes double-stranded molecules. Thus, the inclusion of formamide in hybridization recipes allows a reduction in T_m (and hybridization temperature) of about 0.75°C for each 1% of added formamide. Historically, it has been commonplace to carry out nucleic acid hybridization in buffers consisting of 50% formamide at 42°C; this approximately corresponds to T_m –20°C for a "typical" probe, a parameter that favors maximum hybridization between complementary nucleic acid molecules, with no relaxation of stringency. Conducting RNA analysis in formamide-containing buffers extends the useful life of probe molecules that may be heat labile.

In addition, the reader should be aware that a new generation of nonformamide-based hybridization buffers, which drastically accelerate forward hybridization kinetics, are available. Further, many of these preparations contain a surfactant that prevents the probe from binding to the filter, thereby minimizing background hybridization. Although these qualities make such hybridization reagents desirable, they are also costly and of proprietary formulation.

Hybridization Temperature

In most applications, nucleic acid probes that are at least several hundred bases long can be hybridized at 42°C in a generic hybridization solution, which typically consists of 50% formamide. When more precisely defined hybridization conditions are required or when empirical determi-

---

[2]100× Denhardt's solution = 2% Ficoll 400; 2% polyvinylpyrrolidone, 2% BSA (fraction V).

nation must be made, one may first derive the T_m of the duplex and conduct hybridization at a suitable temperature. With respect to the following equations, the contribution of each of the salient parameters has been discussed above.

T_m for Long Probes

The thermodynamic stability is described as follows. For a double-stranded DNA (DNA:DNA) molecule (Bolton and McCarthy, 1962):

$$T_m = 81.5°C + 16.6 \log[Na^+] + 41(\%G+C)$$
$$- 0.63(\% \text{ formamide}) - (500/L);$$

for DNA:RNA hybrids (Casey and Davidson, 1977):

$$T_m = 79.8°C + 18.5 \log[Na^+] + 58.4(\%G+C) + 11.8(\%G+C)^2$$
$$- 0.5(\%\text{formamide}) - (820/L);$$

for RNA:RNA hybrids (Bodkin and Knudson, 1985)

$$T_m = 79.8° + 18.5 \log[Na^+] + 58.4(\%G+C) + 11.8(\%G+C)^2$$
$$- 0.35(\%\text{formamide}) - (820/L);$$

where

T_m = melting temperature (that temperature at which 50% of the duplex molecules are dissociated into their constituent strands);

$(\%G+C)$ = percentage of guanine + cytosine content, expressed as a mole fraction;

$[Na^+]$ = log of the sodium concentration, expressed in molarity;

L = number of bases that participate in the actual hybridization.

T_m for Oligonucleotide Probes

For shorter probes such as oligonucleotides (Chapter 12), the T_m of the oligomer can be quickly estimated by assigning 2°C for each adenine and thymine and 4°C for each guanine and cytosine (Itakura et al., 1984):

$$T_m = 4(G+C) + 2(A+T).$$

It is important to realize that this relationship holds only for short oligonucleotides (11–27 bases long) and is based on hybridization in 1 M Na^+, in the complete absence of any organic solvents. By definition, at the T_m only 50% of all duplexes are stable; therefore, very stringent hy-

bridization is favored at 5°C below the calculated T_m for short oligonucleotides. At T_m –5°C, all exact match duplexes will form. It is important
to realize that in the design of an oligonucleotide probe there may have
been certain ambiguities pertaining to the sequence of the probe. If the
oligonucleotide probe is not an exact match, then it will be necessary to
lower the stringency of hybridization to tolerate mismatches. If more than
one mismatch is involved, it will be necessary to lower the temperature of
hybridization even more.

When longer oligonucleotides are involved (14–70 bases), the T_m can
be calculated (Sambrook *et al.*, 1989) by

$$T_m = 81.5 - 16.6(\log[Na^+]) + 41(\%GC) - (600/N),$$

where N = the length of the oligonucleotide (assuming a perfect match
between probe and target).

Hybridization and the Northern Analysis

The attempted hybridization between nucleic acid probe and target sequences consists of three major components: prehybridization (also
known as blocking), hybridization, and posthybridization washing (also
known as the stringency washes). The detection of hybridization events,
the subject of the next chapter, is indeed a different art entirely.

The exact hybridization conditions for many of the universally available probes have been defined and are provided by the manufacturers.
Many generic types of hybridization buffers and conditions have been
recommended, combinations that purportedly yield the "perfect blot
every time." Although it is true that the kinetic behavior of some families
of nucleic acid molecules can be predicted with a fair degree of accuracy,
knowledge of the parameters that directly influence hybrid formation is
invaluable in fine-tuning the conditions of hybridization. This becomes a
paramount issue when using probes that are not exactly complementary
to the target(s) (exact match), or perhaps when using a nucleic acid probe
to screen blots or libraries for phylogenetic relatedness. It should be noted
that temperature and salt concentration are the two variables most often
adjusted in order to modify the specificity of hybridization, when such
modifications become necessary.

Prehybridization: Filter Preparation

As described in Chapter 11, nucleic acid samples should be fixed to
filter membranes immediately following transfer, after which the filters

can be stored for extended periods. Before hybridization, filters are subjected to a prehybridization incubation (1) to equilibrate the filter in a buffer identical or very similar to the one to be used for the actual hybridization, and (2) to block the filter paper completely. Blocking (Meinkoth and Wahl, 1984) means covering the entire surface of the filter that is not already occupied by experimental RNA samples with sheared, denatured (heterologous) DNA from a species unrelated to the biological origin of the RNA; usually salmon sperm DNA or calf thymus DNA is selected.[3] Alternatively, newer methods involve blocking filters, especially nylon, with casein or other protein, rather than with heterologous DNA. In general, protein blocking is ideal for isotopic and nonisotopic detection systems alike, yielding outstanding signal to noise, and very low background. Prehybridization, then, is very important so that in the actual hybridization probe molecules do not stick in a nonspecific fashion over the surface of the filter paper, which would cause unacceptable background levels.

Numerous recipes have been concocted for prehybridization, their exact composition being a function of the type of filter and probe used in a specific experimental application. The recommendations contained herein have been used with great success in this and other laboratories.

Protocol: Prehybridization (Long Probes)

1. Pre-wet the filter in 5× SSC or 5× SSPE for 2–3 min.

Note: SSPE is a better buffer than SSC, especially in the presence of formamide (Meinkoth and Wahl, 1984), because the phosphate in SSPE actually mimics the nucleic acid phosphodiester backbone and helps to block the filter and improve the background at the detection stage. This strategy is especially desirable when experimental RNA is immobilized on nylon (+) filters, due to the natural affinity of the membrane for the negatively charged phosphodiester backbone of the nucleic probe. 20× SSPE = 3 M NaCl, 200 mM NaH_2PO_4, 20 mM EDTA, adjust pH to 7.4. 20× SSC = 3 M NaCl, 0.3 M Na_3-citrate, adjust pH to 7.0.

2. Prehybridize the filter(s) for 3–4 h at 42°C in an excess volume (250–1000 μl/cm^2) of prehybridization buffer consisting of 5× SSPE, 5×

---

[3]Heterologous DNA, used as a carrier and a filter membrane block is prepared as a 10 mg/ml stock solution in water. The DNA is boiled to facilitate dissolving and then sheared by drawing it repeatedly through an 18-gauge needle. Stock solutions of heterologous DNA are stored at 4°C. Just prior to use, an appropriate aliquot must be denatured again by boiling for 10 min.

Denhardt's solution, 0.1% SDS, 100 μg/ml denatured salmon sperm DNA, 50% deionized formamide. The following is the proper method for assembling the components of the prehybridization buffer:

a. Prewarm formamide to 45°C.
b. In a separate bottle or tube, mix SSPE, Denhardt's solution, and SDS and place at 45°C. The SDS will go into solution when the remaining ingredients are added.
c. In a fresh polypropylene or glass test tube, mix the required salmon sperm DNA in the volume of sterile water necessary to bring the prehybridization buffer up to final volume. Boil the mixture for 5–7 min.
d. Carefully add the boiled salmon sperm DNA to the SSPE, Denhardt's, SDS mixture, and then add the entire mixture directly to the prewarmed formamide. Swirl carefully to mix. Store this buffer at 42°C until ready to use.

The reason for assembling the buffer in this fashion is to eliminate renaturation of denatured salmon sperm DNA that could occur by plunging a small volume of concentrated stock DNA on ice. Immediate addition of formamide effectively reduces the rate of renaturation, especially when all reagents are prewarmed. After mixing, the buffer is ready to be used immediately. Typically, prepare 100 ml of this buffer to prehybridize two blots and reserve 25 ml, which can be used for hybridization.

Note 1: Routine prehybridization and hybridization are most expediently carried out in a plastic container with a perfectly flat bottom, whose surface area is only slightly greater than the area of the filter. Several configurations of Rubbermaid food storage boxes and Nalge utility boxes work well. Be sure to cover the container with a securely fitting lid to prevent evaporation. This approach precludes introducing air bubbles into heat-sealable plastic bags and avoids spillage of high specific activity hybridization buffer at the conclusion of the hybridization period.

Note 2: Placing the hybridization vessel on a gently moving (30–50 rpm) platform, such as an orbital shaker incubator (LabLine), will circulate the buffer evenly over the entire surface of the filter and yield superior data.

Note 3: Prehybridization with a large (excess) volume of buffer usually favors very low levels of nonspecific interaction between the probe (when added) and the filter. After blocking, the volume of the buffer can be reduced to maintain the probe at a respectable concentration.

3. At the conclusion of the prehybridization period pour off the prehybridization buffer and add a fresh aliquot (optimally 100 μl/cm^2) of hybridization buffer. In most cases, the prehybridization and hybridization buffers will be identical (see Step 2 for preparation).

Probe Denaturation

Double-stranded probes must be denatured prior to use. Because of the enormous variation in labeling approaches in nonisotopic systems, it is best to follow the instructions of the manufacturer for probe denaturation.

C A U T I O N   **Radiolabeled probes should not be boiled to denature because of the dangers associated with radioactive steam. Instead, add 0.1 volume 1 N NaOH to the probe (which is usually in a volume of 50–100 μl) and incubate at 37°C for 10 min or at room temperature for 30 min. The probe can then be added directly to the hybridization solution, without prior neutralization, as long as the total probe volume is less than 150 μl (assumes a minimum of 5 ml hybridization buffer).**

Hybridization

4. Add the probe to the buffer and hybridize for 12–16 h (overnight) at 42°C. Be sure to denature double-stranded probes prior to use. Make certain that the cover of the hybridization chamber is securely in place.

Note 1: For the detection of rare mRNAs or single copy genes, use 5–20 ng of probe per milliliter of hybridization buffer or 1–5 × 10^6 cpm/ml.

Note 2: If the mass of the probe is low, increasing the concentration of the salmon sperm DNA will improve hybridization kinetics.

Note 3: If desired, add a total of 25 ng probe, which will base pair with the molecular weight standards, to produce a visual record along with experimental samples on the same film.

Posthybridization Stringency Washes

Depending on the nature of the probe and the stability of the hybrid, the investigator may need to adjust the exact posthybridization washing conditions for each probe. It is very important to keep in mind that in the context of posthybridization washes, stringency is a function of ionic strength, temperature, and time; thus, it may not be suitable to cut down on the washing time by increasing the temperature and/or lowering the salt.

5. At the conclusion of the hybridization period, pour off the hybridization buffer and wash the filters to remove all of the probe molecules that did not participate in the formation of hybrids.

  a. Wash filters for 30 s, in a room temperature solution of 2× SSPE, 0.1% SDS to remove most of the probe in the box and remaining on the filter membrane.
  b. Wash filters twice for 15 min in 200–300 ml of a room temperature solution of 2× SSPE, 0.1% SDS.
  c. Wash filters twice for 15 min, in 200–300 ml of a solution of 0.1× SSPE, 0.1% SDS at 37°C.

Note: If the background is too high, a final wash in 0.1× SSPE, 0.1% SDS at 42–50°C may be useful.

6. Rinse the filter very briefly in 2× SSPE and place on a piece of Whatman paper just long enough to blot excess buffer from surface of the filter.

Important: Do not allow the filter to dry out to any extent, especially when chemiluminescence is the method chosen for detection.

7a. If using chemiluminescence or chromogenic detection technology, proceed from this step directly to the detection buffers, according to the directions of the manufacturer of the system.

7b. If using isotopes, wrap the filter in plastic wrap and set up autoradiography (Chapter 14).

Note: Be sure to keep filters damp if planning to remove the probe after autoradiography in order to hybridize to another probe(s).

References

Bodkin, D. K., and Knudson, D. L. (1985). Assessment of sequence relatedness of double-stranded RNA genes by RNA–RNA blot hybridization. *J. Virol. Methods* **10**, 45.

Bolton, E. T., and McCarthy, B. J. (1962). A general method for the isolation of RNA complementary to DNA. *Proc. Natl. Acad. Sci. USA* **48**, 1390.

Bonner, T. I., Brenner, D. J., Nenfeld, B. R., and Britten, R. J. (1973). Reduction in the rate of DNA reassociation by sequence divergence. *J. Mol. Biol.* **81**, 123.

Britten, R. J., and Davidson, E. H. (1985). Hybridization strategy. *In* "Nucleic Acid Hybridization: A Practical Approach" (B. D. Hames and S. J. Higgins, Eds.), pp. 3–16. IRL Press, Washington, DC.

Casey, J., and Davidson, N. (1977). Rates of formation and thermal stabilities of RNA:DNA and DNA:DNA duplexes at high concentrations of formamide. *Nucleic Acids Res.* **4**, 1539.

Chang, C. T., Hain, T. C., Hutton, J. R., and Wetmur, J. G. (1974). Effect of microscopic and macroscopic viscosity on the rate of renaturation of DNA. *Biopolymers* **13,** 1847.

Church, G. M., and Gilbert, W. (1984). Genomic sequencing. *Proc. Natl. Acad. Sci.* **81,** 1991.

Denhardt, D. T. (1966). A membrane-filter technique for the detection of complementary DNA. *Biochem. Biophys. Res. Commun.* **23,** 641.

Howley, P. M., Israel, M. A., Law, M.-F., and Martin, M. A. (1979). A rapid method for detecting and mapping homology between heterologous DNAs. *J. Biol. Chem.* **254,** 4876.

Itakura, K., Rossi, J. J., and Wallace, R. (1984). Synthesis and use of synthetic oligonucleotides. *Annu. Rev. Biochem.* **53,** 323.

Lewin, B. (1997). "Genes VI." Oxford University Press, New York.

Marmur, J., and Doty, P. (1961). Determination of the base composition of deoxyribonucleic acid from its thermal denaturation temperature. *J. Mol. Biol.* **3,** 584.

Meinkoth, J., and Wahl, G. (1984). Hybridization of nucleic acids immobilized on solid supports. *Anal. Biochem.* **138,** 267.

Sambrook, J., Fritsch, E. F., and Maniatis, T. (1989). "Molecular Cloning: A Laboratory Manual," 2nd ed. Cold Spring Harbor Laboratory Press, Cold Spring Harbor, NY.

Southern, E. M. (1975). Detection of specific sequences among DNA fragments separated by gel electrophoresis. J. Mol. Biol. **98,** 503.

Subirana, J. A., and Doty, P. (1966). Kinetics of renaturation of denatured DNA. I. Spectrophotometric results. *Biopolymers* **4,** 171.

Thrower, K. J., and Peacocke, A. R. (1968). Kinetic and spectrophotometric studies on the renaturation of deoxyribonucleic acid. *Biochem. J.* **109,** 543.

Principles of Detection

Rationale

The final outcome of an experiment is only as reliable as each step involved in the generation of data. While ensuring the integrity of a nucleic acid sample throughout every experimental manipulation is of paramount importance, another fundamental component in the design of a model system is the methodology for labeling probes and associated method of detecting hybridization events. The selection of labeling and detection techniques is governed primarily by the required degree of sensitivity and resolution. In view of the remarkable refinements in nonisotopic systems, particularly chemiluminescence, one realistic question now is whether to use radioactivity at all.

With its advent, nonisotopic labeling and detection by chemiluminescence was heralded as a realistic alternative to the use of radiolabeled nucleic acid probes in blot analysis and *in situ* hybridization, circumventing many of the traditional difficulties associated with chromogenic (colorimetric) detection. Surprisingly, with its ready availability, chemiluminescence has yet to achieve widespread acceptance, due to a mindset that, fortunately, is slowly changing.

The variety of techniques performed by molecular biologists is no less diverse than the methods used to label and detect nucleic acid probes (antibodies, too). The two goals at the end of every experiment involving hybridization are:

1. **Localize** the probe. It is necessary to determine where on the filter paper, or elsewhere in the assay, the probe managed to base pair stably to its target.
2. **Quantify** the amount of the probe that has been retained in each band or, alternatively, by target molecules found within a particular region of a tissue section.

Broadly, there are two standard detection strategies: those involving radioisotopes, and nonisotopic techniques. Isotopic methods, obviously, involve radiolabeling, coupled with detection by autoradiography (described below) or Cerenkov counting. The nonisotopic methods are becoming increasingly more sophisticated, accommodating detection on filter membranes, in microtiter plate format, directly from gels, and by direct video capture, to name but a few methods. There are four standard types of nonisotopic methods of detection:

1. Chemiluminescence, the enzyme-mediated production of visible light, most often recorded on X-ray film;
2. Fluorescence, the emission of light when certain compounds are ex-

cited with the proper wavelength of light, most often recorded using a fluorescence imaging system;

3. Chemifluorescence, the enzymatic conversion of fluorogenic substrate into a fluorescent product, most often recorded using a fluorescence imaging system;

4. Chromogenic detection, the enzyme-mediated formation of a color precipitate directly on a filter or tissue section. The precipitate itself records the distribution of target sequences.

In the earliest days of nonisotopic labeling and detection, nucleic acids were labeled with biotin and, after hybridization, streptavidin–alkaline phosphatase detection systems, in which a colorless substrate is enzymatically converted into an insoluble color pigment, were used. This method, commonly known as chromogenic or colorimetric detection, failed to gain widespread acceptance due to a severe lack of sensitivity, compared to that achievable with autoradiography. Now, however, many of the labeling techniques originally designed for chromogenic detection methods have been modified to accommodate a detection technology known as chemiluminescence. In many cases the method of probe synthesis to support chemiluminescent detection is identical to labeling for chromogenic detection. The differences in these techniques lie in the posthybridization handling of filters: In addition to differences in detection reagents, stringency washes may differ as well. At the heart of chemiluminescence, the emission of visible light is recorded on a piece of X-ray film.

Autoradiography[1]

Autoradiography is a simple and sensitive photochemical technique used to record the spatial distribution of radiolabeled compounds within a specimen or an object. In the context of nucleic acid-related study, radiolabeled probes are hybridized to target sequences in solution, *in situ*, or are immobilized on a solid support. The capture of ionizing radiation and photons by an emulsion placed in direct contact with a radioactive source provides a relatively permanent record of the decay events associated with unstable radionuclides. Autoradiography is subdivided into two broad groups, commonly referred to as macroautoradiography and microautoradiography, indicative of the type of specimen containing the radioactivity, the type of emulsion necessary for image formation, and the method of examining the results.

[1] Adapted, in part, from Eastman Kodak Company.

The most common applications of autoradiography in molecular biology include the quantitative analysis of nucleic acid hybridization events. If used in conjunction with electrophoresis, the resultant imagery will provide a qualitative aspect as well. For example, in Northern blot analysis of gene expression (macroautoradiography), the extent of hybridization and probe specificity to RNA target molecules is assessed by the magnitude and location of radioisotope emissions from the filter membrane. Microautoradiography, on the other hand, is achieved by coating a thin section of a larger specimen with a light-sensitive emulsion. This technique, also known as *in situ* autoradiography, is usually carried out directly on a microscope slide. Because this technique has the advantage of maintaining natural cellular geometry, the intracellular distribution of radiolabel can be easily determined by light or electron microscopy (i.e., by counting the grains).

In macroautoradiography, a piece of high-speed medical X-ray film is placed in direct contact with a plastic-wrapped filter membrane. This is commonly referred to as direct exposure. In some applications, such as the nuclease protection assays, isotopes can be detected directly in polyacrylamide or agarose gels. The film base is a flexible piece of polyester [Kodak films used for autoradiography are 0.007-in. (7-mil) thick] that has been coated on one or both sides with an emulsion. The radio- and light-sensitive emulsion itself consists partly of sensitive crystals of silver halide (bromides, chlorides, or mixed halides), to which energy is released by ionizing radiation and photons of light. During film exposure, energy is absorbed in the silver halide grains within the film emulsion, with the release of electrons. The resulting negatively charged "sensitivity specks" attract positively charged silver ions, thus forming an atom of metallic silver. Exposure of X-ray film generates the so-called latent image, an invisible precursor to the visible image manifested when the film is immersed in a suitable photographic developing solution. The latent image, consisting only of developable silver grains, is not visible in the darkroom (under safelight conditions) and the emulsion is still exquisitely light sensitive. Formation of the latent image can be improved by preflashing the film (described below) and by low temperature exposure. Once formed, the latent image is not stable for long exposure periods; potential latent image instability is exacerbated by the presence of oxygen and moisture on the film. The final developed image may be assessed by visual inspection or, with greater precision, by digital image analysis.

The formation of a visible autoradiographic image can be carried out manually or by machine. In either case, there are distinct stages in the processing of the film: development, washing, fixation, washing, and finally drying.[2] When immersed in a photographic developer, the exposed silver halide grains in the emulsion of the film become reduced to metallic silver, thus amplifying the latent image enormously. The action of the developer is arrested by immersion of the film in a chemical "stop bath." Although a variety of stop bath solutions may be purchased, a homemade version can by concocted by preparing a solution of 1% acetic acid. Arresting the activity of the developer at precisely the right time is very important; overdevelopment can obscure any image that may have been captured on film. Keep in mind that a piece of X-ray film withdrawn from the developer tray is still coated with the developer, and that the developing process will continue until the action of the developer has been neutralized. In the absence of a suitable stop bath formulation, the film should be immersed in water to dilute and wash off the developer. These actions will not only arrest the developing process but also extend the life of the fixer. Next, the film is immersed in a chemical fixer solution, which dissolves away the unexposed silver halide. A thorough, room temperature washing of the film in water, after the fixation step, is needed to remove processing by-products. Once the film has been completely dried, a relatively stable image is in place. When developing films manually, it is necessary to develop, wash, and fix each film for the same length of time. Without this type of quality control it may be difficult to determine the extent to which observed data variations are due to experimental manipulation or to variation in the mechanics of developing the film. Machine-developed autoradiograms have the advantage of being developed according to a standardized process; however, one is at the mercy of the machine.

The mechanics of an autoradiographic exposure and the choice of emulsion are dependent on the energy emission spectrum of the isotope, the image quality requirements, and the anticipated amount of radioactivity in the system. There are no hard and fast rules about exposure time and exposure strategies; these parameters must be empirically determined with each experiment.

Handling of Filter Membranes

Filter membranes used for Northern analysis, dot-blotting, and similar techniques should not be allowed to dry out completely after the last stringency wash. Instead, these filters should be blotted briefly by resting

---

[2]For manual developing of X-ray films, Kodak D-76, D-19, and GBX developers and standard Kodak fixer work extremely well.

them on a piece of Whatman 3MM paper or the equivalent and then wrapped in plastic. Maintaining filter membranes in a damp state will greatly facilitate the postautoradiography removal of the hybridized probe by an extremely high-stringency wash. Thus, a filter membrane can be screened several times with other probes of interest. Removal of probe from membranes that are allowed to dry completely is significantly more difficult than removal from damp membranes. *Never* place a damp or wet filter membrane in direct contact with a sheet of X-ray film. The two will form a permanent symbiosis, resulting in the ruination of both filter and film.

X-Ray Film

The choice of X-ray film is dependent on whether high-energy or low-energy emitters were used as label and the desired level of resolution. Among the more commonly used types of X-ray film for autoradiography are the Kodak X-Omat AR (XAR), X-Omat LS (XLS), and SB film series. Unopened packages of X-ray film should be stored below 70°F and shielded from penetrating radiation. Once opened, storage below 50% relative humidity is recommended.

The XAR film provides the greatest latitude in terms of sensitivity and applications for most autoradiography procedures. It is a double-coated film on a clear base. The resulting autoradiogram will appear as a dark image on a grey-tinted film. XAR film is available in two configurations: (1) XAR-5 film, provided in alternate interleaved packing that protects each sheet of film during shipment and also prevents static electricity from exposing the film as it is withdrawn from the box; and (2) XAR-2 or Ready Pack film, in which each sheet of film is wrapped in a light-tight envelop.

The XLS film gives greater definition than the XAR films, with less background fog. The film is double-coated on a blue-tinted base. The resulting autoradiogram will appear as a dark image on a blue-tinted film. Although hybridization signals may appear better defined with this film, the average exposure time is about twice that required of the XAR film. For this reason, it may be worthwhile making initial exposures with XAR film, followed by a final autoradiogram with the XLS blue film. Further, autoradiograms on XLS seem to photograph better for publication-quality prints.

SB film has sensitivity similar to XAR film and is particularly well suited for work with low energy isotopes such as ^{35}S, ^{14}C, and ^{3}H. It is a single-coated film on a blue-tinted base. Images generated on single-sided films are often of higher definition than those on double-coated films. This is true because a slightly larger image, formed on a second-side emulsion, is not possible with the SB-type film. Moreover, decay events associated with low-energy isotopes frequently possess insufficient energy to penetrate and expose a second-side emulsion layer. This film is one of the few that should always be developed manually, rather than through an automated developing machine.

Exposure Time

There is no limit as to the number of exposures that can be made from a single radioactive source other than the half-life of the material itself. In the case of high-energy emitters, a Geiger counter probe placed in the vicinity of the filter is good way to assess the activity on a filter membrane. Obviously, a source manifesting 3000 cpm will require a much shorter exposure period than a source registering but a few counts above background. The lack of obvious signal should not be an immediate discouragement; with autoradiography, it is not unusual to have to wait 3 days or longer to observe hybridization signals corresponding to single-copy genes or rare mRNAs.

It is useful to do an overnight exposure at room temperature, with XAR film, to assess the extent of background and nonspecific filter hybridization; subsequent exposure tactics are based on the image derived from this test exposure. In the case of low activity filters one can decrease the exposure time with the use of an intensifying screen or by fluorography. Unlike direct exposure autoradiography, both of these signal amplification techniques are temperature dependent. Low temperature exposure itself, however, will not improve the direct exposure image formation process.

Intensifying Screens

Autoradiographic detection of ionizing radiation can be improved through the use of intensifying screens consisting of a flexible polyester base coated with inorganic phosphors. The predominant type of screen used in conjunction with macroautoradiography is the calcium tungstate ($CaWO_4$) phosphor-coated screen, such as the DuPont Cronex Lightning Plus Intensifying Screen. The organic phosphors convert ionizing radiation to UV or blue light; these screens are blue light emitting, exhibit very low-level noise (background signal), and are extremely compatible with standard autoradiographic use. It is also worthwhile to store intensifying screens in the dark for several hours or overnight before setting up the exposure. This is necessary because "after glow" effects from non-dark-adapted screens can fog the film. Unlike double-coated X-ray films, which have no real sidedness, only one side of an intensifying screen is phosphor-coated, so proper orientation is essential. The $CaWO_4$ phosphor coating gives one side of the screen a white appearance while the uncoated, non-functional side has an off white appearance. In this lab, to minimize confusion and possible error in the dark room, strips of heavy tape have been placed along the edges of the *nonfunctional* side of the intensifying screen so that feeling the tape on one side will help to position the other side facing the film. Low temperature ($-80°C$) autoradiography is necessary for maximum energy conversion efficiency when using an intensifying screen, because film response to low light intensities is improved dramatically at low temperatures. It is also imperative that the sensitivity of the film used match the light emission spectrum of the intensifying screen (reviewed by Laskey and Mills, 1977; Swanstrom and Shank, 1978). The net result, then, is a greater degree of exposure per decay event than in the absence of an intensifying screen. Intensifying screens improve the detection of the high-energy β emitters[3] ^{45}Ca and ^{32}P and γ ray emitters such as ^{125}I (Swanstrom and Shank, 1978; Laskey and Mills, 1977).

The proper juxtaposition of an intensifying screen in an X-ray film cassette is critical if the screen is to be of any use whatsoever during then exposure period. The film is placed between the specimen and intensifying screen so that the film can "see" scintillation events generated by the screen, in addition to being exposed directly by the ionizing radiation associated with isotope emission. The proper location of a second intensifying screen (if used) is behind the specimen. Placement of the film between two intensifying screens with the specimen outside the "sandwich" is not recommended, because it will reduce image sharpness (it may improve sensitivity, however). One should expect a minor decrease in image sharpness when even one intensifying screen is used, due to the spread of β particles from their source as they travel through the film to the screen. In essence, there is considerable bouncing around of β particles within the cassette as they strike the intensifying screen. The additional spread of the energy causes a larger area of the film to be exposed than when only direct exposure to the β energy source is carried out.

---

[3]The autoradiographic characteristics of ^{33}P, used mostly for DNA sequencing, are more like ^{35}S than ^{32}P, precluding the use of an intensifying screen. See Zagursky *et al.* (1991) for details.

Fluorography

Fluorography is an intensification process by which some of the energy associated with the decay process of the isotope is converted to light by the interaction of radio-decay particles with a compound known as a fluor, which exposes the X-ray film. As with intensifying screens, fluorography is an intensification procedure, though it is used mainly when low-energy emitters, especially ^3H, are employed as radiolabels (Bonner and Laskey, 1974; Randerath, 1970). Fluorography is accomplished through the impregnation of a specimen (acrylamide gel, agarose gel, or blot format) with a fluorographic solution. Commercially available fluorographic formulations include mixtures of 2,5-diphenyloxazole or sodium salicylate. While impregnation does not improve detection performance with ^{32}P or ^{125}I, it may enhance detection of ^{14}C and ^{35}S. In this way, fluorographic detection differs from signal amplification through the use of intensifying screens, which may be viewed as a type of external fluor. Fluor-impregnated samples should definitely be exposed to preflashed film (Kodak X-Omat AR) at $-70°C$, to ensure that the response of the film is linear.

Preflashing Film

To increase the sensitivity of film to low intensities of light and thus increase the linear response of the film, an investigator may elect to preflash or hypersensitize a sheet of film just prior to setting up the exposure. By providing a short burst of light, the latent image formation process is initiated; it is then completed by light photons from fluorographic exposure or from intensifying screens. Preflashing is not an effective means of enhancing latent image formation by ionizing radiation alone.

Type of Cassette

The selection of an appropriate light-tight apparatus to accommodate autoradiographic exposures is probably the least involved decision of all. Among the least sophisticated apparatuses are Kodak X-ray exposure holders, consisting of two heavy pieces of cardboard, hinged together and lined with light-tight paper, and a metal clip for locking. It is useful to sandwich this type of exposure holder between two 1-cm-thick sheets of plexiglass for two reasons: First, moisture tends to warp this type of X-ray exposure holder, which could, if not held flat, result in uneven exposure of the film; and second, cardboard provides no shielding of high-energy β particles, posing a potential health hazard to the investigator and others in the vicinity. Cardboard exposure holders may then be sealed in a plastic bag for the duration of the exposure period. Other more deluxe models, such as Kodak X-Omatic cassettes, are constructed of two vinyl-covered aluminum panels mounted in a polyurethane frame, and include permanently mounted intensifying screens. Other types of cassettes, known as vacuum exposure holders, are also available for situations in which oxygen and moisture must be excluded from the system during the exposure period. There is most certainly an exposure apparatus to fit every budget, all of which can also be used to make exposures by chemi-luminescence, should the lab wish to adapt this technology. When assembling the exposure apparatus, it is of paramount importance to immobilize a filter membrane within the cassette; even the slightest movement of the filter at any time during the exposure period will produce a fuzzy image. Immobilization by mounting the plastic-wrapped filter on a piece of Whatman paper cut to the size of the cassette is probably the simplest way. Never tape anything to an intensifying screen.

Development and Fixation

Transformation of the invisible latent image into a visible autoradiographic image occurs by the stepwise treatment of the exposed film with a photographic developer, some type of developer stop bath, a photographic fixer solution, and finally a thorough washing of the film. Each type of X-ray film has a complementary set of developer and fixer solutions recommended by the manufacturer. Films can be developed manually in a dark room, in which case the equipment requirements are minimal. Alternatively, films can be developed automatically by machine, if such an apparatus is available. If autoradiographic exposure was carried out at −80°C, it is worth allowing the film to warm to room temperature before opening the cassette, otherwise the quality of the image and the reproducibility of the assay may be compromised. Films that are developed manually should be washed in running water (16–24°C) for at least 15 min to generate the highest possible quality autoradiogram.

Autoradiography: Suggested Protocol

CAUTION As with all experiments involving isotopes, be sure to follow all guidelines for the proper containment of this material and all safety guidelines to minimize potential health hazards. If possible, it is wise to manipulate radioactive filters behind plexiglass shields.

In order to generate meaningful data, it is critical to standardize both the exposure time and the developing time; evaluation of short- and long-term exposures of the same blot may suggest very different conclusions.

Part I: Setting Up the Exposure

1. At the conclusion of all posthybridization stringency washes (Chapter 13), place the filter on a piece of Whatman 3MM paper just long enough to blot excessive buffer from the filter.

2. Wrap the filter in plastic wrap, taking care to avoid wrinkling on the side of the filter to which the probe has hybridized. Be certain to fold excess plastic wrap behind the filter so that there is no possibility that the X-ray film will get wet.

Note: If a damp or wet filter comes into contact with a sheet of X-ray film, they will form a permanent symbiosis; it will be impossible to separate the two, resulting in loss of both filter and film, and ruination of the experiment.

3. Hold a Geiger counter probe about 6 in. above the surface of the plastic-wrapped filter, to estimate the required time for autoradiography. It is not unusual to have very few "clicks" registering just above background when assaying rare mRNAs by Northern analysis or single-copy genes by Southern analysis. Significantly greater activity may be observed in the region of the filter corresponding to the molecular weight markers if the hybridization cocktail contained an aliquot of probe capable of base pairing to these sequences in addition to the probe used to hybridize to the sequences under investigation. Extremely high activity distributed over the surface of the filter may be indicative of unacceptably high background.

4. Place the plastic-wrapped filter, a sheet of XAR-5 film, and an intensifying screen together in a cassette, under safelight conditions (e.g., GBX-2 Safelight; Kodak No. 141-6627). Store at −80°C overnight.

Part II: Developing the Film

5. At the conclusion of the exposure period allow the cassette to warm to room temperature. If using an automatic X-ray film developing machine, open the the cassette, again under safelight conditions, and process the film according to the protocol for the developing machine, and then proceed to step 15. If manually developing the film, follow Steps 6–14.

6. Photographic developer[4] should be prepared in advance and stored

---

[4]**Caution:** Always avoid skin contact with photographic developer, stop bath, and fixer, and be certain to wear eye protection when working with these and all chemicals.

in a tightly sealed dark bottle because it is light sensitive and oxidizes rapidly. In this lab, Kodak D-76 developer (Kodak No. 146 4809) is used most often. Pour enough developer into a photographic developing tray so that the film is submerged when placed into the developer. The developer should be clear and colorless when poured from its storage bottle; as it ages and oxidizes, it acquires a yellow tint. As long as the developer is clear, it will continue to function, even if slightly yellowed. If the developer is brown, it is completely useless and should be discarded at once.

7. Fill a second photographic developing tray with some type of stop bath solution, or simply water.

8. Fill a third photographic developing tray with photographic fixer (Kodak No. 197 1753), which, like the developer, was prepared in advance and stored in a dark, tightly sealed bottle. Again, the volume of fixer should permit the film(s) to be submerged with ease.

9. Take note of the placement of everything in the darkroom before turning off the lights. Fumbling around in the dark is counterproductive.

10. Turn the safelight on and room lights off. Wearing gloves, open the X-ray exposure cassette and, before removing the film, bend one corner of the film so that the resulting asymmetry will show how the film was resting in the cassette. Do not remove the filter from the cassette or otherwise change its relative position, and be certain that the plastic-wrapped filter is not sticking to the film (occurs frequently).

11. Slide the film into the developer and submerge all at once; do not float the film on the surface of the developer. Begin timing the developing stage, and gently agitate the developing tray. An exposed film will generally begin to manifest bands within 60 s, though the developing time may be extend to as much as 2.5 min. When the bands begin to appear, note the elapsed time, using it as a guideline for subsequent films. This facilitates the determination of a standard developing time and gives greater meaning to data generated in different experiments.

12. At the conclusion of the developing period, quickly remove the film from the developer and place in the stop bath. Rinse the film for 20–30 s.

13. Transfer the film to the third tray containing the photographic fixer and submerge. Gently agitate this tray, and notice how, after a few moments, the film itself begins to clear, rendering the characteristic X-ray film appearance. Fixation is complete when the film has *completely* cleared, usually in less than 5 min.

14. At the conclusion of the fixation step, it is safe to turn on the room lights. Carefully transfer the developed film into a sink filled with warm water and allow it to rinse for at least 15 min in order to remove com-

pletely all of the residual chemistries. This postdeveloping wash will produce superior quality films. Hang the films to dry.

15. If very little is evident on the film after this initial exposure, then the background is likely to be very good. Set up another exposure, store at −80°C for 3 days, and then repeat the steps for developing the film. If background on the initial exposure is high, usually indicated by the appearance of an outline of the filter paper on the film, with no apparent bands, unwrap the filter paper and repeat the most stringent posthybridization wash;[5] for greater stringency, filters may be washed at 55–60°C in 0.1× SSPE, 0.1% SDS for 10–15 min. Repeat steps pertaining to wrapping the filter and setting up the exposure.

16. Follow instructions of the filter manufacturer for the removal of hybridized probe and/or storage of the filter.

Nonisotopic Procedures

At the heart of nonradioactive methodologies for nucleic acid probe synthesis and posthybridization detection is the incorporation of a label into probe molecules, which can subsequently be localized on a filter membrane with reproducible resolution and sensitivity. Among these nonradioactive methodologies are (1) chemiluminescence, a process in which light is generated by enzymatic modification of suitable substrates, and (2) chromogenic (colorimetric) detection, in which an enzymatic activity causes a color precipitate to be deposited directly on the filter at the site of the reaction. In chemiluminescence, light emission is recorded on a piece of X-ray film, and the required exposure period is significantly shorter than is necessary for probe detection by autoradiography. In chromogenic detection, the results of the experiment are assessed by direct examination of the precipitate on the surface of the filter. The obvious advantage of both chemiluminescence and chromogenic techniques is circumvention of the handling and containment of radioactive materials, not to mention the mounting disposal problems associated with their use. Moreover, detection time is often reduced from hours or days to minutes.

Numerous systems that support nonisotopic labeling and detection by chemiluminescence have been developed by many major suppliers of

---

[5]If using chemiluminescence, a follow-up stringency wash will mandate blocking the filter again, along with all of the subsequent steps for hapten recognition, binding of the enzyme conjugate, and delivery of the appropriate substrate. This constitutes a somewhat labor-intensive process, suggesting that it is best to wash the filter stringently the first time, to minimize the likelihood that additional washes will be required.

biotechnology reagents and products. In a very general sense, the mechanics of nonisotopic detection methodologies involve (1) assembly of a "molecular sandwich" by the ordered application of a series of reagents, as in the use of biotin (Bio), digoxigenin (DIG), or fluorescein (Fl), or (2) enzyme linkage directly to the probe molecules [e.g., enhanced chemiluminescence (ECL); Amersham, Arlington Heights, IL]. Further, nonisotopic detection often involves both stringent and nonstringent filter washes in moderate to high concentrations of SDS. Final washes must be performed in a clean container so as not to inhibit the activity of the required enzyme (discussed below).

Biotin

Biotin is a small water-soluble vitamin that can be readily conjugated to a number of biological molecules. For nucleic probe synthesis, biotinylation is accomplished in one of the following ways:

1. Biotinylated nucleotides. Virtually all of the classical methods for enzymatic probe synthesis, including random priming, nick translation, and polymerase chain reaction (PCR), can be performed using biotinylated nucleotides, specifically Bio-11-dUTP, which functions as a dTTP analog. The prefix Bio-11 means that the nucleotide is biotinylated and that there is an 11-atom spacer arm separating the biotin itself from the nucleoside component. Bio-21-dUTP and Bio-21-UTP are also available. By this arrangement there is no steric interference in the labeling reaction itself, and the biotin is more readily accessible to streptavidin (described below).

2. Biotinylated primers. The easiest method of biotinylation is during the actual synthesis of oligonucleotides, which can be used directly as probes or to support the polymerase chain reaction. The investigator indicates the preference for the inclusion of a 5' biotin when ordering the oligonucleotide and need not worry again about the inclusion of the biotin label in the detection component of the assay. The minimal additional cost of biotinylation during oligonucleotide synthesis is easily justified by the enormous efficiency and convenience.

3. Photoactivatable biotin acetate[6] (Forester et al., 1985; McInnes et al., 1987; Denman and Miller, 1989). This material, an aryl azide derivative of biotin, can be used to label single- or double-stranded DNA or RNA. The labeling reaction is driven by visible light and generally requires

---

[6]Photobiotin acetate, N-(5-azido-2-nitrophenyl)-N'-(3-biotinylaminopropyl)-N'-methyl-1,3-propanediamine.

20–30 min to complete. The incorporation of biotin occurs, on average, once in every 50–100 nucleotides, thereby minimizing its usefulness for labeling oligonucleotides. Probes labeled with photoactivatable biotin acetate are as stable as any other biotin-labeled nucleic acid molecules.

4. Psoralen biotin. Psoralens and their derivatives are planar tricyclic compounds that intercalate into double-stranded nucleic acids upon irradiation with 320- to 400-nm light (Schleicher & Schuell references). These molecules become covalently attached to nucleotide bases, preferentially to thymidines. To a lesser extent, single-stranded molecules, such as oligonucleotides, may also become modified. Psoralen biotin derivatives (Schleicher & Schuell) are very useful for the synthesis of biotinylated probes for nucleic acid hybridization. Unincorporated psoralen biotin conjugates are removed by simple extraction with n-butanol.

After being labeled, biotinylated probes are stable for at least 1 year, when stored at –20°C. An aliquot of probe is stored just prior to use and, if double stranded, is boiled and rapidly cooled just prior to addition to the hybridization buffer. Biotinylated probes support both chromogenic detection and chemiluminescence.

Few molecules exhibit the high affinity observed between biotin and streptavidin ($K_d = 10^{-5}\ M^{-1}$). Streptavidin is a tetrameric protein (MW 60,000) isolated from the bacteria *Streptomyces avidinii* and has four biotin binding sites. Unlike avidin (from egg), streptavidin has a neutral isoelectric point at physiological pH, with few charged groups, and contains no carbohydrate. These properties reduce nonspecific binding and background problems that would otherwise be experienced and, in so doing, enhance the sensitivity of many forms of this assay. Many biotin–streptavidin applications pertaining to nucleic acid hybridization and related techniques have been described (Leary *et al.*, 1983; Wilchek and Bayer, 1984; Hofman and Finn, 1985), and the diversity of applications increases with each passing day. Thus, one of the first steps in biotin-based detection systems is the binding of biotin with a preparation of streptavidin. In some systems, streptavidin is modified, existing as a conjugate, most often to alkaline phosphatase (AP) or horseradish peroxidase (HRP). AP dephosphorylates a substrate compatible with the mode of detection, resulting in the emission of light (dioxetane substrates) or the formation of a color precipitate (e.g., BCIP and NBT).[7] HRP, in the presence of H_2O_2, also initiates a set of chemical reactions, resulting in the emission of light (luminol mediated) or the formation of a color precipitate (e.g., 4-chloronapthal mediated).

[7]BCIP, 5-bromo-4-chloro-3-indolyl phosphate; NBT, nitroblue tetrazolium.

Digoxigenin

Digoxigenin is a derivative of the cardiac medication digitalis. It is widely used as nucleic acid label (Boehringer Mannheim, Indianapolis, IN) and supports both chromogenic and chemiluminescence detection applications. DNA probes are most efficiently synthesized by random priming (Chapter 12) with digoxigenin-dUTP, and RNA probes are synthesized by *in vitro* transcription with digoxigenin-UTP. These nucleotides are linked via a spacer arm to the steroid hapten digoxigenin. Alternatively, oligonucleotide primers may be synthesized with a DIG label at the 5^9 end. The DIG-labeled oligonucleotide can be incorporated into a PCR product or used directly as a probe for nucleic acid hybridization. In any event, the resulting DIG-labeled molecules then function as hybridization probes in much the same manner as any other probe. DIG labeling supports posthybridization detection by chemiluminescence or by formation of an insoluble color precipitate directly on the filter membrane or tissue section.

Following the posthybridization stringency washes, DIG-labeled probes are detected by enzyme-linked immunoassay, using an antibody conjugate (anti-DIG-alkaline phosphatase conjugate). As with systems involving biotin, the emission of light or the formation of precipitate is mediated by dephosphorylation of a substrate compatible with the method of detection, either colorimetric or chemiluminescence.

Fluorescein

Fluorescein is a hapten that can be incorporated by standard probe synthesis reactions. Fluorescein-labeled nucleotides, fluorescein-11-dUTP, can be used to generate continuously labeled or end-labeled probes, either by enzymatic labeling reactions or by the 5' end modification of oligonucleotides during their synthesis. Following hybridization, high-affinity antibodies prepared against fluorescein are used to localize hybridization events. The anti-fluorescein antibodies have been modified, existing as a conjugate with alkaline phosphatase or horseradish peroxidase, a chemistry that supports luminol-based chemiluminescence. It is also possible to make use of the intrinsic fluorescence of the hapten to monitor the labeling of the probe, though specialized, expensive equipment is required.

Regarding direct detection of fluorescein probes hybridized on filter papers: Fluorescein has a relatively high photobleaching rate and pH-dependent fluorescence, rendering quantitative measurement problematic. While widely used for a variety of nucleic acid-based applications including DNA sequencing, fluorescence *in situ* hybridization, and general probe synthesis, the photobleaching rate of this fluorophore is a significant disadvantage. Gradually, newer instrumentation and newer dyes are overcoming some of the intrinsic limitations of fluorescein; for now, however, filter-based detection of fluorescein probes is best served by anti-fluorescein antibodies.

Direct Enzyme Labeling

The enhanced chemiluminescence (ECL) system (Amersham, Arlington Heights, IL) features horseradish peroxidase linked directly to probe molecules through a glutaraldehyde linker arm, a linkage that is an integral part of probe synthesis itself. Thus, the enzyme is present throughout the hybridization period as well as the posthybridization stringency washes. This labeling and detection precludes the need for any type of hapten recognition, thereby supporting immediate application of chemiluminescence substrates.

Detection by Chemiluminescence

Chemiluminescence is the conversion of chemical energy into the emission of visible light as the result of an oxidation or hydrolysis reaction. Simply stated, chemiluminescence is luminescence that results from a chemical reaction. This technology provides a very sensitive, cost-effective detection alternative to many radioisotopic and fluorescence techniques, and most chromogenic detection processes (Table 1). Cur-

TABLE 1

Comparison of Attributes of Detection by Chemiluminescence versus Detection of Isotopic Decay by Autoradiography

Parameter	Chemiluminescence	Isotopic detection
Sensitivity	Very good to excellent	Excellent
Resolution	Very sharp banding	Doublets often difficult to resolve
Exposure time	Minutes to hours	Days to weeks
Detection	X-ray film	X-ray film
Potential health hazards	Relatively low	Very high
Probe stability after labeling	Most stable for 1 year	Limited by half-life
Posthybridization handling of filters	Detection buffers required	No buffers needed for exposure

rently, there are an enormous number of adaptations of this technology to conventional molecular biology techniques, including Northern analysis, Southern analysis, library screening, and DNA sequencing. While the underlying principles at work in each system are essentially the same, kits for hybridization detection by chemiluminescence differ mostly in the complexity of the system, that is, in the number of reagents and manipulations required of the investigator in order to generate a signal and, to a lesser extent, in the achievable level of sensitivity.

Localization and quantification of probe molecules by chemiluminescence in no way influence the parameters that govern nucleic acid hybridization. In fact, many of the labeling technologies that support detection by chemiluminescence are identical to those used for chromogenic detection. Differences in the detection systems are a result of alternative substrate formulations that become destabilized in the presence of the activating agent. The mechanics for measuring chemiluminescence are similar to those for measuring isotope decay by autoradiography. The relatively intense, albeit brief, emission of light generates a quantifiable image on X-ray film in a fraction of the time needed for detection by autoradiography.

Substrates for Chemiluminescence

The numerous chemiluminescence detection systems fall into what this author considers the two schools of chemiluminescence: the alkaline phosphatase school and the peroxidase school. The classical substrates used in alkaline phosphatase systems are 1,2-dioxetane derivatives, available as Lumigen PPD[9] (Lumigen, Inc., Southfield, MI), and CDP-Star (Tropix, Inc., Bedford, MA); other alkaline phosphate substrates have been developed since the inception of this technology, all of which are dioxetane derivatives. These patented 1,2-dioxetane formulations undergo rapid conversion into luminescent form in the presence of calf intestine alkaline phosphatase (Bronstein and McGrath, 1989). The resulting emission of blue light ($\approx$ 477 nm), is in direct proportion to the amount of alkaline phosphatase present (Figure 1). Lumigen PPD is available in three formulations that are compatible with the alkaline phosphatase system. The first, Lumi-Phos 530, contains a fluorescent enhancer, resulting in light emission at $\approx$ 525 nm. The second formulation, Lumi-Phos 480, is nearly the same as Lumi-Phos 530, but without the fluorescent en-

---

[8]β-galactosidase and luciferase also support chemiluminescence, though their use is most often reserved for reporter gene assays.

[9]Chemically, Lumigen PPD is 4-methoxy-4-(3-phosphatephenyl)spiro[1,2-dioxetane-3,2′-adamantane].

F I G U R E   1

Enzymatic dephosphorylation of dioxetane substrate by alkaline phosphatase leads to the metastable phenolate anion, which, upon decomposition, emits light at $\approx$ 480 nm. The light emission increases with time until a constant intensity is attained.

hancer. A third, newer compound, Lumi-Phos Plus, produces an enhanced signal, emitting at 480–490 nm. Tropix also produces chemiluminescence enhancers, and all of these formulations are compatible with most molecular biology applications.

The substrate used for peroxidase systems, hydrogen peroxide (H_2O_2), is oxidized to oxygen radicals (O_2^-) and H_2, thereby promoting the luminescent oxidation of luminol,[10] a cyclic hydrazide, or any other of a variety of related substrates (Matthews *et al.*, 1985; Thorpe and Kricka, 1986). In the luminol-based chemistry, the oxidation products are 3-aminophthalate and N_2, with the associated emission of light. The location and magnitude of hybridization is most often recorded by X-ray film exposure, much as in autoradiography.

The detection of hybridization events by chemiluminescence often represents the end of a series of steps that include probe labeling, blocking of filters, posthybridization washes, blocking the filters a second time, hapten localization, washing the filter repeatedly, and the catalysis of suitable substrates. Whereas the actual production of light is enzymat-

---

[10]5-Amino-2,3-dihydro-1,4-phthalazinedione.

ic, the method of detection prior to the introduction of AP or HRP into the system can be quite elaborate. The process begins with the labeling of probe molecules that can be localized with little difficulty (as described above) following the posthybridization stringency washes.

In general, the detection mechanics for chemiluminescence involve creating a molecular sandwich. Biotinylated probe molecules, for example, are localized by reaction with a streptavidin/AP or streptavidin/HRP conjugate. Alternatively, DIG-labeled probes are localized by anti-DIG antibodies, which exist as an enzyme conjugate. The same is true of the anti-Fl antibodies used to detected fluorescein-labeled nucleic acid probes.

Upon incubation with the appropriate substrate, enzyme activation results in the emission of visible light. To generate a record of the intensity and location of light emission, substrate-soaked filter papers[11] are quickly plastic wrapped and exposed to X-ray film, much as in macroautoradiography (see "Autoradiography," this chapter, for details). As there is no isotope involved, the resulting image might best be thought of as a chemilumigram, rather than an autoradiogram.

One common difficulty with chemiluminescence as a detection technology is the incidence of unacceptably high levels of background. This is most often due to (1) incomplete washing of filters posthybridization and during the detection steps, (2) incomplete blocking of the filter prior to addition of the enzyme conjugate, and (3) allowing the filter to dry out to any extent at some point after incubation with detection-related reagents has been initiated. In this lab, stringent posthybridization washes and strict attention to the detection protocol recommended by the manufacturers of the numerous systems we have used have produced remarkably good quality exposures, with reproducibly high levels of sensitivity.

Chromogenic Detection Procedures

Many of the chromogenic detection systems that appeared in the mid-1980s failed to gain widespread acceptance for one reason or another. Chromogenic systems seem to have lost popularity with the debut of many refined systems for solid-phase (nucleic acids immobilized on a

---

[11]Schleicher & Schuell has developed a novel method for delivery of the Lumigen in AP-based detection systems, the notion being the *controlled* delivery of the substrate. Substrate sheets, impregnated with Lumigen, are placed beneath the filter paper such that the Lumigen diffuses upward to react with the AP on the side of the filter opposite that which is in direct contact with the substrate sheet. This approach has broad applicability and is a commonly used technique in this laboratory.

filter) detection by chemiluminescence. One noteworthy exception is the Genius system (Boehringer Mannheim, Indianapolis, IN), which has become very respected as a method for nonisotopic *in situ* hybridization/detection.

Chromogenic and chemiluminescent detection procedures are often similar up to the addition of the very last reagent, the enzyme substrate. In the final step of a typical AP-based chromogenic detection, the enzyme-mediated dephosphorylation of BCIP and subsequent interaction with NBT result in the formation of a blue/purple precipitate, which is deposited directly on the filter membrane. This reaction occurs only when enzyme is present, at sites that correspond to hybridization events. There is no film exposure involved as there is no autoradiography or chemiluminescence. The chromogenic detection process may often require up to several hours to reach completion. It has been reported that one factor responsible for the low sensitivity observed in some chromogenic systems is that the activity of alkaline phosphatase is inhibited as the color precipitate is formed and deposited on the filter (Pitta *et al.*, 1990). Enzyme inhibition in this manner quickly compromises the quantitativeness of the assay, resulting in unreliable, nonrepresentative data. The final step of an HRP-mediated, chromogenic detection involves the application of either H_2O_2 and 4-chloronapthal, or TMB.

In order to reprobe filters that were previously subjected to chromogenic detection methods, very harsh washes of the filter are required, to remove the precipitate (and probe) from the filter. Each type of filter membrane and nonisotopic detection system has a set of recommended guidelines for stripping previously hybridized filters, and preparations for reprobing.

References

Bonner, W. M. (1987). Autoradiograms: ^{35}S and ^{32}P. *Methods Enzymol.* **152**, 55.

Bonner, W. M., and Laskey, R. A. (1974). A film detection method for tritium-labeled proteins and nucleic acids in polyacrylamide gels. *Eur. J. Biochem.* **46**, 83.

Bronstein, J., and McGrath, P. (1989). Chemiluminescence lights up. *Nature* **338**, 599.

Denman, R. B., and Miller, D. L. (1989). Use of photobiotinylated deoxyoligonucleotides to detect cloned DNA. *BioTechniques* **7**, 138.

Forester, A. C., *et al.* (1985). *Nucleic Acids Res.* **13**, 745.

Hofman, K., and Finn, F. M. (1985). Receptor affinity chromatography based on the avidin-biotin interaction. *Ann. N.Y. Acad. Sci.* **447**, 359.

Laskey, R. A., and Mills, A. D. (1977). Enhanced autoradiographic detection of ^{32}P and ^{125}I using intensifying screens and hypersensitized film. *FEBS Lett.* **82**, 314.

Leary, J. J., Brigati, D. J., and Ward, D. C. (1983). Rapid and sensitive colorimetric method for visualizing biotin-labeled DNA probes hybridized to DNA or RNA immobilized on nitrocellulose. *Proc. Natl. Acad. Sci. USA* **80,** 4045.

Matthews, J. A., Batki, A., Hynds, C., and Kricka, L. J. (1985). Enhanced chemiluminescent method for the detection of DNA dot-hybridization assays. *Anal. Biochem.* **151,** 205.

McInnes, J. L., *et al.* (1987). *BioTechnology* **5,** 269.

Pitta, A., Considine, K., and Braman, J. (1990). Flash chemiluminescent system for sensitive nonradioactive detection of nucleic acids. *Strat. Mol. Biol.* **3**(3), 33.

Randerath, K. (1970). An evaluation of film detection methods for weak β-emitters, particularly tritium. *Anal. Biochem.* **34,** 188.

Swanstrom, R., and Shank, P. R. (1978). X-ray intensifying screens greatly enhance the detection by autoradiography of the radioactive isotopes ^{32}P and ^{125}I. *Anal. Biochem.* **86,** 184.

Thorpe, G., and Kricka, L. (1986). Enhanced chemiluminescent reactions catalyzed by horseradish peroxidase. *Methods Enzymol.* **133,** 331.

Wilchek, M., and Bayer, E. A. (1984). The avidin-biotin complex in immunology. *Immunol. Today* **5**(2), 39.

Zagursky, R. J., Conway, P. S., and Kashdan, M. A. (1991). Use of ^{33}P for Sanger DNA sequencing. *BioTechniques* **11,** 36.

RT PCR

Rationale

Without a doubt, the revolutionary polymerase chain reaction (PCR)[1] has influenced profoundly all of molecular biology, even to the most fundamental level of how a particular experimental problem ought to be approached. RNA PCR is that technology by which RNA molecules are converted into their complementary DNA (cDNA) sequences by any one of several reverse transcriptases, followed by the amplification of the newly synthesized cDNA by standard PCR amplification. Because of the classical role of reverse transcriptase (RT) in the synthesis of first strand cDNA, RNA PCR has also become known as RT-PCR. Because of the speed and efficiency of PCR, RT PCR is now the preferred method for the synthesis and analysis of cDNA.

The utility of RT PCR is readily apparent when compared to the traditional methods of RNA analysis, the underlying theme being, of course, quantification of gene expression. A transcript must be present in a cell lysate to support reverse transcription and amplification by PCR; genes that are transcriptionally silent, therefore, are not assayable if no template exists in the reaction tube. Given the extreme sensitivity of PCR, it is now commonplace to detect and quantify transcripts present in exquisitely low abundance.

cDNA Synthesis—An Overview

The synthesis of cDNA is a central, though technically challenging, component of research involving molecular biology techniques. While the classical methods for cDNA propagation have given way to PCR, knowledge of the mechanics of traditional cDNA synthesis and its significance remains central to the proper understanding of the technology and interpretation of data. Thus, a short historical perspective follows, which becomes the foundation for the contemporary approach.

cDNA is enzymatically synthesized *in vitro,* one strand at a time, and high-quality RNA is the first and foremost requirement. Although the selection of poly(A)+ messenger RNA (mRNA) as the starting template for the synthesis of cDNA has historical significance, it is now commonplace to synthesize cDNA directly from total cellular RNA or total cytoplasmic RNA without prior poly(A)+ selection. Moreover, the reader is cautioned that, given the power of PCR, poly(A)+ selection is often neither neces-

---

[1] The polymerase chain reaction is a patented process currently owned by Hoffman-LaRoche. Use of PCR requires a limited license under their patents.

sary nor recommended. The mechanics of the poly(A)$^+$ selection often result in the loss and further underrepresentation of very low abundance transcripts. As a result, this technique, intended to enrich in favor of poly(A)$^+$ mRNA, is often counterproductive. In any event, the selection of poly(A)$^+$ mRNA from previously purified cellular or cytoplasmic RNA may be performed, if desired.

The synthesis of cDNA confers a number of advantages to the investigator wishing to characterize gene structure or expression, develop nucleic acid probes, or express proteins that might otherwise be difficult to purify. First, RNA is a naturally labile single-stranded molecule, and its conversion into more stable double-stranded DNA facilitates long-term storage of these sequences. Second, by synthesizing and cloning the resulting cDNAs from a single source, that is, a **library**, the investigator creates a method for propagating the cDNA as needed; this is greatly facilitated by the huge variety of vector molecules compatible with an equally impressive variety of hosts. Third, very short, non-full-length, nonrepresentative cDNA molecules can be used to screen (hybridize to) members of much more complex genomic DNA libraries. This approach facilitates the isolation of exon and intron sequences from the stuctural portion of genes as well as flanking 5′ and 3′ sequences. Fourth, and perhaps most importantly, the synthesis of cDNA is really the creation of a permanent biochemical record of the cell at the moment of lysis; only those genes that are transcriptionally active are candidates for inclusion in a cDNA synthesis reaction (cDNA cannot be synthesized from mRNA that is absent). In sharp contrast, genomic DNA affords the opportunity to clone all cellular sequences regardless of transcriptional status.

cDNA is synthesized directly from an RNA template; therefore cDNAs present at the end of a synthesis reaction mirror directly the RNA complexity at the beginning of the reverse transcriptase reaction. Because all nucleated cells within an organism have essentially the same genomic material, nearly identical data are expected, for example, from analysis of a genomic splenocyte library and a genomic hepatocyte library from the same organism. cDNAs from these two tissue types are certainly expected to share at least those sequences necessary for cellular viability; however, differential gene induction and repression that confers one cellular phenotype or another is best characterized by examination of cDNA complexity, as at least one component of a more thorough scientific inquiry. The take home lesson: In a given organism cDNAs from various diploid tissues differ, while genomic DNAs essentially do not.

cDNA is synthesized in a stepwise fashion, and very high-quality RNA prepared using guanidinium-based lysis buffers (Chapter 5) favors more efficient first-strand synthesis reactions. Because of the extremely sensitive nature of the cDNA synthesis reactions, it is highly advisable to invest in a cDNA synthesis kit when performing this technique for the first time. A disproportionate number of man hours and other laboratory resources are likely to be wasted trying to optimize several sets of reagents, with no success guaranteed. On a per reaction cost basis, cDNA synthesis kits are easily the most effective means for the synthesis of cDNA. To enhance the understanding of the reader, following is a description of the steps involved in the synthesis of cDNA. For the hearty scientist, a cDNA synthesis protocol follows.

First Strand Considerations

cDNA is synthesized stepwise, one strand at a time, from an aliquot of total RNA or mRNA template material (Chapters 5 and 6). Following heat denaturation to disrupt any RNA secondary structures, a short poly(T) primer, commonly oligo(dT)$_{12-18}$, is annealed to the poly(A) tail of the RNA. Because of the antiparallel base pairing that will occur between the primer and the RNA template, the primer will provide the 3′-OH group required to support 5′ → 3′ synthesis of a molecule complementary to the RNA template, hereafter known as first-strand cDNA:

mRNA

5′_____AAAAAAAAAAAAAAAAAA 3′

HO-TTTTTTTTTTT 5′

Oligo(dT) primers may also be appended at their 5′ end with overhanging sequences that will support the addition of restriction enzyme sites or other useful sequences for directional cloning.[2] These primers are often referred to as adapters:

5′_____AAAAAAAAAAAAAAAAAA 3′

HO-TTTTTTTTTTTTTTCTTAAGCC 5′

↓ reverse transcriptase

5′_____AAAAAAAAAAAAAAAAAAAAGAATTCGG 3′

3′_____TTTTTTTTTTTTTTTTTTTTCTTAAGCC 5′

<hr />

[2]Directional cloning refers to the forced ligation of an insert into a vector in one particular orientation rather than the other. While this type of cloning can be performed any time, it is necessary when the investigator plans to express (DNA→RNA→protein) a DNA sequence.

It is extremely important to realize from the onset that, if primed as described, the resulting cDNA is really poly(A)$^+$-generated cDNA, and not total cDNA, because not all mRNAs are adenylated.[3] To remedy this, an aliquot of random hexamers or random nonmers may be included in the first-strand reaction mixture. Inclusion of these oligomers results in the internal priming of first strand cDNA as well as/instead of priming from the 3' end of the template. Random priming also precludes the possibility that cDNA synthesis is primed from the 3'-most region of the poly(A) tail, which can be as long as 200–250 nucleotides: In such an event, a significant portion of the 5' end of the first-strand cDNA would consist of a lot of T's. In general, random-primed cDNA is usually longer and more representative than cDNA primed by oligo(dT) alone.

The type of enzyme required for the synthesis of first strand cDNA is properly known as an "RNA-dependent DNA polymerase," far more commonly known as reverse transcriptase, the choice of which has become a bit more complicated in the past few years, requiring some forethought. There are several types of reverse transcriptase:

AMV, from avian myeloblastosis virus;

MMLV, from Moloney murine leukemia virus;

Superscript, a modified MMLV from Gibco/BRL;

rTth, from *Thermus thermophilus* (Mn^{2+} dependent);

Taq, from *Thermus aquaticus* (Mn^{2+} dependent).

The classical enzymes are the AMV and MMLV reverse transcriptases. Most investigators prefer the AMV reverse transcriptase because it is a very processive enzyme and is generally accepted to be more efficient than MMLV. One should be aware, however, that many AMV and MMLV preparations contain background RNase H activity. This is a degradative activity that will nick the RNA component of an RNA:DNA hybrid, precisely that which forms when a DNA primer anneals to tem-

---

[3]The net effect of this approach is cDNA synthesis primed from the 3' end of the mRNA. Therefore, cDNA libraries prepared in this fashion do not contain clones corresponding to poly(A)$^-$ mRNA. Scientists should be aware that such libraries persist in circulation. When screening a library that may have been synthesized some years ago, knowledge of the method of cDNA priming may influence interpretation of the outcome of the experiment. To circumvent the potential shortcomings of oligo(dT)-primed cDNA libraries, many newer libraries are prepared by random-primed synthesis of the first-strand cDNA, and in this laboratory generated through the use of random primers *and* oligo(dT) in the same reaction tube.

plate RNA just prior to reverse transcription. The net result is cleavage of the template and concomitant instability of the primer, greatly diminishing the yield of cDNA. Superscript is an engineered enzyme that manifests no detectable RNase H activity, favoring greater efficiency of reverse transcription. One may also be tempted by the Mn^{2+}-dependent reverse transcriptase activities of the thermostable *rTth* and *Taq* polymerases in the popular one-tube, one-reaction buffer format. Although it is true that these enzymes perform very well in many applications where the complexity of RNAs to be reverse transcribed is limited, these Mn-dependent activities lack the "gung-ho" reverse transcriptase activity observed by the other enzymes when the complexity of RNAs is enormous. AMV reverse transcriptase (Promega and Boehringer Mannheim varieties) is routinely used in this laboratory with great success. A comprehensive listing of the enzymes commonly associated with cDNA synthesis and cloning is presented in Table 1.

As with most other molecular biology applications, each enzyme has an optimal set of reaction conditions (ionic strength, pH, monovalent cation, divalent cation), all of which are provided by the 10× or 5× first-strand buffer. This buffer is diluted to a working concentration of 1×, and the reaction is conducted at a temperature compatible with the specific reverse transcriptase being used. The investigator must further supplement the reaction with the primer(s) and a nucleotide cocktail consisting of an equimolar mixture of dATP, dCTP, dGTP, and dTTP, hereafter referred to as dNTP. First-strand cDNA synthesis requires anywhere from 10 min to 1 h. Virtually all of the first-strand reaction components currently used are completely compatible with the chemistries that are subsequently required for amplification of the resulting cDNAs by PCR.

Typical reaction conditions for the synthesis of the first-strand cDNA are as follows: 10 mM Tris, pH 8.3; 50 mM KCl; 5 mM MgCl$_2$; 1 mM dNTP; 1.5 μg primer; 50 U RNasin; 10 U AMV reverse transcriptase; 1 μg RNA; sterile H$_2$O to a final volume of 20 μl. Occasionally, one may wish to supplement a reaction with 10 μg/ml gelatin to stabilize the enzyme. The reverse transcriptase and RNasin are always supplied in a glycerol storage buffer and should never be subjected to repeated freeze/thaw cycles. The other reagents are quite stable at –20°C for extended periods. It is usually to the distinct advantage of the investigator to prepare a cDNA master mix when multiple cDNA synthesis reactions are to be performed. This is very much the same strategy as in the preparation of a PCR master mix, in which all but one components are pipetted together in a large volume, and then aliquoted into the individual reaction

TABLE 1

The Enzymes of cDNA Synthesis and Cloning

Enzyme	Activities	Role in cDNA synthesis
Reverse transcriptase (RNA-dependent DNA polymerase)	$5' \to 3'$ polymerase RNase H (minor)	Enzymatic conversion of RNA into cDNA. Substrates may include RNA, tRNA, rRNA, total RNA, viral RNA
DNA polymerase I (Kornberg enzyme)	$5' \to 3'$ polymerase $5' \to 3'$ exonuclease $3' \to 5'$ exonuclease	Synthesis of second-strand cDNA by the Gubler and Hoffman method ($5' \to 3'$ exonuclease activity removes nicked mRNA)
DNA Polymerase I (Klenow fragment)	$5' \to 3'$ polymerase $3' \to 5'$ exonuclease	Synthesis of second-strand cDNA by the classical method of cDNA synthesis
RNase H	Nicks mRNA component of RNA:DNA hybrid molecules	Nicking activity provides numerous primers ($3'$-OH) for synthesis of second-strand cDNA
S1 nuclease	Single-strand-specific nuclease (RNA and DNA)	Removal of hairpins in classical synthesis (enzyme has specificity for single-stranded domains)
DNA ligase	Ligation of cohesive ends; ligation of blunt ends	Seal phosphodiester backbone ($\ldots 3'$-OH+p5'X $\ldots$)
T4 DNA polymerase	$5' \to 3'$ polymerase $3' \to 5'$ exonuclease	Preparation of blunt ends on ds cDNA; exonuclease activity is 200× greater than DNA pol I
DNA methylase (*Eco*RI methylase)	Methylase	Methylation of internal restriction sites on cDNA; protects cDNA from cleavage by *Eco*RI
Terminal deoxy-nucleotidyl transferase (Terminal Transferase; T.T'ase)	$5' \to 3'$ polymerase	Addition of nucleotides to blunt ended (Co^{2+} dependent) or protruding $3'$-OH groups (Mg^{2+}-dependent)
Thermus thermophilus polymerase (*rTth* polymerase)	Reverse transcriptase $5' \to 3'$ polymerase	High-temp. first-strand synthesis (Mn^{2+}-dependent); second-strand synthesis/amplification by PCR (Mg^{2+})
Thermus aquaticus polymerase (*Taq* polymerase)	Reverse transcriptase $5' \to 3'$ polymerase	Poor RT'ase activity for first-strand synthesis; second-strand synthesis/amplification by PCR (Mg^{2+})

tubes.[4] This approach precludes procedural errors due to pipetting. To quantitatively determine the efficiency of reverse transcription, and as a basis of normalization among various samples, one may wish to consider spiking the first-strand reaction with an aliquot of [α-^{32}P]- or [α-^{3}H]dCTP; the measured amount of label incorporation is directly proportional to the mass of the newly synthesized cDNA. Alternatively, it is also possible, albeit far more cumbersome, to electrophoresis an aliquot of cDNA on a gel, stain it with SYBR Green and, using image analysis software, quantify and compare the resultant electrophoretic smears of cDNA observed among the samples on the gel. The mass and size distribution of radiolabeled cDNAs can be assessed by electrophoresis of an aliquot of the first- and second-strand synthesis reactions, coupled with autoradiography directly from the gel. Ideally, the exposure should show a lot of smearing toward the top of the gel: The closer the smearing is to the wells, the better.

In the days before PCR, at the conclusion of the first-strand synthesis, second-strand cDNA was initiated by the addition of the enzyme DNA polymerase, often without prior removal or inactivation of reverse transcriptase. This approach was useful because DNA polymerase and reverse transcriptase working together have an enhanced ability to plough through secondary structures that can compromise the length of the resulting double-stranded cDNA. Now, however, with the exception of the Mn-dependent *rTth* and *Taq* activities, failure to heat-inactivate AMV or MMLV reverse transcriptase may interfere with subsequent applications (Kawasaki, 1990).

Second Strand Considerations

The mechanics of the synthesis of first-stand cDNA today are virtually the same as they were at the inception of the technology. The greatest advances in cloning cDNA, by far, pertain to the numerous strategies deployed to support the synthesis of the second-strand cDNA. Very early methods involved alkaline hydrolysis of the original mRNA template when the synthesis of the first strand was complete, and actually relied on the transient formation of a hairpin at the 3′ end of the first strand, which, acting as primer, supported the synthesis of the second-strand cDNA, us-

[4]For large-scale investigations in this laboratory, 1–2 ml of cDNA master mix is prepared (excluding the reverse transcriptase and RNasin) and stored at –20°C. Using this strategy, the investigator has access to identical chemistries for follow-up and repeat applications, thereby minimizing at least one potential source of procedural error.

ing the first strand as a template. The resulting double-stranded molecule was joined by a single-stranded hairpin, requiring an S1 nuclease digestion (Chapter 18) to remove the hairpin for subsequent vector insertion (Efstratiadis *et al.*, 1976). This was a terribly barbaric approach by today's standards, resulting in the loss of sequences corresponding to the 5′ end of the mRNA, as well as exposing the newly synthesized cDNA to the extremely aggressive and often uncontrollable degradative activity commonly associated with the S1 enzyme.

Okayama and Berg (1982) subsequently developed a method by which an oligo(dT)-tailed vector, rather than oligo(dT)$_{12-18}$, was used to prime the synthesis of the first strand. The resulting RNA:DNA was circularized by linker ligation. The real innovation in this technique involved the use of a mixture of RNase H and DNA polymerase I to remove the mRNA template and synthesize the second-strand cDNA. In this approach, the RNA component is nicked by RNase H and degraded by the 5′→3′ exonuclease activity of DNA polymerase, and at the same time the 5′→3′ polymerase activity of DNA polymerase displaces the RNA and synthesizes second-strand cDNA, using the 3′-OH groups from the nicked RNA as primers. Finally, as DNA polymerase is unable to seal the nicks made by RNase H, DNA ligase is introduced to repair the backbone of second-strand cDNA to maximize its stability during the subsequent cloning steps. A major improvement to this general approach came with the procedure of Gubler and Hoffman (1983), in which oligo(dT)$_{12-18}$ was once again used to prime the first cDNA strand, followed by the synthesis of the second cDNA strand using the Okayama and Berg RNase H, DNA polymerase I, DNA ligase cocktail:

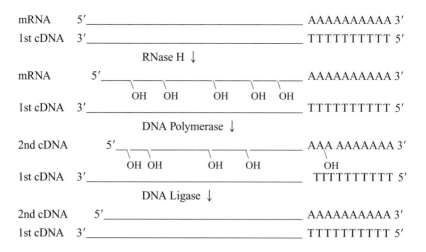

The Gubler and Hoffman method remains the standard format for cDNA synthesis, PCR notwithstanding. Because cDNA synthesis in this manner does not prepare perfectly blunt ends, double-stranded cDNAs are generally incubated for a short time in the presence of T4 DNA polymerase; "polishing" both ends of the molecule favors efficient ligation into a vector.

Upon completion of the first- and second-strand cDNA synthesis reactions, the ends of the resultant molecules are modified to yield cohesive ends compatible with similar end structure in a suitable cloning vector, such as plasmids, cosmids, bacteriophages, or phagemids. The generation of cohesive termini[5] can be accomplished through direct linker ligation

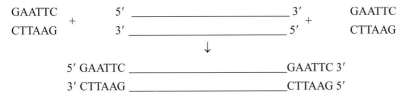

In the example above, the newly added restriction enzyme sites are cleaved with the correct restriction enzyme to yield cohesive (sticky) ends.[6] When directional cloning is desired, it may also be accomplished through the use of adapters, as described above.

Virtually all cDNA synthesis and cloning systems provide the user with linkers and/or adapters to facilitate ligation into a vector of choice. When using PCR, the primers must be designed in advance to provide an end structure compatible with the intended method of cloning.

Ligation Considerations

In every ligation reaction, maximum ligation efficiency is favored when the molecules to be ligated, usually vector and insert, are present in

[5]In addition to linkers and adapters, a very old method for vector:insert ligation known as homopolymerically tailing still appears in the literature, albeit quite infrequently. In this procedure, double-stranded cDNA is tailed with dGTP using terminal deoxynucleotidyl transferase (terminal transferase, for short); the vector, in turn is also tailed, but with dCTP. The resulting poly(dC) and poly(dG) tracts, obviously complementary, are annealed and then ligated together with DNA ligase.

[6]If the restriction enzyme used to generate cohesive ends also cuts an internal recognition site, the cDNA will be diminished in size. To preclude this possibility, one strategy is to methylate the cDNA (e.g., with *Eco*RI methylase) prior to linker ligation and restriction enzyme digestion. In this example, methylated *Eco*RI sites in the body of the cDNA are resistant to cutting by *Eco*RI.

a 1:1 molar ratio. In the synthesis of a library, a near 1:1 ratio of vector to insert will favor completeness in cloning. However, when simply attempting to clone, for example, a single type of purified PCR product, ligation ratios are much less of a concern: Following bacterial transformation and plating on selective media (e.g., 50 µg/ml ampicillin or 12.5 µg/ml tetracycline), a single colony should, at least theoretically, contain the vector + insert when standard cloning methodologies are employed. Since false positives do occur, it is best to have at least a few candidate clones to examine. The take home lesson: A perfect 1:1 ratio is often not a critical parameter in the simple cloning applications described here.

In the event that the generation of recombinant clones is problematic, ligation optimization is calculated by determining the number of picomole (pmol) ends involved in the reaction, and then increasing or decreasing the mass of one or the other to achieve 1:1 vector:insert ends. One must calculate pmol ends for both the insert *and* the vector, and then equalize them in the ligation reaction

$$\text{pmol ends} = \frac{\mu\text{g DNA}}{660 \times \text{No. of bases}} \times 2 \times 10^6,$$

where

> pmol ends = number of molecule ends available to participate in the ligation reaction
>
> µg DNA = mass of the DNA (either the insert or the vector)
>
> 660 = average molecular weight of a nucleotide base pair
>
> No. of bases = the length of the DNA (either insert or vector)
>
> 2×10^6 = conversion factor from µg (10^{-6}) to pmol (10^{-12}) and the fact that each molecule has two ends capable of ligation.

For example, 575 ng of a DNA sequence that is 852 bp long contains 2.045 pmol ends:

$$\text{pmol ends} = \frac{0.575\mu\text{g}}{660 \times 852} \times 2 \times 10^6.$$

Assuming, for cloning purposes, that this DNA is intended to be an insert, the investigator would also need to calculate the number of pmol ends of the intended vector, ensuring an equimolar concentration of each in the ligation reaction.

This strategy is acceptable when an insert of uniform length is intended to be cloned, though it is vastly more complicated when a mixed population of cDNAs, all of which are different sizes, is to be cloned. In this case, the most worthwhile strategy is to vary the insert:vector ratio in a series of ligations, an example of which might be

1:0.25, 1:0.5, 1:0.75, 1:1, 1:1.25, 1:1.5, 1:1.75, 1:2.

At the conclusion of the ligation reaction, the now recombinant molecules are used to transform competent *Escherichia coli* cells,[7] most often using a chemical transformation process and ideally with competent cells that have been purchased (e.g., Promega).

PCR—An Overview

Unquestionably, the widespread implementation of the polymerase chain reaction has fundamentally influenced all aspects of basic research in the life sciences. The popularity of many classical labor-intensive methods of vintage molecular biology (circa 1985) have begun to fade in favor of the faster, more discriminating method of what is precisely defined as a "primer-mediated, enzymatic amplification of specific genomic or cDNA sequences" (Saiki *et al.*, 1985; Mullis *et al.*, 1986; Mullis and Faloona, 1987). That minute quantities of template material can be amplified several millionfold in a matter of a few hours has unleashed enormous potential in the study of infectious disease, gene mutations, ontological relationships among members of gene families, and transcriptional activity in cells and tissues, using RNA as a parameter of gene expression. One should not, however, consider abandoning the classical methods entirely, because of several intrinsic shortcomings of the polymerase chain reaction, namely:

1. In order for the reaction to function properly (at all), the investigator must have direct or indirect knowledge of appropriate primer sequences.

---

[7]cDNA library construction is most efficient when cloning into λ bacteriophage rather than into plasmids. This is because of the huge numbers of clones that must be screened, the much greater efficiency by which screening in performed in λ, and the much higher incidence of false positives when screening bacterial colonies as opposed to bacteriophage plaques. When screening a library, one generally subclones (transfers the insert from one vector into another) after the number of potential clones has been reduced from hundreds of thousands to just a dozen or so. The subsequent manipulations needed to further characterize the insert carried by the vector are performed with much greater ease when working with plasmids.

2. The primers, when base-paired to the template, must not have mismatches at their 3'-OH ends.
3. The template for the reaction (i.e., the target) must be clean enough to facilitate primer annealing and subsequent elongation.
4. The various components of the reaction cocktail must be optimized to yield unambiguous data.

It is the opinion of this author that the polymerase chain reaction should not be viewed as a cure-all for the numerous difficulties that frequently frustrate both seasoned and novice molecular biologists. Further, one should not attempt implementation of PCR to the exclusion of other molecular methods, until one is reasonably proficient with basic isolation, handling, and storage methods for nucleic acids. More than one student of molecular biology has mistakenly believed that the immediate implementation of PCR will solve all problems in the lab. False.

The mechanics of the polymerase chain reaction have been presented and discussed in resources too numerous to list completely; for excellent reviews the reader is directed to Dieffenbach and Dveksler (1995), McPherson and Hames (1995), and Innis *et al.* (1990). For the convenience of the reader, a brief overview of PCR is presented here.

PCR is dependent, first and foremost, on the primers included in the reaction. What then, exactly, is a primer? In this context, a primer is a short, single-stranded DNA sequence, that is, an oligonucleotide, which has been artificially synthesized to the specifications of the investigator. The essential parameters of primers are:

1. They are short, usually less than 30 bases, with 20–25 bases being optimal.
2. They are single stranded.
3. They manifest a free 3'-OH group, available to support the addition of nucleotides by a DNA polymerase. The 5' end of a primer, in contrast, may carry all types of exotic modifications, without compromising its specificity, and need not be perfectly base-paired to the template.

A pair of primers is required to support amplification of a DNA sequence by PCR. The primers must be designed so that they will base pair to opposite strands of the heat-denatured target (template), in such a way that their respective 3' ends face each other. The template is the experimental genomic DNA or cDNA that the investigator wishes to amplify, and it must show at least some degree of complementarity with the primers. Thus, the anticipated PCR product is that which is framed by the 5' termini of a primer pair.

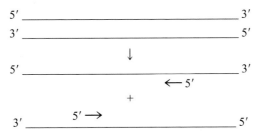

Depending on the way the primers are designed, the polymerase chain reaction will support a variety of amplification formats, including symmetric PCR (Saiki *et al.*, 1985), asymmetric PCR (Gyllensten and Erlich, 1988), inverse PCR (Ochman *et al.*, 1988; Silver and Keerikatte, 1989; Triglia *et al.*, 1988), RNA PCR (Veres *et al.*, 1987), and rapid amplification of cDNA ends (RACE) PCR (Frohman *et al.*, 1988), to name but a few variations on this theme.

The polymerase chain reaction is typically divided into cycles, each cycle consisting of three components, sample denaturation, primer annealing, and primer extension. The successive application of these functionally distinct components, typically in the form of 25–30 cycles, results in the exponential amplification of template molecules. The three distinct components of each cycle are as follows.

1. **Denaturation,** the goal of which is strand separation to facilitate hybridization of primers to the template in the next step. The polymerase chain reaction begins with high-temperature denaturation of all sequences contained within a reaction tube. This includes denaturation of double-stranded template material as well as any secondary structures that may have formed within/between primers.[8] Complete denaturation is an absolute requirement if any PCR product is to be observed:

5'_____3'
3'_____5'

↓ heat

5'_____3'

+

3'_____5'

The temperature and time required for denaturation is a direct function of the complexity of the target as well as the G+C content, if known. In

---

[8]See "Primer Design" for suggestions on how to avoid intra- and intermolecular base pairing of primers.

general, genomic DNA samples require prolonged denaturation to ensure complete dissociation of the template into its constituent single strands. Failure to produce PCR product from a genomic template is very often the result of incomplete target denaturation from the start, which greatly impedes the efficiency of subsequent cycles. Typically, an initially denaturation is conducted at 94°C for as little as 1 min for short, linear DNA sequences, and as long as 10 min for higher complexity genomic DNA.

2. **Primer annealing,** the goal of which is to hybridize primers to complementary sites on the template to provide a 3′-OH substrate, needed to support DNA synthesis in the next step. The second component of each PCR cycle is the annealing or hybridization of a set of primers chosen by the investigator, the 5′ ends of which frame or define the size the size of the PCR product in Cycle 1 (Cycle 2 for RT PCR). For annealing to occur, the temperature inside the reaction tube cannot exceed the melting temperature (T_m) of either of the oligonucleotide primers. For this reason, annealing is usually attempted at least 1–2°C below the lowest calculated T_m of the primers involved:

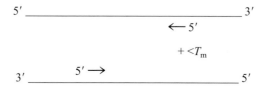

Moreover, primer annealing will not take place unless the region of the target material containing sequences complementary to the primers has been sufficiently denatured.[9] For these reasons, the annealing temperature is the most variable cycling component in PCR, and is a direct function of the primer sequences. Typical annealing temperatures range from 42°C to 65°C for 0.5–2 min.

3. **Primer extension,** the goal of which is to synthesize the PCR products by primer extension along the template molecule to which the primer are base-paired. This is the heart of PCR product synthesis. The inclusion of a thermostable DNA polymerase, for example, *Taq,* from *Thermus aquaticus,* supports primer extension at elevated temperatures. These enzymes can withstand the high temperatures associated with the denaturation component, as well as the primer extension component, most often 68–72°C:

[9]See "Optimization Procedures," for suggestions.

Further, thermostable polymerases make fewer mistakes at elevated temperatures; not all thermostable enzymes, however, have a proofreading activity. Because the error rate of the *Taq* polymerase can be significant, many of the newer protocols call for the use of enzyme blends, that is, more than one thermostable enzyme, functioning in a permissive manner, to increase the yield and fidelity of the amplification process.

In summary, the polymerase chain reaction is little more than the repeated application of the three steps described above: denature, anneal, extend. While the PCR product yield is quite unimpressive for the first several cycles, the later cycles (Cycle 20 and beyond) are responsible for generating the bulk of the usable product (Table 2). Regardless of the name given to a PCR-based technique, the fundamental replication of these steps will always apply.

RT PCR—General Approach

It is easy to see how PCR amplification is accomplished starting with genomic DNA, cDNA, or vector DNA as a template. The same principles apply with respect to the amplification of RNA, with an additional step required to convert RNA "templates" into cDNA first. Because of the extremely complex nature of any assay designed to provide a clearer understanding of gene regulation, conditions governing the success or failure of RNA-based PCR, more than any other technique in molecular biology, require careful forethought coincident with thoughtful empirical determination.

Historically, the synthesis of cDNA has been greatly concerned with the synthesis of full-length or "representative" cDNA molecules, and the extent to which this can be accomplished is directly related to the quality (and quantity) of RNA starting material and, to a lesser extent, to the ability of reverse transcriptase to plough through secondary structures or RNA hairpins. Given the unsurpassed sensitivity and amplification potential of the polymerase chain reaction, the quantity of starting RNA and its degree of purification are less important in terms of its ability to be amplified and identified by PCR than is so absolutely necessary for more traditional cloning methods. Moreover, because PCR primers "frame" the

PCR Theoretical Exponential Amplification

Cycle number	Relative product mass	Base 2 exponent (2^n)
1	2	1
2	4	2
3	8	3
4	16	4
5	32	5
6	64	6
7	128	7
8	256	8
9	512	9
10	1,024	10
11	2,048	11
12	4,096	12
13	8,192	13
14	16,384	14
15	32,768	15
16	65,536	16
17	131,072	17
18	262,144	18
19	524,288	19
20	1,048,576	20
21	2,097,152	21
22	4,194,304	22
23	8,388,608	23
24	16,777,216	24
25	33,554,432	25
26	67,108,864	26
27	134,217,728	27
28	268,435,456	28
29	536,870,912	29
30	1,073,741,824	30
31	2,147,483,648	31
32	4,294,967,296	32
33	8,589,934,592	33
34	17,179,869,184	34
35	34,359,738,368	35

Note. The theoretical accumulation of PCR product for any number of cycles is 2^n, where n = number of cycles completed. In reality, however, the actual amplification of template is much less, due to the depletion of available primer and dNTPs, as well as the cumulative negative effect of repeated sample denaturation (92–95°C) on the polymerase. Product accumulation is exponential in the earlier cycles and subsequently demonstrates what is known as a plateau effect in later cycles, meaning that product accumulation continues, though in a nonexponential manner, as long as functional polymerase remains.

domain that will be amplified, this cloning method is far more tolerant of partially degraded RNA than traditional cDNA synthesis and Northern blot analysis.

Clearly, RT PCR is a two-step process. First, purified RNA is incubated with reverse transcriptase (MMLV, AMV, Superscript, or Mn^{2+}-dependent *rTth*) and an appropriate primer.[10] This reaction results in the production of an RNA:DNA hybrid and is virtually identical to methods used for traditional first-strand cDNA synthesis since the days of yore. In the second reaction, the mechanics of the polymerase chain reaction are responsible for the displacement of the original RNA template, as well as the synthesis and amplification of the second strand: At the onset of PCR Cycle 1 the heat-denaturation step and quasi-alkaline environment produced by the PCR reaction buffer cause RNA:DNA strand denaturation and RNA alkaline hydrolysis, respectively.[11] This clears the way for sequence-specific primer annealing in Cycle 1. Of course, because only one strand of cDNA has been synthesized thus far, only one of the PCR primers will find a template with which to base pair. The primer extension component of Cycle 1 will, in turn, generate the second-strand cDNA. After Cycle 1 is completed, the amplification of both strands of the newly synthesized cDNA will proceed in a symmetric fashion.

Because of the additional steps involved in RT PCR, there is greater opportunity for amplicon contamination of the stock solutions and downstream reactions, compared to the normal peril associated with non-RNA PCR. In order to address this issue, the investigator is well advised to perform the reverse transcriptase reaction in the sample preparation area, as opposed to the PCR amplification area, if such a designation has been made in the lab.

---

[10]In this reaction, which is first-strand cDNA synthesis, one may utilize traditional downstream primers, such as oligo(dT)$_{12-18}$, random primers, RNA-specific sequences, or nontraditional primers, such as those described in Chapters 16 and 17. The net result is, of course, the synthesis of very stable, single-stranded DNA molecules capable of acting as a template for PCR.

[11]RT PCR offers several advantages over traditional methods of cDNA synthesis, the most noteworthy of which are speed and sensitivity. Unlike the still popular RNase H-mediated method of second-strand cDNA synthesis (Gubler and Hoffman, 1983), alkaline hydrolysis of the original RNA template and immediate synthesis of the second-strand cDNA eliminates various precipitation steps and other tedious manipulations. By designing fairly nonspecific primers, e.g., oligo(dT) and combinations of random hexamers or nonomers, entire libraries can be synthesized using PCR; these can then be archived for years for gradual, complete characterization. In contrast, a single sequence-specific primer of adequate length is all that is necessary to support the amplification of a unique PCR product using a method known as anchor PCR, described later in this chapter.

Primer Design

As stated above, PCR requires a pair of properly designed primers. One common source of confusion pertains to the nomenclature used to distinguish the two primers required in the reaction. The **downstream primer** is that which base pairs to the 3′ end of the RNA template, thereby supporting the synthesis of first-strand cDNA; this primer is also referred to as the 3′ primer and as the reverse primer. The **upstream primer** base pairs to the newly synthesized first-strand cDNA; this primer is also referred to as the 5′ primer and as the forward primer. By convention, only one strand of the DNA, the coding strand, written 5′ → 3′, is published. When designing primers from a published DNA sequence, the reader is expected to know that there is indeed another strand base-paired and antiparallel to that which is published. The rules for designing the primers are:

1. The upstream primer is the same sequence as the strand that is published.
2. The downstream primer is the inverted complement of the strand that is published.

To ensure that the correct primer sequences have been selected, it is useful to write out and label both strands of the published sequence and ensure that the primers base pair (antiparallel) to opposite strands and that their respective 3′-OH groups point toward each other, thereby framing the sequence intended for amplification.

Similar rules apply in the design of primers for RT PCR: If an mRNA sequence is known, recall that it is the same sequence as the coding strand of the DNA from which it was transcribed, except for the inclusion of uridine, rather than thymidine, in the transcript. Thus, the downstream primer for RT PCR, if not oligo(dT)$_{12-18}$, should be the inverted complement of a region of the RNA 5′ (upstream) from the poly(A) tail; the further upstream that this primer base pairs, the greater the likelihood of cloning sequences that correspond to the 5′ end of the mRNA. The upstream primer will be the same sequence as the 5′ region of the mRNA, except that DNA primers will contain thymine instead of uracil.

Oligonucleotide primer sequence information can come from previously published sequencing data, knowledge of the nucleotide sequence of a related gene in either the same or a different species, or knowledge of the amino acid sequence of the protein encoded by the gene of interest. The problem with the latter approach is the degeneracy of the genetic code: Some amino acids have multiple codons.

When designing primers based on peptide sequence information, one strategy is to examine the peptide for clusters of amino acids with only one or two possible codons. The single-codon amino acids are methionine and tryptophan. The other amino acids all have two or more possible codons (Table 3). In the case of multiple codons, the investigator may have knowledge of preferential usage of one codon over another in a particular species. One must realize, however, that the occurrence of five or six single-codon amino acids is quite rare, and genes so endowed have, in all probability, already been cloned! Without direct nucleic acid sequence information, the likelihood of designing a definitive oligonucleotide, for use as a primer or as a probe, is slim. It is possible, though, to use several oligonucleotides simultaneously for annealing to a template or target (a technique known as multiplexing), or to use only one oligonucleotide sequence at a lower stringency, for instance, by lowering the hybridization temperature. While these strategies reduce the concentration of the correct (complementary) oligonucleotide sequence and favor semi-nonspecific hybridization, respectively, the important point is that the desired sequence is amplified, albeit among false positives, which can be eliminated in the later stages of sequence identification. At lower stringency, not all of the bases of a primer will necessarily base pair with the template; mismatches (individual bases that are not hydrogen bonded to the template) may abound at lower temperatures, though the primer may continue to anneal enough to support elongation.

The characteristics of useful primers for PCR-associated applications follow.

1. Ideally, primers should be 20–25 bases long in the region that base pairs to the template. Longer primers may be used, though their thermodynamic behavior becomes less predictable. With shorter primers, one runs the great risk of promoting nonspecific primer annealing, because as a primer gets shorter, its discriminatory abilities diminish. A good rule for determining minimum primer length while maintaining primer specificity is to solve for the exponent,

$$4^X > Y,$$

where

4 = the number of possible nucleotides to make up the primer

X = the length of the primer

Y = the genome size of the organism for which the primers are being designed.

Codon Utilization Table for Degenerate Primer Design

Amino acid	DNA codons	RNA codons	Abbreviation	Symbol
Methionine	ATG	AUG	Met	M
Tryptophan	TGG	UGG	Trp	W
Asparagine	AAT AAC	AAU AAC	Asn	N
Aspartic acid	GAT GAC	GAU GAC	Asp	D
Cysteine	TGT TGC	UGU UGC	Cys	C
Glutamine	CAA CAG	CAA CAG	Gln	Q
Glutamic acid	GAA GAG	GAA GAG	Glu	E
Histidine	CAT CAC	CAU CAC	His	H
Lysine	AAA AAG	AAA AAG	Lys	L
Phenylalanine	TTT TTC	UUU UUC	Phe	F
Tyrosine	TAT TAC	UAU UAC	Tyr	Y
Isoleucine	ATA ATC ATT	AUA AUC AUU	 Ile	 I
Valine	GTA GTC GTG GTT	GUA GUC GUG GUU	 Val	 V
Proline	CCA CCC CCG CCT	CCA CCC CCG CCU	 Pro	 P
Threonine	ACA ACC ACG ACT	ACA ACC ACG ACU	 Thr	 T
Alanine	GCA GCC GCG GCT	GCA GCC GCG GCU	 Ala	 A
Glycine	GGA GGC GGG GGT	GGA GGC GGG GGU	 Gly	 G
Arginine	AGA AGG CGA CGC CGG CGT	AGA AGG CGA CGC CGG CGU	 Arg	 R
Leucine	CTA CTC CTG CTT TTA TTG	CUA CUC CUG CUU UUA UUG	 Leu	 L
Serine	TCA TCC TCG TCT AGT AGC	UCA UCC UCG UCU AGU AGC	 Ser	 S
Nonsense codons	TAA TAG TGA	UAA (Ochre) UAG (Amber) UGA (Umber)	Stop Stop Stop	

For example, the size of the human genome is about 3×10^9 bp. Thus, $4^X > 3 \times 10^9$. By solving for X, it is apparent, at least theoretically, that a 17-base oligonucleotide (a 17-mer) is long enough to find only the DNA sequence to which it is targeted, and not another due to the random occurrence of bases; after all, 3×10^9 is a lot of sequence. In practice, a 17-mer remains in what should be thought of as the gray area, where a primer may or may not impart the necessary level of sensitivity. Therefore, a bit longer is better.

2. In a PCR reaction, it is far more important to match primers by T_m than by length or, to an extent, by sequence. T_m differences of more than a few degrees, especially among shorter primers, are likely to have a profound negative effect on the outcome of the reaction in terms of specificity and yield.

3. Mismatches at the 3' end are fatal. Mismatches that involve the 3'-most nucleotide and often the penultimate nucleotide will not support primer extension. Polymerases require a 3'-OH perfectly base-paired to the template, as a substrate for the addition of nucleotides. When primers are designed by trial and error, there is often no way of knowing if the 3' end is perfectly base-paired to the template.

4. The 5' end of a primer can have rather exotic overhangs when the investigator wishes to generate a PCR product with new sequences on one or both ends. Useful sequences to add include, but are not limited to, DNA sequencing promoters, RNA transcription promoters, restriction enzyme sites, and/or sequences that allow the PCR product to be recognized and amplified by an unrelated primer pair for a downstream application, such as competitive PCR (Chapter 17).

5. Each primer sequence should be examined to ensure that it does not exhibit intramolecular complementarity; if it does, then the primer could actually base pair to itself. The formation of what is best thought of as a primer hairpin causes a marked reduction in free primer concentration, thereby drastically reducing the product yield. This is more of a problem than most people might be led to believe. Due to the extremely rapid hybridization kinetics associated with oligonucleotides, it should be intuitive that the primer will base pair to the spatially closest sequence to which it is complementary; if the 5' end of a primer is complementary to its own 3' end, then hybridization is sure to occur. Moreover, intramolecular base pairing can promote nonspecific base pairing to the true template due, in part, to the presence of a few free nucleotides at the 5' and 3' ends of the primer.

6. Each primer pair should be examined to ensure that the primers are not complementary to each other near their 3' ends. When primers base

pair to each other, the synthesis of primer–dimer is favored, detracting greatly from the synthesis of the intended PCR product.

Primer 1   5'_____ 3'

3'_____ 5'   Primer 2

↓

5'_____ 3'
                                    Primer–Dimer
3'_____ 5'

The incidence of primer–dimer is greatest when the primers are GC-rich at their 3' ends. If primer scrutiny suggests that primer–dimer may occur, changing the length of one or both primers by as few as one or two nucleotides will usually rectify this problem.

7. Primers should be designed with average GC content. In humans, for example, the GC content of the genome is approximately 41%. In this laboratory, primers with 40–55% G+C are considered average. This parameter is important because guanine and cytosine base pair with three hydrogen bonds between them, while adenine and thymine base pair with only two. This will preclude having one primer interact with the template more efficiently than the other primer.

8. A particular primer should be designed with an average distribution of nucleotides, as opposed to having a disproportionate density of guanosine and cytidine at one end of the primer, with adenosine and thymidine at the other. This will preclude having one end of the primer base pair with and/or melt from the template with a significantly different efficiency than the other end.

T_m Considerations

It should be clear from the preceding suggestions that the molecular nature of each primer alone is, obviously, very important; how the primers behave together in a reaction tube is of equal importance. One of the most critical aspects of designing useful primers is promoting a proper balance between template specificity, thermodynamic stability when base-paired to the template, and capacity of one primer to function with the other(s) to support PCR. The probable collaborative behavior of one or a pair of oligonucleotide primers is best described in terms of the T_m of each primer involved. The T_m is that temperature at which 50% of the possible annealing events between primer and template have occurred and 50% have yet to occur; thus the T_m is best thought of as an equilibrium temperature at which one primer base pairs to the template for every primer that melts or dissociates from the template.

T_m is a direct function of the length of the primer, the base sequence of the primer, and the concentration of salt (Na^+) in the reaction. One rapid method for determining T_m is to assign 2°C for each adenine and thymine and 4°C for each cytosine and guanine:

$$T_m = 2(A+T) + 4(G+C).$$

This method is reliable for 11-mer to ≈28-mer primers. When the T_m is determined to be greater than 68°C, this calculation is no longer reliable. This method is very widely used to estimate the probable T_m for each of a pair of primers. Conveniently, when primers are synthesized commercially, the specification sheet that accompanies them generally includes the T_m calculation, usually at 1 M NaCl and at 50 mM NaCl. This gives the investigator some flexibility in PCR itself as well as the potential to use either or both of the primers as hybridization probes for library screening, Southern analysis, and the like. When designing primers, it is best to match them closely based on T_m rather than by length.

The T_m defines an annealing equilibrium point; it should be fairly obvious though, that one does not wish to saturate *only* 50% of all complementary target sites, but rather all of them. Thus, knowledge of the T_m of each primer allows the calculation of the temperature of annealing (T_a) in the second step of each PCR cycle. To promote the greatest specificity of primer annealing, one may wish to use an annealing temperature of only a few degrees below the T_m; for example, $T_a = T_m - 5°C$. If annealing 5°C below the T_m fails to generate product, one should consider repeating the experiment at $T_m - 10°C$, and so forth. Incrementally decreasing the annealing temperature is most often required when the one or more mismatches exist between primer and template.

Optimization Procedures

The polymerase chain reaction has become as commonplace in the molecular biology laboratory as a set a micropipettors. Without a doubt, it has revolutionized the way many questions pertaining to cell biology are approached. The reproducibility of PCR-based assays is highly dependent on optimization of the reaction; optimization is frequently required whenever a new set of primers is utilized.

There are numerous potential difficulties, some obvious, others not so obvious, that the investigator must adddress. Several of these are listed below, and are best taken to heart. Succinctly, the efficiency and specificity of PCR are influenced by each component of the reaction.

1. Master mix. The standard method for sample preparation involves the preparation of a large volume of reaction mix, containing all but one of the reaction components. This master mix is then aliquoted into individual reaction tubes, into each of which the one variable component is added. Master mix preparation minimizes tube to tube variations attributable to pipetting technique. A typical PCR reaction consists of 10 mM Tris–Cl, pH 8.3 (at 20°C); 1.5 mM MgCl$_2$; 50 mM KCl; 200 μM dNTP; 0.5 μM of each oligonucleotide primer; 2.5 U *Taq* polymerase/100 μl reaction; 10^2–10^5 copies (< 1 μg)/100 μl reaction. Each of these components is subject to change, though the parameters listed here represent a good place to begin.

2. Divalent cation. Pay particular attention to the final magnesium concentration when using a 10× PCR reaction buffer without MgCl$_2$, requiring addition by the investigator. Mg^{2+} binds the dNTPs and the template (phosphodiester backbone) quantitatively, thereby exerting a measurable effect on the efficiency and specificity of the reaction.

3. Template quality. PCR amplifies segments of DNA that lie between the primers. Thus, the reaction is a bit more tolerant of partially degraded DNA and RNA than some of the more classical techniques in molecular biology. Templates that are severely degraded are unlikely to support the amplification of anything. Wherever possible, always check template quality by minigel electrophoresis before proceeding to PCR.

4. Primers. If newly designed primers fail to generate a product when annealing close to the T_m, reduce the annealing temperature incrementally, and do so at least twice. If the primers continue to fail, do not waste time with primers that do not work: Select alternative primer sequences.

5. Reaction components. Test the components of each new PCR kit before using it with precious template. Many PCR kits contain a test template and control primers to show that the reagents are functional. Failure to generate a product can just as easily be due to compromised reagents as to poor primer design or degraded template.

6. Sample preparation areas. While it is clearly desirable to have adequate laboratory square footage available to maintain separate rooms for template preparation, reagent preparation, and PCR cycling, this is not always possible in smaller laboratories. Instead, to minimize the inherent dangers of amplicon contamination (PCR carryover), it is best to at least define areas within the lab where template alone will be handled, as well as areas exclusively for reagent preparation and PCR cyling. Further, investigators should be very cognizant of which tubes are moved into each area, and in which direction. In general, three areas that pertain exclusively to the preparation and handling of PCR-related items should be desig-

nated. The movement or flow of materials should always be from the area with the cleanest or most highly purified materials (reagent preparation), to a lesser purity area (template preparation), to the final area containing the least pure or dirtiest materials (the location of the thermal cycler and the gel boxes/plate readers for product analysis). Unidirectional movement of sample components related to PCR as suggested by the flowchart in Figure 1 is likely to minimize the incidence of carryover contamination. The investigator is well advised that equipment such as microcentrifuges, micropipettors, and even lab coats should not leave their designated areas. Failure to pay careful attention to this aspect of PCR technology, both for DNA and RNA templates, is likely to result in horrifying contamination problems, if not sooner, then later.

7. Positive controls. Limit the number of positive controls; these can easily contaminate stock solutions and cause unmitigated havoc in various downstream applications.

8. Negative controls. Each set of reactions and each gel run should contain at least one negative control, the nature of which is dependent on the experiments being performed.

9. Aerosol contamination. The use of positive displacement micropipet tips (Figure 2) is strongly recommended to prevent micropipettor-mediated contamination. Alternatively, one may wish to consider the implementation of aerosol-resistant barrier tips. Aerosol-borne template can easily get into stock solutions and nonintended reactions simply because a contaminated instrument was used. Always practice safe pipetting.

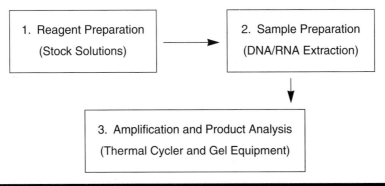

FIGURE 1

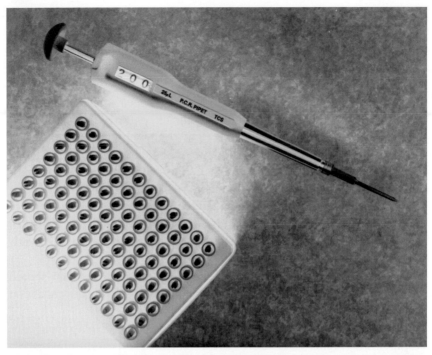

FIGURE 2

Positive displacement micropipette tips. a. The microsyringe-like nature of positive displacement tips precludes aerosol formation which might otherwise contaminate the micropipettor itself, resulting in amplicon chaos. Courtesy of Tri-Continent Scientific, Inc. (Grass Valley, CA).

10. Thermal cycler. Always optimize PCR conditions and perform subsequent reactions in the same thermal cycler. Because of the enormous variation in efficiency among the different models, reactions thought to be optimized in the older machines suddenly manifest new PCR products in the newer machines. These "new" bands were, in fact, probably being generated in the lesser efficiency machines, but their relatively low mass on the gel put them below the level of detection.

11. Uracil-*N*-glycosylase (UNG). This has been used with great success for the inhibition of contaminating carryover template. Succinctly, PCR reactions include dUTP, which functions as a dTTP analog. The amplified PCR product contains uracil, rather than the thymine that would normally be expected. Prior to amplification, the UNG enzyme is used to facilitate breakage of any carryover uracil-containing template

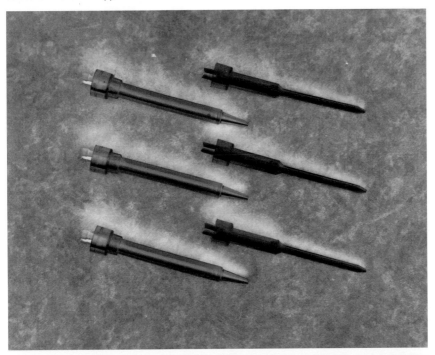

b. Positive displacement tips are available in more than one size in order to facilitate the preparation of stock solutions and master mixes.

from previous reactions. Fresh, thymine-containing template added to the reaction mix by the investigator is not a substrate for the UNG enzyme, and is therefore unaffected by this incubation. The inclusion of dUTP in the reaction does not influence the specificity or efficiency of product amplification, and the UNG itself can be subsequently destroyed by heating.

12. Hot start PCR. Reactions assembled at room temperature or on ice and then raised to 94°C to inititate the cycling obviously must pass 72°C on the way up to 94°C. As 72°C is the optimal temperature for primer elongation (at least for *Taq*), nonspecific PCR product can be generated if the primers are imperfectly base paired to sites on the template where they would not ordinarily base-pair under stringent conditions. To preclude these difficulties, numerous strategies have been employed to inhibit the synthesis of any product at all until both the primers and the template have been completely denatured prior to the onset of

Cycle 1. These strategies include, but are not limited to, JumpStart Taq (Sigma); TaqStart (antibody) (Sigma; Clontech), AmpliWax (Perkin Elmer), and HotWax Mg^{2+} Beads (Invitrogen), AmpliTaq Gold DNA Polymerase (PE Applied Biosystems), or simply preheating the reaction above 72°C and then adding the missing component(s) directly into the tube(s).

13. Enzyme blends. Many of the historical difficulties associated with the synthesis of representative cDNA carry over into the realm of RT PCR. One current trend involves the use of various combinations of thermostable polymerases, that is, enzyme blends, which are currently being used with great success for the production of genomic-length PCR products. Because of the efficiency of this technology, known as long-range PCR, it is exploited frequently in this laboratory for the efficient amplification of cDNAs that correspond to exquisitely low abundance mRNAs, as well as those longer mRNAs in the cytoplasm and much longer unspliced hnRNA. A typical enzyme blend, for example, might consist of *Taq* polymerase with another thermostable polymerase. *Taq* is an efficient (processive) enzyme with respect to its intrinsic $5'{\rightarrow}3'$ polymerase activity, though the half-life of this enzyme is about 60 min at 95°C, and it lacks $3'{\rightarrow}5'$ exonuclease activity (proofreading). Thus, if *Taq*, or any other nonproofreading enzyme, makes an error, the error is stably incorporated into the PCR product. Other thermostable enzymes, such as those from the genus *Pyrococcus* exhibit greatly enhanced thermostable qualities and also exhibit a proofreading capability. *Taq* combined with one of these other enzymes is much more efficient and amplifies with higher fidelity than either alone. Moreover, in instances where the number of cycles is extended beyond 30, to perhaps 35 or even 40, PCR product can continue to accumulate, because the second enzyme more or less takes over after the *Taq* has been, literally, burned out.

14. Optimization kits. Consider investing in a commercially available kit containing combinations of chemistries known to enhance the PCR process (e.g., Sigma No. OPT-1). This system, for example, contains several 10× buffer formulations and six adjuncts known to enhance PCR: *E. coli* single-strand binding protein (Chou, 1992), formamide (Sarkar *et al.*, 1990), ammonium sulfate (Olive *et al.*, 1989), bovine serum albumin (Paabo *et al.*, 1988), dimethyl sulfoxide (Cheng *et al.*, 1994; Winship *et al.*, 1989), and glycerol (Cheng *et al.*, 1994). The influence of each of these adjuncts is summarized in Table 4. One worthwhile strategy is to select the best 10× PCR buffer, based on product yield and absence of nonspecific product, and then embellish it with the various adjuncts to optimize the reaction for a particular template and primer set.

TABLE 4

The Effects of Adjuncts that Enhance PCR

PCR adjunct	Working concentration	Effect
Ammonium sulfate	15–30 mM	Influences denaturation and annealing profiles of the template and primers
Bovine serum albumin	10 –100 μg/ml	Stabilizes *Taq* (and other enzymes, too)
Dimethyl sulfoxide	1.0–10.0%	Decreases incidence of template depurination; may also facilitate strand renaturation
Formamide	1.25–10.0%	Increases specificity in GC-rich template domains
Glycerol	15–20%	Increases *Taq* stability; lowers template T_m
Single-strand binding protein	0.7–1.5 μg/assay	Accelerates primer annealing; inhibits extension of mismatched primers

This will save time and money whenever a new set of reaction conditions is used.

Analysis of PCR Products

In most settings, agarose gels represent the matrix of choice for the evaluation of PCR products. Because of the ease of gel preparation, 1–3% agarose gels are quickly cast to accommodate any size PCR product. These gels may be prepared with or without ethidium bromide added to the gel solution after it has been melted and cooled to about 55°C. Alternatively, gels may be stained with SYBR Green *after* electrophoresis, for superior sensitivity.

A variety of agaroses are available from many vendors (e.g., FMC BioProducts). One may elect to use standard or low melting temperature agarose, as well as an agarose grade that favors separation of low molecular PCR products. In this lab, SeaKem GTG (FMC BioProducts) is routinely prepared as a 2.5–3.0% solution in 1× TAE. This recipe favors exceptional resolution of PCR products ranging from 200 to 1000 bp.

CAUTION When melting high agarose concentrations such as these in a microwave, the tendency for boilover, both in the microwave, and immediately after heating, is quite high. Great care must be taken when preparing these high concentration gels to prevent injury to the investigator and to avoid burning large clumps of agarose in the flask or beaker (especially when using a hot plate).

RT PCR Quality Control Points

Among the more notorious difficulties encountered when preparing for an experiment is the integrity of the template material, a facet that takes on an entirely new meaning when working with RNA. For those laboratories actively investigating gene expression, it is likely that PCR is being used to amplify one or another transcript[12] with primers and parameters that have long since been optimized. The frustration index rises when one is unable to generate a product using a freshly prepared sample of RNA, or an RNA sample that "used to work." The following should be pondered, along with the optimization items discussed above.

1. Assess the quality of the RNA just prior to use by electrophoresis. The minigel format is more than adequate to simply show the presence of the 18S and 28S ribosomal RNAs (rRNAs). RNA that has been stored in aqueous buffer, rather than as an ethanol precipitate, degrades rapidly. This is exacerbated when the sample has been subjected to repeated freeze/thaw cycles. In this lab, previously purified RNA is heated to 65°C for 5 min and then electrophoresed in a minigel (1% agarose in $1\times$ TAE; no denaturants needed), the purpose being to assess the integrity of the sample (no blotting). As long as there is reasonable definition to the 28S and 18S rRNAs, the investigator may proceed confidently.

2. Test each primer pair without template. This is necessary to show that the stock solutions are not contaminated with an amplicon and to give any indication of primer–dimer.

3. Test each primer alone, with template. This is the only true way to show that self-priming is not a possibility and that the primer pair is specific.

4. Test each sample of RNA by omitting the reverse transcriptase from the first-strand synthesis reaction. Then use the contents of this tube as template material for PCR. In this sample, known as the "no RT control" or the "RT⁻" control, no products should be observed. In the absence of reverse transcription of the RNA template, the only bands that could possibly be generated must have come from a genomic DNA template contaminating the sample. The standard protocol in most labs performing any type of RT PCR is to DNase the samples at the time of isolation, after which the RT⁻ control is performed.

---

[12]Related RNA-based PCR techniques are described in detail in Chapters 16 and 17.

5. Test difficult RNA samples with primers that may not be of direct experimental interest, but that are at least known to generate a PCR product each time RNA is isolated, intact and clean, from a particular biological source. This can be performed using very boring primers, such as those for ß-actin or GAPDH, just to show that one can amplify something (anything) from the RNA in question. If a PCR product is evident with control primers, then the reason for failure to generate a product from experimental primers can be investigated more intelligently. In summary: No signal can mean no transcription *or* that the RNA sample is poor and unable to support amplification of anything. This latter possibility will have been validated, or ruled out, through the use of control primers.

Related Techniques

5' RACE PCR

One of the key drawbacks of traditional cDNA synthesis is the failure to generate a cDNA sequence that corresponds to the 5' end of the mRNA. This loss of sequence makes it impossible to clone full-length sequences or map the 5' end of specific transcripts. A method known as RACE (rapid amplification of cDNA ends; Frohman *et al.*, 1988) was developed mainly for cloning sequences that characterize transcript 5' and 3' ends. In 5' RACE, the synthesis of first-strand cDNA is accomplished with a downstream primer, either oligo(dT) or, ideally, a gene-specific primer that base pairs as close to the 5' end of the transcript as possible. First-strand synthesis is followed by the placement of extra nucleotides of known sequence at the 3'-OH end of the newly synthesized cDNA:

GGGGGGGGG _____5' 1st-strand cDNA

This can be accomplished by terminal transferase-mediated homopolymeric tailing, or by T4 RNA ligase-mediated attachment of a defined oligomer to the 3' end of first-strand cDNA. Many investigators prefer the ligation of what is best thought of as an anchor,[13] the sequence of which is predetermined to approximate the calculated T_m of the gene-specific downstream primer that will be used to support amplification by

---

[13]This cloning approach is also known as anchor PCR (Loh *et al.*, 1989).

PCR. In this strategy, the 3′ end of the anchor is best blocked with an amino group to preclude ligation to both ends of the cDNA:

NH$_2$-XXXXXXXXXXp 5′ + 3′ HO_____p 5′ 1st-strand cDNA

Next, an oligonucleotide complementary to the anchor (the anchor primer) functions as the upstream primer and, in conjunction with an RNA-specific downstream primer, supports the amplification of one species of cDNA of which the 5′ end is favored. This is only one of several versions of this cloning strategy; newer methods (the New RACE) suggest a dephosphorylation step prior to begining the RACE protocol to eliminate any partially degraded RNAs from the reaction, and the annealing of the anchor primer to the anchor prior to reverse transcription (see Frohman (1995) for details):

    5′ YYYYYYYYYY 3′ →

    NH$_2$-XXXXXXXXXX_____5′ 1st-strand cDNA

3′ RACE PCR

In order to favor cloning sequences that correspond to the 3′ end of the transcript, one may wish to exploit a method known as 3′ RACE. In this approach, mRNAs are primed for first-strand cDNA synthesis using a downstream primer with the generic structure 5′-anchor-TTTTTTTTTTTTTTTTTT-OH 3′, where the anchor is a unique sequence located 5′ to oligo(dT) used for standard cDNA synthesis. Amplification of cDNAs synthesized with this downstream primer is favored through the use of an anchor primer, complementary to the anchor, and the use of a gene-specific primer. While one might envision 3′ RACE by simply using oligo(dT) as the downstream primer, experience strongly advises against the use of any AT-rich sequence: oligo(dT) is 100%!

Nested PCR

Occasionally, the failure to generate a PCR product may be related to the exquisitely low abundance of the template or, perhaps more commonly, to the purity of the template. In one strategy, an aliquot of a PCR reaction, which manifests no product, is diluted 1:100 into a fresh reaction mix and subjected to an additional 30 cycles. However, a technique involving two sets of primers, known as nested PCR, is widely accepted to render a more satisfactory outcome, compared to simply running through more cycles; this is especially true when there is a purity issue to be re-solved.

In nested PCR, one of the two sets of primers (the inner primer set) is designed to base pair completely within the region amplified by the other set of primers (the outer primer set). Succinctly, the sample is amplified using the outer primers for 30 cycles, after which an aliquot (dilution op-tional) is amplified for an additional 25–30 cycles using the inner primers. If properly designed, the outcome of the nested PCR approach is usually most gratifying. One must be certain to examine all four of the proposed primer sequences to ensure that they conform to the standard rules for primer design, as discussed above.

Protocol: First-Strand cDNA Synthesis

In advance:

1. Assess the quality of the RNA template by minigel electrophoresis, as described above.

2. Prepare cDNA 10× first-strand buffer (500 mM KCl; 100 mM Tris, pH 8.3; 15 mM MgCl$_2$).

cDNA synthesis:

1. Prepare cDNA master mix. For each 25-µl reaction, mix 2.5 µl 10× first-strand buffer; 2.5 µl 1 mM dNTP stock; 1.0 µg oligo(dT)$_{12-18}$ *or* 2 µg random primers *or* 1 µM sequence specific primer (volume vari-able); 1 µl RNasin; 1 µl 20 U/ml AMV reverse transcriptase; water to 23 µl.

2. Aliquot the cDNA master mix into the required number of tubes. Unused master mix (without RNasin or reverse transcriptase) may be stored frozen at –20°C for extended periods.

3. Heat each sample of RNA to 65°C for 3 min to disrupt secondary structures that may interfere with reverse transcription. Immediately cool RNA on ice for 3 min.

4. Add an aliquot of each RNA sample to each aliquot of master mix. Ideally, plan to use 0.5–1.0 µg RNA per reaction. The RNA should be at a concentration of about 500 ng/µl. If not, compensate by adjusting the amount of water added to the master mix prepared in Step 1.

Note: If only one type of RNA template is to be run in duplicate or trip-licate, the RNA can be heat denatured and added directly to the cDNA master mix, which is then aliquoted. This approach precludes variation due to template mass in each reaction.

5. Incubate first-strand synthesis reactions at room temperature for 5 min to facilitate primer annealing and then at 42°C for 60 min.

6. At the conclusion of the cDNA synthesis reaction, heat the sample to 95°C for 5 min to destroy the reverse transcriptase; if not heat inactivated, residual reverse transcriptase can have a negative impact on subsequent applications. Store newly synthesized cDNA at 4°C until ready to use. If the PCR reactions will be set up on another day, then store the cDNA at –20°C.

Protocol: PCR Amplification of cDNA

1. Prepare PCR master mix using PCR reagents purchased from a licensed source. Typically, for each 50-μl reaction to be performed, mix 5 μl 10× PCR buffer (without Mg); 2 μl 25 mM MgCl$_2$ (subject to change if optimization is required); 0.1 μM upstream primer (volume variable); 0.1 μM downstream primer (volume variable); 0.25 μl *Taq* DNA polymerase; water to 45 μl.

Note 1: If different sequences are to be amplified in each reaction tube, then omit the primers from the master mix, adding each pair to the proper reaction tube directly. If the cDNA is shared among all samples, it may be added directly to the PCR master mix.

Note 2: Never add all of cDNA to the master mix. By reserving at least a small aliquot, it will be possible to amplify several sequences from the same experimental source, if not today, then perhaps for some unforeseen purpose in the future.

2. Aliquot 45 μl of the master mix into the appropriate number of microfuge tubes.

3. Add 5 μl of cDNA first strand products from each experimental source into each tube.

4. If using a thermal cycler without a heated lid, overlay 30 μl of mineral oil onto each reaction. Thermal cyclers with heated lids do not required mineral oil-overlaid reactions.

5. Cycle the samples according to the following suggested profile. These parameters have been optimized for a Perkin Elmer 480 thermal cycler. Remember that cycling parameters are subject to optimization for

each new set of primers, changes in the reaction cocktail, and model of thermal cycler.

Initial melt	94°C for 3 min
Cycling	
Denaturation	94°C for 1 min
Annealing[14]	55°C for 1 min
Extension	72°C for 2 min
Final extension	72°C for 10 min
Soak	4°C until ready to use

6. Remove an aliquot (5 μl from a 50-μl reaction) and analyze size and variety of products by minigel electrophoresis. In general, PCR products are commonly analyzed in 2–3% agarose gels prepared in 1× TAE buffer, as described earlier in this chapter.

Cloning PCR Products

The ability to amplify and reamplify DNA sequences by PCR has, in most instances, rendered traditional insertion into a cloning vector rather obsolete. After all, if one requires more of the same product, it can be generated again by PCR using the same primers. Cloning of PCR products is needed to (a) facilitate sequencing, (b) facilitate *in vitro* transcription, (c) physically separate a mixture of PCR products, and (d) support the expression of recombinant proteins. Of these, sequencing and *in vitro* transcription can be supported without cloning: Reamplification of PCR products with modified up- and downstream primers consisting of appropriate 5' promoter sequences precludes the requirement for a vector.

Should cloning be deemed necessary, there are several options. *Taq* DNA polymerase exhibits a very useful habit of adding an extra nucleotide, most often an A, at the 3' end of each strand of the product that it generated. This is the basis for a convenient method known as TA cloning, in which the investigator anneals and then ligates the *Taq* A-tailed product into a vector that exhibits a 3' T overhang on each strand. These vectors may be purchased from any of a number of sources or sim-

---

[14]The correct temperature at which to anneal (T_a) is a direct function of the T_m of the oligonucleotides being used. Consult the specification sheet that accompanied the primers when they were manufactured. If uncertain where to begin, use $T_a = T_m - 5°C$. Recall that the T_m may be quickly estimated: $T_m = 2°(A+T) + 4°(G+C)$.

ply prepared in-house and should possess a selectable marker (e.g., Amp^r or Kan^r) as well as a *lacZ* gene for blue/clear color selection. The ligation itself is mediated by T4 DNA ligase in a reaction requiring several hours. Note that thermostable enzymes used for PCR that demonstrate a $3' \rightarrow 5'$ exonuclease (proofreading) activity prepare blunt ended PCR products, which cannot be used for TA cloning unless they are tailed. This is easily accomplished by incubation of the completed PCR products in *Taq* polymerase for 10 min at 72°C, followed by standard phenol:chloroform extraction to remove the *Taq* and the proofreading enzymes, and salt and alcohol precipitation.

There are several advantages to working with A-tailed products:

- precludes having any prior knowledge of the PCR product sequence;
- eliminates the need to synthesize primers carrying 5′-overhanging restrictions enzyme or other sequence(s);
- purification of PCR products is usually not required;
- most T-tailed cloning vectors carry the *lacZ* gene, enabling blue/clear color selection;
- directional cloning is possible when hemi-phosphorylated PCR products are generated (5′ phosphorylate one of the primers);
- vector cloning sites are generally located in a polylinker region, facilitating excision, if desired, using any of a number of restriction enzymes.

Protocol: Rapid TA Cloning of PCR Products

1. Prepare three ligation reactions for each PCR product to be cloned. In each tube, the ratio of vector to insert will be varied to favor maximum ligation efficiency. Note that multiple ligations are suggested only for first-time users; it is generally necessary to perform only one ligation reaction because large numbers of transformants (i.e., maximum ligation efficiency) is not required when only a single type of insert is expected.

Reaction 1: 1 μl 10× ligase buffer; 10 ng T-tailed vector (volume variable); 1 μl PCR product; H$_2$O to 9 μl; 1 μl T4 DNA ligase.

Reaction 2: 1 μl 10× ligase buffer; 5 ng T-tailed vector (volume variable); 2.5 μl PCR product; H$_2$O to 9 μl; 1 μl T4 DNA ligase.

Reaction 3: 1 μl 10× ligase buffer; 2.5 ng T-tailed vector (volume variable); 5 μl PCR product; H$_2$O to 9 μl; 1 μl T4 DNA ligase.

2. Mix the reaction(s) gently by stirring it with the micropipet tip following the addition of the ligase.

3. Incubate for a minimum of 3 h at 16°C.

Note: Ideally, the ligation should continue overnight.

4. Thaw an aliquot of frozen (–80°C) competent cells on ice.

Note: These cells may be purchased or prepared in advance. Be sure that the bacterial host strain is compatible with blue/clear color selection if using this selection scheme. E. coli strain JM109, which may be purchased competent from Promega, works quite well in this regard.

5. Pipet 20-μl aliquots of competent cells into prechilled microfuge tubes.

6. Add 1 μl ligation reaction to each tube. Stir gently with pipet tip.

7. Incubate tubes on ice for 20 min.

8. Place the tubes in a 42°C water bath for *exactly* 45 s. Do not shake these tubes.

Note: This heat-shock step is the most critical of all, often determining the success or failure of the transformation. Do not heat at a high temperature or for a longer period than recommended here.

9. Immediately return the tubes to the ice for 2 min.

10. Add 400 μl room temperature LB medium or room temperature SOC medium.[15]

11. Incubate at 37°C for 1 h with shaking (200 rpm).

12. Plate an aliquot (10–100 μl) of each transformation reaction onto LB agar plates that have been supplemented with the appropriate antibiotic. Reserve the unused portion of the transformation by storing it at 4°C.

13. Spread the transformation reaction aliquot over the surface of the agar as completely as possible using a sterile bent glass rod or similar implement.

14. Allow all of the liquid to absorb into the agar; 5 min is usually satisfactory.

15. Invert the plates and incubate at 37°C overnight.

---

[15]LB medium (per liter): 10 g tryptone, 5 g yeast extract, 10 g NaCl; autoclave. SOC medium (per liter): 20 g tryptone, 5 g yeast extract, 0.6 g NaCl, 0.19 g KCl, 2 g $MgCl_2$, 2.5 g $MgSO_4$, 3.6 g glucose; autoclave.

16. The following morning, remove the plates from the incubator as soon as possible. This will minimize the formation of satellite colonies.

17. To test whether a colony on the dish is a true transformant, pick a portion of the colony from the plate, using a sterile toothpick, and inoculate the live bacteria directly into a 100-μl PCR reaction mix: 10.0 μl 10× PCR reaction buffer (containing Mg); 2 μl 10 mM dXTP; 0.5 μM downstream primer (volume variable); 0.5 μM upstream primer (volume variable); 1 μl Taq polymerase.

Note: The heating steps associated with PCR will break open the bacterial cells added to the reaction and, in so doing, make the plasmid-borne template (i.e., the insert) accessible to the primers.

18. Overlay the reaction with mineral oil, if necessary, and cycle with the same amplification parameters used to generate the PCR product.

19. At the conclusion of the amplification, remove a representative aliquot of the reaction (10 μl) and analyze by agarose gel electrophoresis. Colonies that contain a true insert will manifest a band identical in size to that observed for the PCR product prior to cloning.

20. Return to any of the original colonies that generated a PCR product. Use the remainder of the colony to streak a new agar plate and to prepare a glycerol stock for long-term storage.

TopoCloning

In a second approach to cloning PCR products, the vaccinia enzyme topoisomerase I is used to ligate an A-tailed PCR product into a T-tailed vector. Known as TopoCloning (Invitrogen), this method of cloning is rapid and efficient, often rendering >90% cloning efficiency in as few as 5 min and at room temperature. The vectors that support TopoCloning are linearized and "activated" with the topoisomerase I enzyme (Shuman, 1994), meaning that the enzyme has been directly linked to the vector. Many of the utilities associated with standard TA cloning, described above, also apply to TopoCloning and blunt end ligations as well. Upon ligation, the topoisomerase I enzyme is released from the now recombinant plasmid construction. Bacterial transformation proceeds as usual.

References

Chenchik, A., Diachenko, L., Moqadam, F., Tarabykin, V., Lukyanov, S., and Siebert, P. D. (1996). Full-length cDNA cloning and determination of mRNA 5' and 3' ends by amplification of adaptor-ligated cDNA. *BioTechniques* **21**(3), 526.

Cheng, S., *et al.* (1994). *Proc. Natl. Acad. Sci. USA* **91,** 5695.

Chou, Q. (1992). *Nucleic Acids Res.* **20,** 4371

Dieffenbach, C., and Dveksler, G. S. (1995). "PCR Primer." Cold Spring Harbor Laboratory Press, Cold Spring Harbor, NY.

Efstratiadis, A., Kafatos, F. C., Maxam, A. M., and Maniatis, T. (1976). Enzymatic *in vitro* synthesis of globin genes. *Cell* **7,** 279.

Frohman, M. A. (1995). Rapid amplification of cDNA ends. *In* "PCR Primer" (C. W. Dieffenbach and G. S. Dveksler, Eds.). Cold Spring Harbor Laboratory Press, Cold Spring Harbor, NY.

Frohman, M. A., Dush, M. K., and Martin, G. R. (1988). Rapid production of full-length cDNAs from rare transcripts: Amplification using a single gene-specific oligonucleotide primer. *Proc. Natl. Acad. Sci. USA* **85,** 8998.

Gubler, U., and Hoffman, B. J. (1983). A simple and very efficient method for generating cDNA libraries. *Gene* **25,** 263.

Gyllensten, U. B., and Erlich, H. A. (1988). Generation of single-stranded DNA by the polymerase chain reaction and its application to direct sequencing of the HLA-DQA locus. *Proc. Natl. Acad. Sci. USA* **85,** 7652.

Innis, M. A., Gelfand, D. H., Sninsky, J. J., and White, T. J. (Eds.). (1990). "PCR Protocols: A Guide to Methods and Applications" Academic Press, San Diego, CA.

Kawasaki, E.S. 1990. Amplification of RNA. *In* "PCR Protocols: A Guide to Methods and Applications" (M. A. Innis, D. H. Gelfand, J. J. Sninsky, and T. J. White, Eds.). Academic Press, San Diego, CA.

Loh, E. L., Elliott, J. F., Cwirla, S., Lanier, L. L., and Davis, M. M. (1989). Polymerase chain reaction with single sided specificity: Analysis of T cell receptor delta chain. *Science* **243,** 217

McPherson, M. J., and Hames, B. D. (Eds.). (1995). "PCR 2: A Practical Approach." IRL Press/Oxford University Press, Oxford, England.

McPherson, M. J., Quirle, P., and Taylor, G. R. (Eds.). (1991). "PCR: A Practical Approach." IRL Press, Oxford, England.

Mullis, K. B., and Faloona, F. A. (1987). Specific synthesis of DNA *in vitro* via a polymerase-catalyzed chain reaction. *Methods Enzymol.* 155, 335–350.

Mullis, K. B., Faloona, F. A., Scharf, S., Saiki, R., Horn, G., and Erlich, H. (1986). Specific enzymatic amplification of DNA *in vitro*: The polymerase chain reaction. *Cold Spring Harbor Symp. Quant. Biol.* **51,** 263.

Ochman, H., Gerber, A. S., and Hartl, D. L. (1988). Genetic applications of an inverse polymerase chain reaction. *Genetics* **120,** 621.

Okayama, H., and Berg, P. (1982). High efficiency cloning of full-length cDNA. *Mol. Cell. Biol.* **2,** 161.

Olive, D., *et al.* (1989). *J. Clin. Microbiol.* **27,** 1238.

Paabo, S., *et al.* (1988). *Nucleic Acids Res.* **16,** 9775.

Saiki, R. K., Scharf, S., Faloona, F., Mullis, K. B., Horn, G. T., Erlich, H. A., and Arnheim, N. (1985). Enzymatic amplification of β-globin genomic sequences and restriction site analysis for diagnosis of sickle cell anemia. *Science* **230,** 1350.

Sambrook, J., Fritsch, E. F., and Maniatis, T. (1989). "Molecular Cloning: A Laboratory Manual," 2nd ed. Cold Spring Harbor Laboratory Press, Cold Spring Harbor, NY.

Sarkar, G. *et al.* (1990). *Nucleic Acids Res.* **18,** 7465.

Shuman, S. (1994). Novel approach to molecular cloning and polynucleotide synthesis using vaccinia DNA topoisomerase. *J. Biol. Chem.* **269,** 32, 678.

Silver, J., and Keerikatte, V. (1989). *J. Cell. Biochem. (Suppl.)* **13E,** 306.

Triglia, T., Peterson, M. G., and Kemp, D. J. (1988). A procedure to *in vitro* amplification of DNA segments that lie outside the boundaries of known sequences. *Nucleic Acids Res.* **16,** 8186.

Veres, G., Gibbs, R. A., Scherer, S. E., and Caskey, C. T. (1987). The molecular basis of the sparse fur mouse mutation. *Science* **237,** 415.

Winship, P. R., *et al.* (1989). *Nucleic Acids Res.* **17,** 1266.

Messenger RNA Differential Display

Rationale

The exquisite orchestration of gene expression at the transcriptional level and beyond is responsible for the differentiation of cells and tissues and, ultimately, for conferring the phenotype of the organism. It has long been the goal of many a molecular biologist to be able to determine which genes are expressed in common among two or more different cell types and, more importantly, which genes are expressed differentially when comparing these cell populations side by side.

That RNA is an excellent parameter of gene expression is a recurrent theme in this volume. Prior to the emergence of the polymerase chain reaction, now a mainstream tool of the molecular biologist, several variations of a method known generically as subtraction hybridization (Sargent and Dawid, 1983; Sargent, 1987; Lisitsyn et al., 1993; Hedrick et al., 1984) were utilized as a means of subtracting away sequences shared by two populations of RNA/complementary DNA. Methods that support the physical separation of differentially expressed sequences include chromatography through hydroxyapatite[1] (Bernardi, 1971; Sargent, 1987), biotinylation coupled with streptavidin precipitation, biotinylation coupled with magnetic bead separation[2] technology (Rodriguez and Chader, 1992; Lambert and Williamson, 1993; Coche et al., 1994), biotinylation coupled with phenol extraction (Travis and Sutcliffe, 1988), and cross-hybridization of entire libraries (Kulesh et al., 1987). Interestingly, some of these methods are now enjoying a resurgence. In general, each of these methods has its shortcomings, with respect to both efficiency and sensitivity, and none of them alone renders a comprehensive snapshot of gene expression at the moment of cell lysis. The result of physical subtraction (Figure 1) of commonly expressed sequences is a population of differentially expressed (transcribed) molecules suitable for cloning, restriction mapping, sequencing, and expression studies. Although some of these methods have been widely used, in general they favor the selection of differentially expressed sequences of moderate to high abundance and not necessarily those sequences that are underrepresented from the start.

Many investigators now attempt to achieve these same experimental ends by taking advantage of the efficiency and exquisite sensitivity of the polymerase chain reaction (PCR). In short, the strategy involves coupling

---

[1]Hydroxyapatite $(Ca_5(PO_4)_3OH)_2$ is a form of calcium phosphate. Historically, it has been used for the purification of a variety of macromolecules, including nucleic acids. The unique nature of this matrix permits the physical partitioning of supercoiled DNA from linear molecules, and single-stranded from double-stranded DNA.
[2]See Chapter 6 for magnetic bead purification of poly(A)$^+$ mRNA.

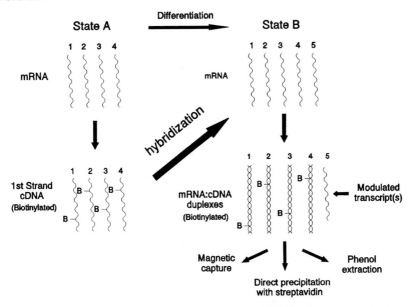

State A → Differentiation → State B

FIGURE 1

General strategy for the identification of differentially expressed sequences by subtraction hybridization. mRNA from State A is reverse transcribed, rendering single-stranded first-strand cDNAs. In one version of this approach, the cDNA is biotinylated either by the incorporation Bio-21-dUTP or by photobiotinylation. These first-strand cDNAs are then used as probes to base pair with the mRNAs that State A cells and State B cells have transcribed in common. The resulting RNA:cDNA hybrids are then processed for physical separation. This same approach can be used to screen entire libraries.

standard methods for complementary DNA (cDNA) synthesis to PCR-based amplification, using large combinations of short primers, for the purpose of evaluating two or more experimental populations in a side-by-side comparison by polyacrylamide gel electrophoresis (PAGE). In so doing, transcripts common to the RNA populations under investigation are manifested as PCR products of identical size, side by side, in adjacent lanes on a gel, when identical primer sets are used. More importantly, this method facilitates the identification of bands in one lane that are *absent* or of *different abundance* in the corresponding location of the adjacent lane(s).

Several variations on this standard method appear ubiquitously and are described by a variety of names. Perhaps the most scientifically correct terminology to describe this technique is messenger RNA (mRNA) differential display (Liang and Pardee, 1992); hence, the title of this chapter.

More commonly used, and instantly recognizable, however, is Differential Display PCR. One may shorthand this method as DDPCR,[3] or use the more cumbersome DDRTPCR (Bauer *et al.*, 1993). Other nomenclature that has evolved to describe this general approach, but with a subtle variation in primer design, is RNA fingerprinting and RNA arbitrarily primed PCR (RAP-PCR) (Welsh *et al.*, 1992; McClelland and Welsh, 1994). Although subtly different from one another, each of these techniques essentially accomplishes the same end, albeit achieving varying levels of completeness or "representativeness" with respect to the complexity of transcripts identified. In in this volume the technique will hereafter be referred to as differential display PCR or DDPCR, for the sake of simplicity.

General Approach

DDPCR is all about the speed, specificity, and sensitivity of PCR by which at least a section of each of the various transcripts being produced by the cell is amplified. The strategy behind DDPCR is quite simple, incorporating traditional synthesis of cDNA from purified RNA, followed by amplification of the products of the first-strand cDNA reaction. Unlike traditional cDNA synthesis, several cDNA reactions are performed for each sample, each of which is supported by a unique combination of downstream primers and upstream primers, which, in any given reaction tube, will support the amplification of a subset only of all possible PCR products. These products are subsequently analyzed by electrophoresis and compared directly to the other sample(s) under investigation (e.g., control vs. treated). Taken together, these steps constitute a structured method for simultaneous examination of all transcribed sequences. The major steps of DDPCR are as follows.

1. RNA Isolation

Identification of modulated sequences by DDPCR begins with the isolation of extremely high-quality RNA. In particular, the isolation of RNA by guanidinium–acid–phenol is one of the better, more widely accepted, purification methods for analysis by DDPCR. Investigators may wish to explore the protocol found in Chapter 5, or use one of the numerous commercial guanidinium-containing reagents to accomplish this isolation. Among the advantages of the DDPCR method (Table 1) is the require-

---

[3]When using this shorthand, be certain to use uppercase letters. The shorthand ddPCR strongly suggests that the writer is referring to dideoxy (dd-) sequencing by PCR.

TABLE 1

Advantages and Disadvantages of Differential Display PCR

Advantages
- simplicity
- very sensitive method for detecting specific transcripts
- versatility
- speed
- systematic approach for the identification of differential gene expression
- semi-quantitative method
- poly(A)$^+$ selection neither required nor recommended
- simultaneous display of all up- and down-regulated genes
- reproducible data
- small μg quantities of RNA required (often less than 3 μg)

Disadvantages
- high incidence of false positives (20–40%) after amplification
- extremely pure samples of RNA required
- exquisitely sensitive to contaminants, especially genomic DNA
- false positives also observed after reamplification
- suboptimal for use as a probe without extensive validation
- labor intensive generation/analysis of PCR products after isolation of RNA populations
- large amounts of PCR reagents required to *fully* characterize samples
- purified cDNAs are often contaminated with unrelated cDNA sequences
- technique is highly sensitive to template quality and other variables that are often not major points for concern in standard PCR reactions
- reproducible data?

ment for small input RNA mass. It is strongly recommended, however, that a larger mass of RNA be purified and stored as an ethanol precipitate, thereby ensuring that precisely the same RNA template, from the same biochemical source, isolated on the same day by the same investigator, will be available when the time comes to repeat the experiment and, it is hoped, reproduce the data. Moreover, the selection of poly(A)$^+$ RNA is neither required nor recommended. There are four reasons for this:

1. Selection of poly(A)$^+$ RNA confers no specific advantage with respect to the synthesis of first-strand cDNA.
2. Attempted selection of poly(A)$^+$ RNA may actually be responsible for the loss of some of the template material, resulting in the further underrepresentation of low abundance (rare) transcripts.
3. There exists a remote possibility of poly(dT) contaminating the sample.

4. Poly(A)$^+$ selection is yet another opportunity for the accidental introduction of RNA contamination, thereby reducing the usefulness of the sample.

Last, but certainly of profound importance, is the requirement for removing all traces of contaminating genomic DNA by DNase-treating the sample. Remember that primers will base to any complementary molecule, DNA or RNA. The efficient removal of DNA is validated by the no-reverse transcriptase control reaction, described below.

2. cDNA Downstream Primer Design

The standard approach to cDNA synthesis intended to support analysis by DDPCR is to perform several cDNA synthesis reactions for each unique RNA sample, reverse transcribing only a fraction of all possible messages into cDNA in any one reaction tube. Thus, each reverse transcriptase reaction is *intended* to generate subpopulations of cDNA so as to resolve discrete bands upon electrophoresis, rather than a glorious smear. The key to accomplishing this lies with the primers: Downstream primers support the conversion of RNA into cDNA by means of a modified oligodeoxythymidylate 5′ TTTTTTTTTVV (below), where V = A, C, or G. The 3′ dinucleotide sequence degeneracy ensures the reverse transcription of all possible mRNAs; further, the 3′ dinucleotide functions as an anchor, ensuring that the primer will base pair at the 5′-most end of the poly(A)$^+$ tail. Thus, during the actual cDNA synthesis reaction, valuable time will not be lost reverse transcribing the poly(A) tail: An oligo(dT) primer (5′ TTTTTTTTTTTT) can base pair well into the body of the poly(A) tail, hundreds of nucleotides away from the coding portion of the transcript.

There are two prevailing views pertaining to the design of the downstream primers. The basis for generating a subset of cDNA is the downstream primer recognition of the one or two nucleotides located immediately 5′ to the poly(A) tail of adenylated RNA. In one approach, the generic structure of the downstream primers is the standard oligo(dT) structure followed by a single additional nucleotide at the 3′ end (Liang *et al.*, 1994), for example, $T_{12}V$ (5′ TTTTTTTTTTTTV; where V = A, C, or G). Thus, a different reverse transcriptase reaction for each of the three downstream primers is expected to generate three different pools of cDNA, each containing about one-third of all possible cDNAs. In the second approach, the generic structure of the downstream primers is $T_{11}VV$ (5′ TTTTTTTTTTTVV; where V = A, C, G). The degenerate dinucleotide approach at the 3′ end of the downstream primers thus favors annealing to and synthesis of cDNA only from those RNA transcripts characterized at their 3′ end by 3′ $A(A)_n$WW, where W is complementary to V. Many investigators who perform differential display PCR routinely will agree that downstream primers that manifest degenerate 3′ dinucleotides correlate with more representative PCR products, that is, a virtually complete pattern of all mRNAs, as well as reproducibility of data.

Consideration of the possible dinucleotide combinations reveals 4^2 or 16 permutations:

AA	CA	GA	TA
AC	CC	GC	TC
AG	CG	GG	TG
AT	CT	GT	TT

Of these, only nine are frequently used, at least at the onset of an investigation involving this method:

AA	CA	GA
AC	CC	GC
AG	CG	GG

Why the differential? First, elimination of a downstream primer 5′ T_{11}TT is clearly in order, because such a primer confers no first-strand cDNA priming specificity. Using such a primer would support first-strand cDNA synthesis from all adenylated transcripts, an older method for the synthesis of whole cDNA libraries![4] Next, TA, TC, and TG, may be eliminated because using these three dinucleotides renders a downstream primer with the generic structure T_{11}TV, which is the same as having only a single degenerate nucleotide at the 3′ end. Further, the exclusion of the AT, CT, and GT 3′ dinucleotide primers also is suggested, because (a) it is the 3′-most nucleotide that confers the most specificity of the primer; (b) the short nature of these primers favors mispriming[5] (to an extent) and it is quite likely that RNAs consisting of the sequence

5′ . . . . AWAAAAAAAAAAAAAAAAAAA 3′

---

[4] A primer T_{11}TT, essentially oligo(dT), can also be used to generate cDNA for library construction by PCR. Amplification of the library by PCR is supported by the judicious selection of the upstream primer(s).

[5] Mismatches can be tolerated as long as the nucleotide at the 3′ end of the primer is not involved. cDNA synthesis, PCR, and a variety of other molecular biology techniques *require* 3′ base pairing of the nucleotide that holds the 3′-OH substrate to the template.

will be converted into first-strand cDNA; and (c) the use of these primers correlates with large amounts of rather small PCR products that are difficult to resolve and that contribute heavily to background within the gel.

In this lab, the nine commonly used downstream primers are:

5′ TTTTTTTTTTTAA

5′ TTTTTTTTTTTAC

5′ TTTTTTTTTTTAG

5′ TTTTTTTTTTTCA

5′ TTTTTTTTTTTCC

5′ TTTTTTTTTTTCG

5′ TTTTTTTTTTTGA

5′ TTTTTTTTTTTGC

5′ TTTTTTTTTTTGG

If, after careful introspection, it is deemed necessary to utilize downstream primers with a 3′ AT, CT, and/or GT dinucleotide, then these may certainly be added to the primer regimen.

3. cDNA Synthesis

The first enzymatic manipulation of RNA destined for differential display is the synthesis of first-strand cDNA mediated, as usual, by reverse transcriptase. This can be performed using any of the widely available systems for cDNA synthesis; one only need invest in a first-strand cDNA synthesis system (e.g., Boehringer Mannheim No. 1483 188), allowing immediate introduction of the newly synthesized cDNA into a PCR reaction. Due to the intrinsically labile nature of RNA, the small mass of RNA used for each cDNA synthesis reaction, and the peculiarities of this technique, strong consideration should be given to the use of a non-RNase H-exhibiting reverse transcriptase (see Chapter 15 for details). Further, one-tube reverse transcriptase PCR systems, which are growing in popularity and which are excellent for many applications, ought to be avoided for DDPCR.

4. Upstream Primer Design

The design of the upstream primers is, generally speaking, less structured than that of the downstream primers. In this laboratory, a collection of 24 semi-random 10-mers are routinely used for differential display with a great deal of success. The upstream primers should be thought of as semi-random because it is not cost effective, necessary, or sane to consider the use of the 4^{10} possible upstream primers. The literature is full of "Materials and Methods" sections containing the didactically useless disclaimer that ". . . upstream primers were rationally designed . . .". This means that the upstream primers selected consist of average GC content (usually 50%) and are unlikely to anneal in an intramolecular fashion or to support primer–dimer formation. Given the relatively relaxed primer annealing conditions, and the fact that the primers are so short (nondiscriminating), only seven or eight of the bases in any given upstream primer are expected to anneal. While this relatively relaxed stringency may at first be unsettling, it is a necessary component of the DDPCR approach: The idea is to be able to amplify *any* message that nature can generate.

In one system for DDPCR (Differential Display Kit; Display Systems Biotech), the following well-designed upstream primers are provided; they are routinely used in this laboratory:

5′ GATCATAGCC	5′ CTGCTTGATG
5′ GATCCAGTAC	5′ GATCGCATTG
5′ AAACTCCGTC	5′ TGGTAAAGGG
5′ GATCATGGTC	5′ TTTTGGCTCC
5′ GTTTTCGCAG	5′ TACCTAAGCG
5′ GATCTGACAC	5′ GATCTAACCG
5′ TGGATTGGTC	5′ TCGGTCATAG
5′ GGAACCAATC	5′ GATCTGACTG
5′ GATCAATCGC	5′ TCGATACAGG
5′ TACAACGAGG	5′ GATCAAGTCC
5′ GATCTCAGAC	5′ GGTACTAAGG
5′ GATCACGTAC	5′ CTTTCTACCC

These upstream primers support the subsequent synthesis of the second-strand cDNA and, in conjunction with the downstream primer, they will support the amplification of the cDNA by symmetric PCR.

5. Round 1 PCR Amplification

It should be clear from the preceding information that there exists the potential for a very large number of PCR reactions: 9 downstream primers × 24 upstream primers × 2 different cDNA populations (minimum). It would not be reasonable to consider the use of all of these simultaneously. Instead, it is best to store 8 of the 9 cDNA synthesis reactions at −80°C; only 1 cDNA reaction at a time is thus amplified by PCR. Should interesting bands become manifest by DDPCR with this first sample, these bands can be more fully explored, and the remaining cDNA synthesis reactions investigated at a later date.

a. Reaction profile. The PCR cycling profile for differential display PCR differs somewhat from that of standard symmetric PCR amplifications. For DDPCR, the optimal reaction conditions feature short denaturation, annealing, and primer extension components; these short cycles are the key to reproducibility and representative data. In addition, it is not unusual to perform routinely as many as 40 cycles, followed by the fairly standard final primer extension at 72°C for 5–10 min. The large number of cycles is often needed to amplify the "very low" abundance RNAs (Chapter 6) to the level of detection. In so doing, these parameters favor maximum flexibility with respect to the particular upstream and downstream primer combination within any given tube; at the same time the reaction conditions are tolerant of mismatches, maximizing the likelihood that all first-strand cDNA templates will be capable of supporting amplification by PCR. Further, not all instrumentation is suitable for running DDPCR. Modification of the cycling parameters is often necessary, the exact nature of which must be determined empirically for each thermal cycler. A suggested cycling protocol is:

Initial melt	94°C – 3 min
Cycling parameters	94°C – 20 s ⎤
	40°C – 30 s ⎬ 40 cycles
	72°C – 30 s ⎦
Final extension	72°C – 5 min
Soak	4°C

b. Mineral oil. The use of mineral oil in differential display PCR is to be avoided. It is recommended whole-heartedly that these reactions be conducted in a thermal cycler with a heated lid such as the Perkin Elmer Model 9600 or Model 9700, or other authorized thermal cycler.[6] While differential display PCR can certainly be performed with a mineral oil

---

[6]A list of authorized thermal cyclers is available by contacting P.E./Applied Biosystems, Director of Licensing, 850 Lincoln Center Drive, Foster City, CA 94404. Tel: 650/570-6667.

overlay, the use of mineral oil is often responsible for decreased numbers of bands and correlates well with nonreproducibility of the banding pattern, which, for this technique especially, can have catastrophic consequences. If a thermal cycler with a heated lid is not available in the laboratory, it is in the best interests of the investigator to make friends with someone who has one.

c. Enzyme blends. A powerful emerging technology known as long-range PCR exploits the combined abilities of two different enzymes to generate very large PCR products, and do so with high fidelity. This approach has been applied in this laboratory for DDPCR. Although the use of these enzymes may not be intended to generate genomic-length PCR products, the *efficiency* of two different thermostable enzymes working concertedly results in the amplification of cDNAs that correspond to very low abundance mRNAs; these sequences might not otherwise be detectable.

Succintly, enzyme blends often consist of *Taq* polymerase with some other thermostable polymerase. There are many different combinations of enzymes that can be used, one common example of which is *Taq* used in conjunction with the *Pwo* polymerase (Table 2). *Taq* polymerase lacks a proofreading activity (3′→5′ exonuclease) and can routinely generate PCR product up to 2–3 kilobases with minimal fanfare. *Pwo* polymerase has a proofreading activity and can thus reduce the error frequency associated with *Taq* severalfold. Together in the same reaction tube, these enzymes overcome the length limitation associated with *Taq* alone, and with high fidelity. Because of the enhanced stability that *Pwo* exhibits, it can continue to synthesize product long after *Taq* has been heat inactivated. This translates into an opportunity for amplification of rare cDNAs that might otherwise remain below the level of detection. Further, be-

TABLE 2

Collaborative Efforts of Thermostable Enzyme Blends

	Taq Polymerase	*Pwo* Polymerase
Half Life	> 1 h at 95°C	> 2 h at 95°C
5′ → 3′ polymerase	+++	+
5′ → 3′ exonclease	+	−
3′ → 5′ exonuclease (proofreading activity)	−	+
Source	*Thermus aquaticus*	*Pyrococcus woesei*

cause of the proofreading capacity of the enzyme combination, data derived upon sequencing are much more reliable.

d. Appropriate controls. It is of critical importance to be able to document the reliability of data, and in a technique as prone to error as DDPCR, suitable controls are all the more invaluable.

First, always run a no-reverse transcriptase (RT⁻) control whenever RNA is to be reverse transcribed. The failure to generate product will support the notion that the RNA prep is free of contaminating genomic DNA. If PCR product is generated in the control reaction, the investigator must DNase-treat the sample (with RNase-free DNase I) before moving forward. It is a tremendous time saver to automatically DNase-treat all RNA samples during the isolation protocol, regardless of the technique used to harvest the RNA, rather than having to back-track, DNase the samples, and then reprecipitate the sample. One should assume that all RNA samples are tainted with genomic DNA, even samples purified in density gradients (e.g., CsCl), until treated with RNase-free DNase I.

Second, be certain to run control reactions with only one primer, to show that the primers are not self-priming. One method that is useful in this regard is to kinase (5′ end label) the downstream primers. Thus, the only PCR products visible by autoradiography are those generated with an upstream *and* downstream primer, as opposed to upstream primers alone.

Third, a simple way to ensure authenticity of a differentially amplified product is to assay two concentrations of the same RNA sample. A true positive, one which will generate a signal with a least two starting concentrations of the same RNA, and no signal at any concentration of the other RNA, can be further investigated. This approach precludes wasting time investigating artifactual variations, that is, differences in signal intensity due to the way the samples were pipetted together or the way the gel was loaded.

Fourth, it is wise to identify internal markers within a given set of reactions electrophoresed on the same gel; that is, look for the superabundant sequences in the same relative position on the gel. A very short exposure to X-ray film will show these strong signals only, and not the lesser strength signals: This is a good way to make a fingerprint of the reaction and is way to check efficiency each time a set of samples is assayed because these abundant "landmark" samples should show up each time an experiment is repeated.[7] Differential display PCR is at least semi-quanti-

---

[7]This is analogous to the identification of internal markers in peptide maps generated by 2 D gel electrophoresis.

tative[8] at this level, reflecting the complexity of the RNA starting material: Signal intensities of electrophoretically resolved products are in proportion to the abundance of the template material that supported the synthesis of these PCR products.

e. Pilot study. In view of the many inherent difficulties associated with differential display PCR in general, it is strongly recommended here that a pilot study, or at least a pilot set of reactions, be developed before actually investing labor and materials on samples that may well be irreplaceable. It often becomes apparent early on that reproducibility is a key concern. Precisely because of the nature of the primers involved, hot start PCR may also be helpful.

6. Round 1 Product Analysis

The conclusion of the first round of PCR is a very exciting time in the DDPCR protocol. The observation of gene modulation, perhaps involving never before observed sequences, is imminent. Autoradiography is the surest method for seeing everything there is to see, though other methods may, in some instances, render satisfactory observation of the myriad PCR products as well. The following are methods for visualizing DDPCR products from most to least sensitive:

^{35}S dATP

^{33}P dCTP

Silver staining

SYBR Green I

Ethidium bromide

For any preparative method, ^{35}S or ^{33}P is indispensable. However, silver staining has been used successfully in the identification of modulated sequences by DDPCR (Lohman *et al.*, 1995); while it does not offer the level of sensitivity afforded by radiolabeling as of this writing, it certainly shows great promise. SYBR Green I can be used to simply show the high and medium abundance products and, on occasion, one *might* observe a difference in expression. Ethidium bromide is useful only for showing that the reactions worked; in the beginning, it might be helpful to elec-

---

[8]There is a limit to the quantitativeness of DDPCR, and one should realistically expect data generated by this technique to be only semi-quantitative at best. Subsequent analyses, including nuclease protection (Chapter 18) and competitive PCR (Chapter 17) are used to measure changes in gene expression in a much more quantitative manner.

trophorese a small aliquot through 3% agarose and then stain it with SYBR Green or ethidium bromide before making the effort to run and analyze the DDPCR products on a polyacrylamide gel.

Whereas the literature is ripe with protocols calling for the use of denaturing gels (6% acrylamide, 8.3 M urea) for the analysis of DDPCR products, nondenaturing gels have also been used with great success. If absolutely necessary, the DDPCR reactions can be concentrated with a Speed-Vac for no more than 10 min prior to loading the gel, though it will not improve resolution or sensitivity. This vacuum-concentrating approach should be avoided altogether when using denaturing gels because it causes volume variations among the samples. If the nondenaturing approach fails to generate easily resolvable bands, then repeat the electrophoresis using a standard DNA sequencing gel. Most often, gels are run at constant power of 50 W until bromophenol blue reaches the bottom of the gel. For autoradiography, the gel is then dried down under vacuum onto Whatman paper (80°C for 2 h), and exposed to film overnight. Regarding the recovery of bands of interest: A most worthwhile strategy is to make two X-ray film exposures of the dried-down gel. The first is reserved as a record of the experiment, and the second is used to physically overlay the gel after it has been rehydrated to facilitate band recovery. To confirm that the correct band has been recovered, dry down the gel again, and do a third exposure to show that the DNA is no longer there. Last, the reamplification of the cDNA from the nondenaturing gels is much easier; with denaturing gels, overnight dialysis is strongly suggested to remove the urea, followed by ethanol precipitation of the cDNA fragments prior to reamplification. When purified from nondenaturing gels, one may use the gel slice directly for PCR reamplification.

7. Round 2 PCR Amplification → Reamplification

After DNA products believed to be differentially expressed have been recoved from the Round 1 gel, they must be reamplified for subsequent characterization. The cycle parameters should be identical to those used to generate the product in the first round amplification, after which the size of the product can be estimated by agarose gel electrophoresis (2–3%).

The polyacrylamide gel slice is generally placed in 100 μl sterile water for 30–60 min, during which time enough of the DNA should leach out of the gel slice such that, upon using an aliquot of the water in a second amplification, the PCR product should be regenerated, often in a 50- to 100-μl reaction. If reamplification is problematic, try reducing the amount of water into which the gel slice is placed and then heating the sample to

90°C for 10 min. Although some protocols have suggested boiling the gel slice, the amount of usable template which results is often greatly diminished, compared to heating alone. It may also be helpful to increase the concentration of the primers two- to fivefold. Further, it may be helpful to practice the reamplification protocol using nonmodulated cDNAs to streamline the mechanics of reamplification.

Product Variety: What to Expect

Phenotypic change unquestionably involves several genes, some of which may, or may not, be suspected by the investigator as being differentially expressed. A key advantage of differential display PCR is that it favors the simultaneous display of all transcripts present in the cell at the moment of cell lysis. The speed and sensitivity of PCR facilitates the detection of very low abundance transcripts in a relatively short time frame. A most important concept here is that differences in gene expression are manifested in the form of PCR products, though the investigator has no idea what these sequences are until they are subsequently characterized.

Typical DDPCR reactions generate usable products that range from 50 to 800 bp, though it is not uncommon to clone sequences in excess of 1 kb. Upon electrophoresis, DDPCR products manifest themselves as being highly abundant, not-at-all abundant, and everywhere in between. By running the bromophenol blue dye front to the bottom of a polyacrylamide gel, unincorporated nucleotides will run off the gel, products that range up to 700–800 bp are usually resolvable, and products larger than 800 bp begin to show signs of compression. Clearer definition may be given to the larger molecular weight products by double loading the gel, as is often done for DNA sequencing purposes. This may be especially valuable when enzyme blends are used to generate the PCR products.

The greatest sense is made from the data when all of the PCR products generated using the same downstream primer are electrophoresed on a single gel. "State A" PCR reactions are loaded adjacent to "State B" PCR reactions, in order of upstream primer used; an example of this loading strategy is shown in Table 3. This permits the direct side-by-side comparison of the products from each pair of reactions. Products of identical molecular weight will be observed when cells in State A and cells in State B are producing the same transcript (Figure 2); the appearance of band without a corresponding band in the adjacent lane suggests differential expression (Figure 3). Conveniently, 48-tooth combs used for DNA sequencing permit the loading of all primer combinations, using a single

TABLE   3

Gel Loading Strategy for DDPCR

Lane	cDNA source	Downstream primer	Upstream primer
1	State A	3	1
2	State B	3	1
3	State A	3	2
4	State B	3	2
5	State A	3	3
6	State B	3	3
7	State A	3	4
8	State B	3	4
9	State A	3	5
10	State B	3	5
11–48	*Same strategy*		

Note. PCR products generated using identical primer combinations are loaded into adjacent lanes, permitting the direct side-by-side comparison of the variety of cDNAs amplified. Bands of equal molecular weight in adjacent lanes correspond to sequences transcribed in both cell types. Differences in intensity, including the complete absence of a band in one of the lanes, suggest differential modulation. Standard size DNA sequencing gels favor excellent resolution, and a 48-tooth comb permits loading all of the upstream primer combinations, with a single downstream primer, for both starting cell types.

downstream primer, on the same gel (2 cell samples × 1 downstream primer × 24 upstream primers). The use of a molecular weight marker is not required in Round 1 amplification, the purpose of which is merely to highlight the regulated sequences. Molecular weight standards are more valuable in the Round 2 screening.

In all probability, the RNA was originally prepared by guanidinium–acid–phenol extraction, a perfectly legitimate method. Recall, however, that such chaotropic reagents yield mRNA from the cytoplasm mixed with unspliced heterogeneous nuclear RNA (hnRNA). Thus, there exists the potential that a sequenced PCR product may show both exon and intron sequences. In addition, it is often the case that two different size cDNAs, corresponding to the same transcript, are identified on the same gel. This occurs because in both cases all of the molecules of a specific transcript are reverse transcribed by the same downstream primer, though two or more upstream primers (different reaction tubes) find complementary sites on the cDNA, thereby supporting amplification by PCR. Although this seems to frustrate many, the phenomenon should be viewed as a means of maximizing the chances that any given transcript will be amplified.

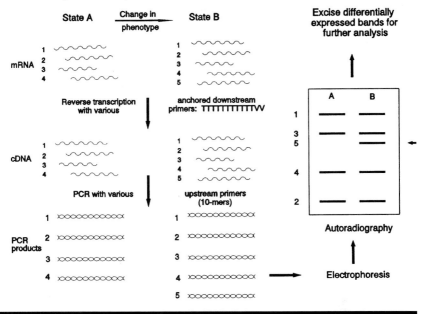

FIGURE 2

Schematic showing the general method used for mRNA differential display. Transcripts produced in both cell types are displayed simultaneously as PCR products when identical primer combinations are used. The presence of a PCR product from only one sample strongly suggests differential transcriptional activity of the corresponding gene.

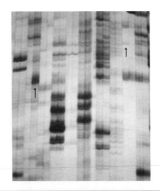

FIGURE 3

mRNA differential display. Typical appearance of DDPCR products following electrophoresis and autoradiography of the first-round amplification. Identical bands appearing in adjacent lanes represent genes transcribed in common. Bands unique to one lane in a set are potentially differentially expressed sequences and are, therefore, candidates for further characterization. Notice the differing intensities of the various product pairs, directly reflecting the relative abundance of the corresponding transcripts.

Protocol: mRNA Differential Display

The following method is based on the use of the 9 downstream primers and 24 upstream primers described above. The method is systematic and thorough, allowing the investigator to stop at any of several points in the protocol without loss of sensitivity or resolution. The anchored downstream primers should be prepared as a 25 μM stock solution, and the upstream primers prepared as a 2 μM stock solution (e.g., Display Systems Biotech).

I. Synthesis of cDNA

1. Ensure that high-quality RNA is available to support the synthesis of first-strand cDNA. This is ascertained by minigel electrophoresis and visualization of the 28S and 18S ribosomal RNA (rRNA), with minimal smearing below the 18S rRNA. For these reactions, dilute the RNA to a concentration of 100 ng/μl in sterile H_2O and store on ice. It is best not to rehydrate an RNA ethanol precipitate until just prior to use.

2. Prepare the cDNA master mix (sufficient for 18 reactions):

38.0 μl 10× first-strand reaction buffer (100 mM Tris, 500 mM KCl; pH 8.3)

76.0 μl 25 mM MgCl$_2$

9.5 μl 10 mM dNTP mix

19.0 μl RNase inhibitor (50 U/μl)

13.3 μl H_2O

15.2 μl AMV reverse transcriptase

Set aside on ice.

Note: The required volume is dependent on the number of downstream primers to be used and the number of replicates for each. Scale up or down accordingly.

3. Label a set of microfuge tubes for each sample of RNA to be reverse transcribed. For convenience, label tubes for RNA from cell Type A A1–A9, and those for RNA from cell Type B B1–B9. Add 3.0 μl of each 25 μM downstream primer stock solution to a different tube (only one primer in each tube).

4. Add 3.0 μl (300 ng) of RNA from cell Type A to tubes A1–A9. Add 3.0 μl of RNA from cell Type B to tubes B1–B9.

5. Add 5.0 μl of DEPC-H_2O to each tube.

6. Heat tubes to 70°C for 10 min. Immediately cool on ice.

7. *Briefly* pulse-centrifuge tubes to collect sample volume at the bottom of the tube.

8. Add 9.0 µl of the cDNA master mix (Step 2) to each tube.

9. Incubate cDNA synthesis reactions at 42°C for 1 h.

Note: MMLV reserve transcriptase may favor slightly larger cDNA products. If using the MMLV enzyme, incubate at 37°C for 1 h (be sure to use MMLV first-strand reaction buffer).

10. Heat tubes to 95°C for 5 min to heat inactivate the reverse transcriptase. cDNA samples to be used immediately can be placed on ice. Other cDNA samples should be pulse-centrifuged and stored at –80°C.

Note: Failure to heat inactivate the reverse transcriptase will almost certainly have a negative influence on the efficiency of downstream applications. Failure to store the newly synthesized cDNA properly will compromise its stability.

II. Amplification of cDNA Subpopulations by PCR

At the conclusion of cDNA synthesis, the samples are ready for immediate amplification by PCR. It is not reasonable to work with more than one downstream primer at a time, because of the large number of upstream primers, and reactions, that should be tested. One of the nine cDNA synthesis reactions should be selected, and the other eight stored frozen. Accurate pipetting is essential for reproducibility. If the detection method will be autoradiography, the master mix must include [α-^{33}P]dCTP (10 mCi/ml) or [^{35}S]dATP.

1. Prepare DDPCR master mix as follows (sufficient for 48 reactions):

    100.0 µl 10× PCR buffer (without $MgCl_2$)

    120.0 µl 25 mM $MgCl_2$

    5.0 µl [^{35}S]dATP (10 mCi/ml)

    10.0 µl 500 µM dNTP

    10.0 µl *Taq* polymerase

    361.0 µl DEPC-H_2O

Note 1: It is most judicious to prepare a single, large volume master mix for all of the PCR reactions that will be run, excluding the Taq polymerase and the radiolabel. The volume would be 9 times larger than shown above (9 downstream primers). This master mix is aliquoted into 9 microfuge tubes, which are stored at –20°C.

Note 2: The reactions are optimized at 3 mM MgCl₂.

Note 3: For PCR products larger than 450 bp, increase dNTP concentration from 5 μM, as written, to as much as 15 μ*M.*

2. Add 108 μl downstream primer (use the same primer as in the cDNA synthesis reaction, above) to the master mix.

3. Split this DDPCR master mix into two different tubes, one for Type A cDNA, and the other for Type B cDNA.

4. Add 18 μl of each cDNA synthesis reaction to a master mix aliquot, maintaining one master mix for Type A cDNA and one for Type B cDNA.

5. Distribute 15 μl of the premixed components (this protocol, Steps 1–4) into 24 microfuge tubes.

6. Add 5 μl of upstream primers to each microfuge tube (only one upstream primer per tube).

7. Pulse centrifuge briefly to collect all components at the bottom of the tube.

8. If using a thermal cycler without a heated lid, overlay each reaction with only 20 μl mineral oil. Otherwise, skip this step.

9. Amplification protocol:

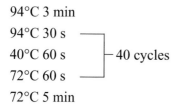

    94°C 3 min
    94°C 30 s
    40°C 60 s   ⎫ 40 cycles
    72°C 60 s   ⎭
    72°C 5 min

III. PCR Product Analysis

Without a doubt, optimum resolution is favored by PAGE analysis of the PCR products. It is not necessary to run a denaturing (i.e., urea-containing) polyacrylamide gel; most of the time nondenaturing gels work just as well.

A. Optional: Preliminary Analysis. To determine the relative efficiency of a set of DDPCR samples, an aliquot of each can be briefly electrophoresed through a 3% agarose gel made up in 1× TBE or 1× TAE buffer, followed by staining with SYBR Green or even ethidium bromide. After 20 min at 100 V, the gel can be inspected for the presence of DDPCR products (Figure 4). Failure to observe anything most likely means that the reaction efficiencies were poor and that a long autoradi-

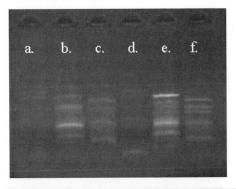

FIGURE 4

Appearance of DDPCR products after first-round amplification when stained with ethidium bromide. Aliquots of six samples were electrophoresed in a 3% agarose gel supplemented with 1 μg/ml ethidium bromide. This diagnostic can be used as a preliminary indicator of the efficiency of the reactions. Notice that, even with this very insensitive method of visualization, some differences are obvious among the samples.

ographic exposure will be required to fully visualize the sample. Staining with fluorescent dyes, while certainly an insensitive detection method, almost always gives the investigator something to look at when the reaction is robust.

1. Melt 3.0 g agarose in 105 ml 1× TAE buffer. This will be enough agarose solution to prepare two 12 × 14-cm gels, each consisting of a double row of wells.

CAUTION   **Agarose at this high concentration tends to superheat and boil over quite readily. Handle with great care.**

2. After boiling, do not wait too long for the gel to cool because at this high concentration, gel polymerization occurs rapidly. If staining with ethidium bromide, add 20 μl of 10 mg/ml ethidium bromide directly to the hot agarose and cast the gel. If staining with SYBR Green I, cast the gel without prior addition of this dye. Gels are stained with SYBR Green after electrophoresis.

3. Remove no more than 5 μl from each DDPCR sample, and add 1 μl gel loading buffer. For easier pipetting, add 5–10 μl of water to each 5-μl DDPCR aliquot and 2 μl of gel loading buffer.

4. Electrophorese at 100 V until dye has migrated 2–3 cm into the gel.

5. Gels containing ethidium bromide may be viewed directly by UV transillumination. Otherwise, submerge gels in a solution of 2× SYBR

Green I prepared in 1× TAE buffer. After 15 min, observe gels. Longer staining times in SYBR Green may enhance staining of weak bands. Photograph.

CAUTION   Be sure to observe all safety precautions when working with or near a UV transilluminator, when working with mutagenic fluorescent dyes, and when working with radioactivity.

B. Nondenaturing polyacrylamide gel electrophoresis. The preparation of polyacrylamide, denaturing and nondenaturing, is reserved for individuals experienced in the handling of monomeric acrylamide, a rather nasty neurotoxin. Unless working in a lab where the *proper* handling of such reagents is commonplace, it is quite prudent to buy the required polyacrylamide gels precast; these may be purchased to fit virtually any electrophoresis configuration and of any composition. Routinely used is a 6% polyacrylamide gel, electrophoresed in 1× TBE.

The resolution of DDPCR products by nondenaturing PAGE is usually excellent. One need not switch to a denaturing gel system unless smearing in the lanes is a chronic difficulty.

1. **Wearing gloves!** Carefully unwrap precast gel, and rinse wells with distilled H_2O. Keep gloves on at all times, and never assume that a polymerized gel is not a potential health hazard.

2. Assemble PAGE apparatus, clamping gel into place.

3. Pre-run the gel for approximately 20–30 min.

4. Transfer one-half of each DDPCR reaction (10 μl) to a fresh microfuge tube, and add 10 μl of nondenaturing gel loading buffer (0.2% bromophenol blue; 10 mM EDTA, pH 8.0; 20% ficoll). The remaining 10 μl of each sample should be stored at –80°C until needed again.

5. After mixing with loading buffer, apply 5 μl of each sample to wells of the gel.

6. Run gels at a constant power of no more than 50–60 W, until the bromophenol blue dye has reached the bottom of the gel.

7. Carefully remove the gel from the electrophoresis chamber. Dry gel down onto a sheet of Whatman filter paper, for 2 h at 80°C. Do not fix the gel in methanol/acetic acid.

CAUTION   Keep in mind that DDPCR reactions containing radiolabel will contaminate the gel box and electrophoresis buffers with radioactivity. Handle and dispose of contaminated equipment and reagents safely, and in accordance with in-house policy.

8. Set up autoradiography as usual and make at least one overnight exposure.

Note: A second exposure is ideal, should the first film become damaged during the process of band excision from the gel.

9. Develop film(s) and examine for the presence of differentially expressed sequences.

Note: One of the more time-consuming aspects of this technique is the analysis of the data and the identification of bands that are of some interest. Some of the better quality image analysis software can scan digitized images of these gels and pick out the bands that are clearly present in one sample and clearly absent in the other, as well as being able to determine which band in a pair differs by more than a predetermined amount, that is, a five-fold or better induction or repression.

Protocol: Identification and Selection of Differentially Expressed Sequences

As suggested above, nondenaturing gels are recommended rather than traditional denaturing gels. This is because denaturing gels often afford too much resolution for the analysis of DDPCR products, resulting in the appearance of two or more bands on a denaturing gel for every one band on a nondenaturing gel. The reasons for this include (a) DNA strand separation and (b) the addition of an extra nucleotide (usually adenosine) by the *Taq* polymerase at the 3′ end of the PCR product.[9] The complexity of the autoradiogram, therefore, makes recovery of discrete bands difficult.

1. Using the autoradiogram as a guide, cut the bands of interest from the filter paper, and transfer each to a separate microfuge tube containing 100 μl sterile H_2O.

2. Allow the tubes to sit at room temperature for 30 min. Heat each tube to 90°C for 10 min and then centrifuge to collect the residual gel and filter paper at the bottom of the tube.

3. Carefully transfer the supernatant to a fresh tube. The eluted DNA is contained in this supernatant.

4. Perform reamplification of eluted DNA using the same upstream and downstream primers that generated the band in the Round 1 amplification.

25.0 μl eluted DNA (H_2O from Step 3)

5.0 μl 10× PCR buffer (no Mg^{2+})

---

[9]The addition of a single nucleotide 3′ overhang does not occur when using a proofreading thermostable polymerase, alone or in combination with *Taq*.

6.0 μl 25 mM MgCl$_2$

5.0 μl 500 μM dNTP

5.0 μl 2 μM downstream primer (dilute 25 μM stock)

5.0 μl 2 μM upstream primer

1.0 μl *Taq* polymerase (5 U/μl)

Note: In the event that reamplification difficulties are encountered, reduce the MgCl$_2$ concentration to 1.5 mM, increase the concentration of each primer to 1 μM, and increase the number of reamplification cycles (below) to 40.

5. Reamplify the eluted DNA using the same PCR amplification profile used to generate the original products:

Initial melt:	94°C 3 min	
Cycling profile:	94°C 30 sec	
	40°C 60 sec	30 cycles
	72°C 60 sec	
Final extension:	72°C 5 min	

6. At the conclusion of the second round amplification, remove a 10-μl aliquot, mix with 2 μl loading buffer, and electrophorese on a 2.5% agarose gel (1× TBE or 1× TAE), along with PCR-range molecular weight standards (e.g., Promega PCR Marker, Cat. No. G-3161).

7. Stain gel with SYBR Green or ethidium bromide. Photograph.

Recovery of Differentially Expressed Sequences by Affinity Capture

The amplification and reamplification of DDPCR products, while a milestone in the quest to identify differentially modulated genes, is often not sufficient to guarantee the homogeneity of the product. For example, in Northern analysis, it is not uncommon to see multiple bands if a non-purified DDPCR product is used as a nucleic acid probe. What appears to be a single band even after reamplification is often two or more different cDNAs "hiding" under the appearance of a single band; these are related to one another only by the sequences at their respective ends, which are recognized by the primers. Unless these products are physically separated, there is no guarantee as to the reliability of the data from subsequent applications.

In Northern analysis, probe sequences base pair with the full-length

RNA that originally supported reverse transcription into cDNA. If the probe is heterogeneous, then each unique sequence will hybridize to a different full-length transcript. Moreover, when total cellular RNA is used as target material in the Northern analysis validation of DDPCR products, it is not unreasonable to expect multiple bands that represent unspliced, precursor forms of the same transcript (hnRNA).

There are two standard approaches for separating DDPCR-generated products. First, the investigator may wish to clone the sequences directly into a vector suitable for downstream applications, as described below. Direct ligation is perhaps the most expedient, though unsophisticated approach. In some cases, however, it is possible to use the DDPCR band (consisting of one or more sequences) as a probe for Northern analysis. Succinctly, aliquots of the same, or at least nearly identical, RNA used to perform DDPCR are electrophoresed under standard denaturing conditions, and then blotted onto a nylon filter (as described in Chapter 11). The products of DDPCR reamplification are then labeled (radiolabeling or chemiluminescence) and used as a probe for Northern analysis. The notion behind this approach is that one of the two strands of DDPCR-generated cDNAs will base pair to the same RNA species that originally supported its synthesis. If more than one cDNA is present, then these will find and base pair with their respective parental RNA transcripts. Upon detection, the signal indicates the location(s) of the hybridized probe. Moreover, a quantifiable signal in one lane (e.g., State A RNA), and absence thereof in the adjacent lanes (e.g., State B RNA), is very convincing proof that one is dealing with a truly differentially regulated sequence.

When this approach works, it is quite a time-saver. The bad news, however, is that many cDNAs generated by PCR often come from RNA transcripts that are well below the level of detection by traditional Northern analysis, even in cases of poly(A)$^+$ selection.

How, then, is the physical separation or "capture" of these different cDNAs accomplished? Upon appearance of a signal by Northern analysis, the region of the nylon filter paper containing the hybridized cDNA is literally cut out, using the X-ray film as a guide, in order to sequester each observable band. This small piece of nylon, containing the immobilized RNA target and hydrogen-bonded cDNA probe, is placed in a microfuge tube containing 100 μl sterile H_2O. The tube is then heated to 95°C for 5 min, a high-stringency environment in which the cDNA will be eluted from the filter. The water in the tube is then used to assemble a PCR mix in which (a) the missing strand, that is, second-strand cDNA, will be synthesized and (b) the now homogeneous product will be amplified nicely for further characterization. This method is known as affinity capture, and perhaps might best be thought of as a reverse Northern blot (Zhang *et al.*, 1996). Not only is this a method for the separation of a mixture of identical length sequences, it is potentially a method for confirming the differential status of purified cDNA.

Cloning PCR Products

In general, the same methods used for cloning PCR products generated in any other PCR-based assay may be applied here. Insertion into a vector is another method by which a mixture of DDPCR-generated cDNAs (same size, different sequences) can be separated. Currently, the most popular methods include TA cloning, in which adenosine-tailed PCR products are ligated into a thymidine-tailed vector; and TopoCloning, in which the "rejoining" activity of DNA topoisomerase I is exploited for rapid PCR product–vector ligation (Chapter 15).

Cloning of PCR products may facilitate sequence and propagation of these clones, though it is not absolutely necessary. Because it is unlikely that more than one insert will be ligated into a single vector molecule, the bacterial colonies that result when the ligation mix is plated into selective media following transformation represent vectors containing a single insert. In order to authenticate a colony as a true transformant, 10 colonies (or another reasonable number) are picked randomly with sterile toothpicks. The bacteria on the toothpick are used to streak a fresh agar plate, for archiving the clone, and are also used to inoculate a fresh PCR mix containing the same upstream and downstream primers used to generate the DDPCR products. Agarose gel analysis of these products should manifest a band that is precisely the same size as the insert.

Confirmation of Differential Expression

After DDPCR is run once using all primer combinations, a second run is conducted using only those primer pairs that show differences in the first run. To an extent, this approach favors the elimination of false positives. Of course, the exact experimental milieu that defined the first run should be duplicated as closely as possible, to afford an opportunity to reproduce the data. One should not expect, however, that any two runs will be precisely reproducible because (a) cell culture conditions are impossible to duplicate (gene expression is dependent on cell density, number of cell doublings, conditioning of media, etc.) and (b) tissue samples from different subjects are exactly that: biochemically different subjects. While many similarities are to be expected, many differences should not be un-expected. To maximize the reproducibity of different runs from the same sample, the investigator should aliquot the RNA during the initial isola-tion procedure such that the RNA is precipitated in several aliquots, only one of which is centrifuged and used as cDNA template material. Upon completion of the entire DDPCR protocol, the investigator will then have access to a fresh aliquot of RNA that was isolated on the same day, by the same person, and that represents precisely the same biochemical status of the cells.

At present, there are four reliable methods for assessing whether a se-quence purported to be differentially expressed, using the differential dis-play method, truly represents a biochemical change in cell: Northern analysis (Chapter 11), ribonuclease protection assay (RPA; Chapter 18), nuclear runoff assay (Chapter 19), and single-strand conformation poly-morphism (SSCP; Orita *et al.*, 1989). SSCP is a simple technique, based on the fact that the migration of single-stranded DNA in polyacrylamide gels is dependent on the actual nucleotide composition of the molecule. For review and protocols, see Fujita and Silver (1995).

Subsequent Characterization

Northern analysis, RPA, and nuclear runoff assay are used to further characterize the transcriptional behavior of a gene whose cDNA has been purified. These methods, while excellent from a gene expression perspec-tive, do little to identify the actual gene(s) under investigation. Because of the ready availability of DNA sequencing, there is no reason the identity of clones cannot be quickly ascertained. In view of the relatively short na-ture of DDPCR-generated sequences, it is probably not worth the time and effort to generate a restriction map; the distribution of restriction sites will be known when the clone is sequenced!

The primers used for DDPCR are too short for standard sequencing protocols. To facilitate sequencing of DDPCR-generated products, ex-tended primers can be synthesized and used during the reamplification step. These extended primers feature 5′ sequencing promoters, such as T3 and SP6; it may be especially convenient to design the primers with desirable restriction enzyme sites as well. Thus, reamplification is sup-ported using Primers A′ and B′, rather than A and B, will support the di-rect sequencing of DDPCR products (Reeves *et al.*, 1995; Wang and Feuerstein, 1995).

The isolation of differentially modulated sequences opens many possibilities for gaining a clearer understanding of this aspect of gene regulation in the cell. Isolated DNA sequences can be reamplified repeatedly, constituting a limitless source of nucleic acid probe; probe synthesis by merely 15 cycles, in the presence of [α-^{32}P]dNTP, followed by 5 min at 72°C, is all that is required. Probe sequences can be used to perform gene expression assays using previously untested RNA samples, or to screen libraries in order to identify the larger portions of the gene as well as flanking regulatory elements.

Applications of Differential Display

Virtually any model in which experimental manipulation is likely to cause any changes in the cellular biochemistry is a suitable candidate for evaluation using DDPCR technology. mRNA differential display affords a side-by-side comparison of mRNA from different sources and is excellent for discerning both the induction and repression of genes. Examples include, but certainly are not limited to, the study of gene modulation

- in cell culture in the presence of pharmacological concentrations of growth factors or drugs;
- when comparing quiescent to senescent cells to log-phase cultures;
- as a function of seasonal changes in fruits and vegetables;
- in response to the transfection of foreign genes;
- as new markers for diagnostic purposes;
- as a function of differentiation;
- for the isolation of new members of multigene families;
- to discern transcriptionally regulated sequences versus those that are posttranscriptionally regulated (relates to method of RNA preparation; refer to Figure 2 in Chapter 3).

Trouble-Shooting mRNA Differential Display

Among the more common complaints induced by the use of DDPCR is the seemingly nonreproducible nature of the reactions, and there appear to be limitless possibilities to explain this. Among the peculiarities of this method is a particular sensitivity to primer quality; the reactions work best when they are HPLC purified. For this reason it is strongly recommended that the first-time user invest in a kit for DDPCR (Table 4), mostly because of quality control, which is often not possible when primers are synthesized in a typical in-house core facility.

		TABLE 4

DDPCR Kits

Company	Kit	Catalog No.
Display Systems	Diffential Display PCR Kit	62-6088-02 (PGC Scientifics)
GenHunter	RNAimage Kit	G501
Clontech	Delta Differential Display Kit	K-1810-1
Stratagene	AP-PCR Kits	200440 - 200443
Genomyx	Hieroglyph	GX418-1

The other peculiarity is the exquisite sensitivity of DDPCR to the precise concentration of all reaction components; thus, accurate pipetting and the use of master mixes for both cDNA synthesis and PCR are essential. The items below outline what may be done to minimize aggravation and maximize reproducibility and productivity.

1. Nonnegotiables

A. High-quality RNA

- good RNase-free technique
- guanidinium thiocyanate- or guanidinium HCl-based lysis buffers[10]
- acid–phenol extraction
- double precipitation with guanidinium lysis buffer
- DNase treatment; must be RNase-free
- wash RNA pellets thoroughly/extensively with 70% ethanol prepared in DEPC-treated H_2O[11]
- proper storage of purified RNA[12]

B. Reverse transcription into cDNA and PCR:

- equal mass of RNA per cDNA reaction (200–300 ng)
- enzymes lacking endogenous RNase H activity may be helpful
- avoid Mn^{2+}-dependent reverse transcriptase activity

---

[10]NP-40 lysis buffers can also be used, though considerably greater skill is necessary to generate representative DDPCR products beginning with RNA so purified.

[11]Recall that cationic salts bind dNTPs quantitatively and can cause havoc in the cDNA synthesis and PCR reactions.

[12]DDPCR is less tolerant of partially degraded RNA than other PCR-based applications.

- always work with a cDNA master mix
- always work with a PCR master mix

2. Common Problems

A. Heavy smearing in one or more lanes:

- optimize Mg^{2+} concentration (usually a reduction; decrease incrementally by 0.1 mM)
- reduce the number of cycles to as few as 30
- decrease the amount of enzyme in the reaction
- slightly increase annealing temperature

Note: In some cases, smearing may be reduced if Cycles 1–5 are annealed at 40°C, and Cycles 6–40 are annealed at 45°C. Be aware, however, that this strategy may reduce the variety of bands by as much as 25–30%.

B. No product:

- check the RNA!
- add an RNase inhibitor (e.g., RNasin)
- test the components of the RT and PCR reactions, especially the enzymes
- check off each component as the reactions are pipetted together
- check PCR cycle parameters

C. Low yield:

- incomplete homogenization or lysis of samples
- final RNA pellet incompletely redissolved
- $A_{260}/A_{280} > 1.65$ (suggests protein or other contaminants)
- RNA degradation:
  - tissue was not immediately processed/frozen after removing from the animal
  - purified RNA was stored at –20°C, instead of –70°C
  - aqueous solutions or tubes were not RNase-free

D. DNA contamination:

- sample was not DNase-treated

E. Nonreproducibility of data:

- use thin-walled tubes (200 μl work well); for thick walled tubes, extend cycle parameters
- consider hot start PCR (described in Chapter 15)
- prepare single, large volume master mix per cell type and aliquot for each reaction
- match T_m of upstream and downstream primer pairs more closely
- consider enzyme mixtures designed for long PCR (e.g., *Taq* + *Pwo*)
- consider elimination of mineral oil: use a thermal cycler with a heated lid
- RNA not isolated by the same method
- RNA not isolated on the same day
- cells/tissue not in the same state (different parts of the cell cycle; the tissue was isolated from biochemically different individuals; late log vs. early log; senescent vs. quiescent)
- different primer pairs used/misused
- different master mix used
- optimize and run reactions in the same thermal cycler (not model number)

3. General Suggestions

- Keep denaturation steps as short as possible
- Keep denaturation temperature as low as possible (90–92°C)
- Elongation temperature can be reduced to 68°C
- Extend elongation time for each successive cycle (5–20 s per cycle)
- Accurate pipetting is critical for reproducibility. Have micropipettors recalibrated before you begin, and at regular intervals thereafter. Positive displacement pipettors may be especially useful for the elimination of aerosol carryover contamination.
- The mRNA differential display technique is especially sensitive to the exact input mass of RNA. As with most RNA PCR reactions "less is more," meaning that it is counterproductive to overload the reaction with starting material (200–300 ng of RNA is optimal).
- Unlike many PCR-associated techniques, differential display seems to be quite sensitive to the quality of the starting material. The high-

est quality RNA is generally prepared in guanidinium-containing buffers.

- It is well worth investing in a kit that contains a selection of primers, rather than trying to have primers made to order or trying to synthesize them in-house.

- Silanized microcentrifuge tubes may be helpful in minimizing losses during the various manipulations because cDNA/PCR products are present in picogram quantities (or less) prior to reamplification (see Appendix F for protocol).

References

Bauer, D. *et al.* (1993). Identification of differentially expressed mRNA species by an improved display technique (DDRT-PCR). *Nucleic Acids Res.* **21,** 4272–4280.

Bernardi, G. (1971). *Proc. Nucleic Acid Res.* **2,** 455.

Callard, D., Lescure, B., and Mazzolini, L. (1994). A method for the elimination of false positives by the mRNA differential display technique. *BioTechniques* **16,** 1096.

Coche, T., Dewez, M., and Beckers, M. C. (1994). Generation of an unlimited supply of subtracted probe using magnetic beads and PCR. *Nucleic Acids Res.* **22,** 1322–1323.

Fujita, K., and Silver, J. (1995). Single-strand conformational polymorphism. *In* PCR Primer: A Laboratory Manual (C. W. Diefenbach and G. S. Dveksler, eds). Cold Spring Harbor Laboratory Press. CSH, NY.

Hedrick S. M., Cohen, D. I., Nielsen, E. A., Davis, M. M. (1984). Isolation of cDNA clones encoding T-cell specific membrane-associated proteins. *Nature* **308,** 149–153.

Kulesh, D. A. *et al.* (1987). Identification of interferon-modulated proliferation-related cDNA sequences. *Proc. Natl. Acad. Sci. USA* **84,** 8453.

Lambert, K. N., and Williamson, V. M. (1993). DNA library construction from small amounts of RNA using paramagnetic beads and PCR. *Nucleic Acids Res.* **21,** 775–776.

Liang, P., Averboukh, L., Keyomarsi, K., Sager, R., and Pardee, A. B. (1992). Differential display and cloning of messenger RNAs from human breast cancer versus mammary epithelial cells. *Cancer Res.* **52,** 6966–6968.

Liang, P., Averboukh, L., and Pardee, A. B. (1993). Distribution and cloning of eukaryotic mRNAs by means of differential display: Refinements and optimization. *Nucleic Acids Res.* **21,** 3269–3275.

Liang, P., Bauer, D., Averboukh, L. *et al.* (1995). *Methods Enzymol.* **254,** 304–321.

Liang, P., *et al.* (1994). Differential display using one-base anchored oligo-dT primers. *Nucleic Acids Res.* **22,** 5763–5764.

Liang, P., and Pardee, A. B. (1992). Differential display of eukaryotic messenger RNA by means of the polymerase chain reaction. *Science* **257,** 967–971.

Lisitsyn, N., Lisitsyn, N., and Wigler, M. (1993). Cloning the differences between two complex genomes. *Science* **259,** 946–951.

Lohman, J., Schickle, H, and Bosch, T. (1995). REN Display, a rapid and efficient method for nonradioactive differential display and mRNA isolation. *BioTechniques* **18,** 200–202.

McClelland, M., and Welsh, J. (1994). RNA fingerprinting by arbitrarily primed PCR. *PCR Methods Appl.* **4**, S66–S81.

Muller, H., *et al.* (1994). Cyclin D1 expression is regulated by the retinoblastoma protein. *Proc. Natl. Acad. Sci. USA* **91**, 2945–2949.

Orita, M., Suzuki, Y., Sekiya, T., and Hayashi, K. (1989). Rapid and sensitive detection of point mutations and DNA polymorphisms using the polymerase chain reaction. *Genomics* **5**, 874–879.

Reeves, S. A., Rubio, M., and Louis, D. N. (1995). General method for PCR amplification and direct sequencing of mRNA differential display products. *BioTechniques* **18**(1), 18–20.

Rodriguez, I. R., and Chader, G. J. (1992). A novel method for the isolation of tissue-specific genes. *Nucleic Acids Res.* **20**, 3528.

Sager, R., Anisowicz, A., Neveu, M., Liang, P., and Sotiropoulou, G. (1993). Identification by differential display of alpha 6 integrin as a candidate tumor suppressor gene. *FASEB J.* **7**, 964–970.

Sargent, T. D. (1987). Isolation of differentially expressed genes. *Methods Enzymol.* **152**, 423–432.

Sargent, T. D. and Dawid, I. B. (1983). Differential gene expression in the gastrula of *Xenopus laevis. Science* **222**, 135–139.

Travis, G. H., and Sutcliffe, J. G. (1988). Phenol emulsion-enhanced DNA-driven subtractive cDNA cloning: Isolation of low-abundance monkey cortex-specific mRNAs. *Proc. Natl. Acad. Sci. USA* 1696–1700.

Wang, X., and Feuerstein, G. Z. (1995). Direct sequencing of DNA isolated from mRNA by differential display. *BioTechniques* **18**(3), 448–452.

Welsh, J., Chada, K., Dalal, S. S., Cheng, R., Ralph, D., and McClelland, M. (1992). Arbitrarily primed PCR fingerprinting of RNA. *Nucleic Acids Res.* **20**, 4965–4970.

Zhang, H. (1995). *Cancer Res.* **55**, 2537.

Zhang, H., Zhang, R, and Liang, P. (1996). *Nucleic Acids Res.* **24**, 2454–2455.

Zimmermann, J. W., and Schultz, R. M. (1994). Analysis of gene expression in the preimplantation mouse embryo: Use of mRNA differential display. *Proc. Natl. Acad. Sci. USA* **91**, 5456–5460.

Quantitative Polymerase Chain Reaction Techniques

Rationale

The polymerase chain reaction (PCR)[1] is a widespread molecular technique that has revolutionized every aspect of biotechnology since its inception. In short, it is a primer-mediated, enzymatic amplification of specific genomic or complementary DNA (cDNA) sequences. PCR is divided into cycles, each of which typically consists of three essential components: sample denaturation, primer annealing, and primer extension. The successive application of these functionally distinct components, typically in the form of 25–30 cycles, results in the exponential amplification of target molecules. Readers who are not yet familiar with the mechanics of RNA-based PCR are strongly encouraged to read Chapter 15 in its entirety before moving forward.

Historically, the major difficulties associated with the synthesis of cDNA, both for cloning purposes and as an indirect measure of transcript abundance, have been the inability to generate near full-length, *representative* cDNA molecules and the overall low efficiency of the process of reverse transcription itself. Currently, the extent to which these can be accomplished is related directly to the

1. quality of the RNA template,
2. quantity of the RNA template, and
3. cDNA synthesis methodology itself (reviewed in Chapter 15).

Given the unsurpassed amplification potential that PCR offers, the input mass and purity of RNA are less important simply in terms of its ability to be amplified by PCR, compared to the much more stringent requirements for purity and integrity demanded by more traditional cloning methods. This is true partly because PCR primers "frame" the domain that will be amplified, making PCR more tolerant of partially degraded RNA than are traditional cDNA synthesis and Northern analysis. For quantitation, however, input mass of RNA and overall amplification efficiency are key parameters.

Quantitative Approaches

Quantitative approaches of all persuasions have been developed for the purposes of (a) quantifying transcript abundance and (b) determining gene copy number, the former being the focus of this chapter. Historically, quantification has been something of a challenge because of sample-

---

[1]The polymerase chain reaction is a patented process currently owned by Hoffman-LaRoche. Use of PCR requires a limited license under their patents.

to-sample variation. A primary concern among investigators wishing to measure transcript abundance is realization that the PCR product mass is dependent on the efficiency of the reverse transcriptase *and* the efficiency of the thermostable enzyme used to support PCR. Thus, any reliable method for quantitation must factor in the relative efficiencies of both enzyme-mediated reactions because the overall efficiency of a technique is the product of the efficiencies of the individual steps involved.

With respect to the reverse transcriptase reaction (cDNA synthesis), normalization of samples in the past has been based on sample input mass, performed most often by preparing uniform dilutions of a single aliquot of the cDNA. Alternatively, cDNA synthesis reactions run in the presence of an isotopic tracer permit normalization based upon incorporated cpm. Although these methods offer a reasonably accurate glimpse of the cellular biochemistry, a more reliable method is to measure some type of control or reference sequence, comparing its behavior to that of the experimental sequences under investigation. The inclusion of such a control is of prime importance because of the variegated nature of cDNA synthesis, with efficiencies commonly ranging from 10 to 90%.

Because PCR is an exponential process, the degree to which template is amplified is described by the equation

$$N = N_0(1 + \text{eff})^n,$$

where

N is the degree of amplification of template

N_0 is the initial amount of template

eff is the efficiency of the reaction

n is the number of cycles performed.

The efficiency of any PCR reaction or set of reactions is subject to profound change, according to the influence of many of the parameters addressed in Chapter 15. Under ideal conditions, PCR amplification should influence equally both control and experimental templates, when the reactions are properly formed.

Types of Internal Controls

Investigators should be aware that currently there is no all-purpose control (i.e., reference) transcript applicable to every system and every circumstance. This is a fact of life. Happily, one is usually able to identify a gene or two for each model system with little difficulty. The general idea is to identify a transcript whose abundance shows minimal modulation during the course of the experimental manipulation(s); these are often referred to as housekeeping genes. One big problem, however, is that the sensitivity of PCR has shown that these same sequences, once thought to be relatively invariant, can change dramatically in terms of transcript abundance in response to experimental challenge, requiring the suitability of each to be determined empirically. Housekeeping sequences often include glyceraldehyde phosphate dehydrogenase (GAPDH), transferrin receptor, hypoxanthine–guanosine phosphoribosyl transferase (HPRT), histone, fibronectin, β actin, dihydrofolate reductase (DHFR), transfer RNA (tRNA) and ribosomal RNA (rRNA). While this list is by no means intended to be exhaustive, it does underscore those genes whose transcripts are most frequently given consideration as controls for the purpose of quantification. The most important point with respect to the selection of an internal control transcript is 100% dependent on the biochemistry of the system under investigation. When identified, primers are designed for both the control and experimental transcripts, and used in one of the following ways.

1. Use of a second set of primers in a different tube with an aliquot of the cDNA from the same experimental source. This is an early form of transcript quantification using PCR. In this approach, the problems are severe:

   a. mRNAs for housekeeping genes often do not remain constant throughout the cell cycle;
   b. a second set of primers will not have the same thermodynamic character as the first set and will support the synthesis of a second PCR product with a different efficiency;
   c. quantification in this manner requires that the mRNAs for the target and control genes be present at similar levels to permit meaningful interpretation;
   d. both mRNAs must be assayed during exponential amplification;
   e. the control and experimental samples are in different tubes, introducing the distinct possibility of "tube effects."

2. Use of a second set of primers in the same PCR tube (Murphy *et al,* 1990; Gaudette and Crain, 1991). This is also known as multiplex PCR. Although this is an improvement over the two tube approach, the potential for significant inaccuracy persists because:

   a. mRNA for housekeeping genes often do not remain constant throughout the cell cycle;

   b. a second set of primers will not have the same thermodynamic character as the first set and will support the synthesis of a second PCR product with a different efficiency;

   c. quantification in this manner requires that the mRNAs for the target and control genes be present at similar levels to permit meaningful interpretation;

   d. both mRNAs must be assayed during exponential amplification.

At least both primer sets are in the same tube.

3. Competitive PCR, in which an "exogenous" internal standard, recognized by the *same* set of primers, is used. Succinctly, the cDNA under investigation *and* what is best thought of as a secondary target compete for the same primers (Becker-André and Hahlbrock, 1989; Gilliand *et al.*, 1990) and other reaction components.

   There are two forms of competitive PCR. In one approach, an artificial transcript is produced *in vitro* and then used to spike the experimental sample (Wang, *et al.,* 1989; Ikonen *et al.,* 1992; Vanden Heuvel *et al,* 1993). The rationale behind the *in vitro* transcription approach is that an artificial transcript mixed directly with the experimental material will be subjected to the same physical parameters; therefore, the extent of tran-

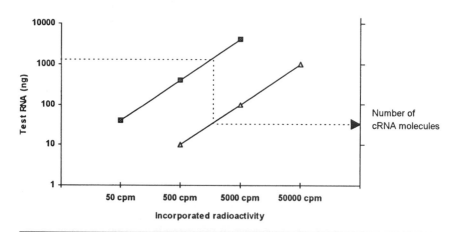

FIGURE 1

Interpolation of transcript abundance with a standard curve. In one approach, PCR is performed in the presence of ^{32}P to label products. Following electrophoresis, bands are cut out of the gel and label incorporation determined by Cerenkov counting. In this example ■ represents the mass of cellular RNA from a sample under investigation and △ corresponds to the number of competitor RNA molecules used to spike the reaction tube. The data might then be expressed as so many molecules of competitor (corresponding to an identical number of gene-specific molecules) per ng or per µg of test RNA.

scription may be determined by assaying the amount of cDNA from the artificial transcript, whose input mass is known precisely, and using the data to construct a standard curve (Figure 1). *In vitro* transcription of an artificial RNA requires the ligation of a DNA template next to a suitable transcription promoter (Figure 2); alternatively, a transcription template can be constructed by PCR (Vanden Heuvel *et al.*, 1993) in which a bacteriophage RNA polymerase promter is appended 5' to a DNA sequence intended for transcription (Figure 3). In either case, the resulting control transcript generally exhibits a deletion, compared to the naturally occurring sequence. Because experimental and control transcripts essentially share the same sequence, they are expected to be reverse transcribed and amplified by PCR with the same efficiency, and will be distinguishable by their difference in size.

    In the second approach, what might best be thought of as a nonhomologous template, sharing only the primer recognition sequences with the endogenous sequence(s) under investigation, is constructed. This is accomplished by first using PCR to place primer recognition sites on the end of the competitor DNA. Composite primers, the 5' end of which is sequence (gene) specific and the 3' end of which is competitor specific, are used to generate the competitor (Figure 4). When constructed in this

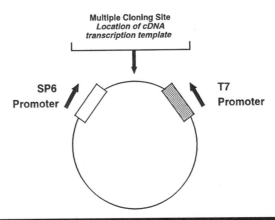

FIGURE 2

Template construction for *in vitro* transcription. The ligation of a template DNA adjacent to one or more transcription promoters, such as T7 (shown), SP6, or T3, allows the synthesis of large quantities of single-stranded RNA molecules of uniform length (the vector must be linearized prior to transcription). These transcripts can be used to spike an experimental sample, as a means of determining the efficiency of reverse transcription into cDNA. Such transcripts are useful also as nucleic acid probes, details pertaining to which are found in Chapter 12.

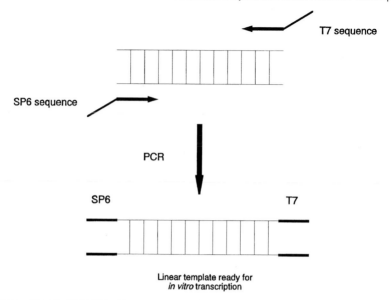

T7 sequence

SP6 sequence

PCR

SP6   T7

Linear template ready for
in vitro transcription

F I G U R E   3

Synthesis of a transcription template by PCR. Antisense RNA (cRNA) sequences are produced by *in vitro* transcription. For quantitative purposes, the transcription template can be synthesized using one primer with a 5' T7 overhang and the other primer with a 5' oligo(dT) overhang. The thymidylate tract is necessary for producing an artificial poly(A) tail, which is required to support some types of reverse transcription.

manner, both competitor and the corresponding cDNA will be amplified using a single set of primers. Often, nonhomologous competitor sequences are used for assaying gene-specific cDNA abundance. If spiked into the cDNA, rather than into the RNA, quantitation in this manner does not factor in the efficiency of reverse transcription. Depending on the context of the experiments being performed, this issue may or may not be of importance.

Competitive PCR: Key Considerations

Competitive PCR is exactly what the name implies: a competition reaction in which two templates compete for amplification. This is possible only if both the experimental cDNA and the competitor have identical primer recognition sites. The success that either or both of these potential templates will experience in an given reaction tube is a function of their molar ratios: While it might be possible to amplify both of these templates, the template represented by the larger number of molecules will yield more product, at the expense of the other template. It follows, there-

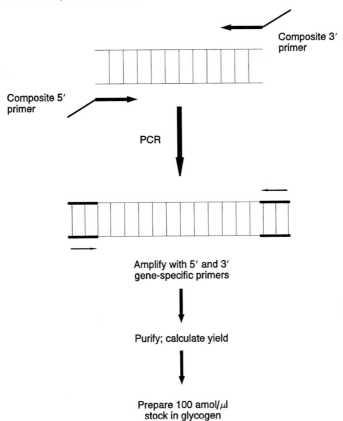

Composite 3′
primer

Composite 5′
primer

PCR

Amplify with 5′ and 3′
gene-specific primers

Purify; calculate yield

Prepare 100 amol/μl
stock in glycogen

F I G U R E   4

Construction of a nonhomologous competitor DNA template. Both the upstream and downstream composite primers contain 5′ gene-specific sequences, which, following PCR, will become part of the competitor. The addition of sequence in this manner is analogous to the method of adding a restriction enzyme site, or anything else, to a PCR product. Subsequently, in the competitive PCR reactions, shorter primers, consisting only of gene-specific sequence, are used to amplify both the competitor and the cDNA under investigation. The extent of amplification of each of these template species is a direct function of their molar ratio.

fore, that if both templates are present in equimolar amounts, they should be amplified with equal efficiency. Therein lies the quantitative beauty of this technique; in order to determine mass of an unknown cDNA in an experimental sample, dilutions of the competitor template are made until an equimolar concentration is identified (Figure 5). That's it. Plain and simple. Or is it? For the convenience of the reader the advantages and disadvantages of quantitation by standard RNA PCR and competitive PCR are compared in Tables 1 and 2.

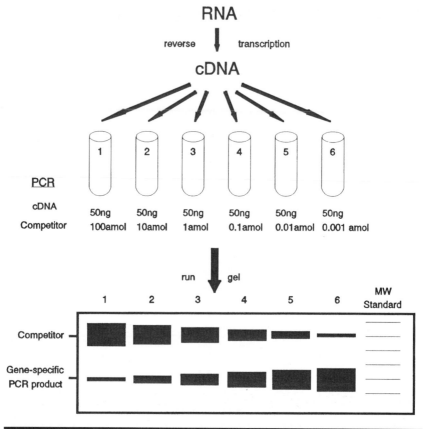

Transcript quantification by competitive PCR. Coamplification of both test cDNA and competitor, when properly designed, produces two distinct bands upon electrophoresis. In each reaction tube, both products were generated using the same set of primers and under the influence of the same physical parameters. The objective is to identify, by visual inspection or through the use of image analysis software, the reaction in which both bands are of equal intensity, meaning that the mass of each was equivalent at the onset of amplification. The abundance of the test template under investigation can be quickly determined by relating the lane where $IOD_{sample}/OD_{competitor} = 1.0$ to the mass of competitor that was spiked into the corresponding reaction tube along with the experimental cDNA.

The following are the key considerations in the design of a competitive PCR experiment.

1. Quantity of the Competitor Must Be Kown

The competitor is synthesized by reverse transcription, PCR, or both, after which it must be purified and its concentration determined either by UV absorption or by digital image analysis (Chapter 5). Typically, the

RT-PCR Quantification: Advantages and Disadvantages

Advantages
- Simplicity
- Speed
- Versatility
- Detection of very low abundance mRNAs
- Small numbers of cells/tissue mass needed
- Poly(A)$^+$ selection not necessary, nor recommended
- Plateau-associated difficulties in the PCR reaction can be overcome by the inclusion of internal PCR reaction controls
- Precludes classical difficulties of blot-associated analysis
- Reproducible data

Disadvantages
- Reverse transcriptase reaction is the major source of variability among samples
- Quantification can be difficult due to exponential nature of PCR
- Plateau phase is problematic without reference transcripts
- Small variations in pipetting or technique can cause drastic changes in product yield and variety
- Sensitivity to contaminants, especially genomic DNA
- Reproducible data?

amount of competitor synthesized will be the same as for most other sim-

Competitive PCR: Advantages and Disadvantages

Advantages
- One set of primers recognizes both the competitor and experimental templates
- Control transcripts, produced by *in vitro* transcription, are useful indicators of the efficiency of reverse transcription
- Identification of exponential PCR cycles not required
- PCR plateau effects influence competitor and cDNA equally
- Highly reproducible
- Products are easily distinguishable by electrophoresis
- Products can be quantified using SYBR Green or ethidium bromide
- No isotope required

Disadvantages
- Several dilutions required to identify the correct ratio of target with competitor
- Unique competitor required for each new target to be assayed
- A second set of primers must be synthesized to construct the competitor
- Nonhomologous DNA competitors do not address efficiency of reverse transcription
- Must confirm that target and competitor are amplified with equivalent efficiencies before using the competitor as quantitative tool

ilar synthesis reactions, on the order of 1–3 μg. For maximum accuracy, the exact number of nucleotides of the competitor must be known, from which the concentration of the newly synthesized competitor can be expressed in micromoles (μmol). When working with homologous RNA competitor transcripts, dilutions ranging from 1 to 10,000 femtograms (fg) of competitor are spiked into 100 ng of cellular RNA. Nonhomologous competitor templates are generally diluted down to a new stock concentration in the attomole (amol) range.[2] For quantitative analyses, the competitor is further diluted along with identical aliquots of the sample. When the mass of the experimental and competitor templates is identical, so too should be the mass of each PCR product.

2. Amplification of the Competitor and the Target Must Be Identical

One set of primers must recognize both types of template and amplify them with the same efficiency. To accomplish this, the investigator adds sequence complementary to the primers to the ends of an unrelated DNA molecule, which will become the competitor. Although homology is always preferred, the intended DNA competitor sequence does not have to be related to the experimental cDNA under investigation as long as:

a. the sequence of the intended competitor is known;
b. the intended competitor has average GC content;
c. the intended competitor does not have AT- or GC-rich domains;
d. the size of the intended competitor is ±200 bp of the size of the experimental cDNA.

3. Reverse Transcription Efficiency of the Template and the Competitor Must Be Identical

To correlate data from any RNA-based PCR assay with the conditions *in vivo,* one must factor in the efficiencies of both the reverse transcriptase reaction and PCR. The coamplification of the experimental cDNA and the competitor addresses the issues surrounding the efficiency of PCR: Whatever influences are exerted on the cDNA are also exerted on the competitor. By spiking the RNA sample with a transcript produced *in vitro,* a standard curve can be generated to address the efficiency issues surrounding reverse transcription. To do this, the control transcript must be designed so that it will be recognized by whatever primer is used to support reverse transcription, for example, a poly(A) tract recognized by oligo(dT).

---

[2]For the benefit of investigators unaccustomed to working with such low concentrations, a review of the units involved is presented in Table 3.

Unit Analysis

Unit	Mass	No. of molecules
1 mole	10^0 mole	6.02×10^{23}
m, millimole	10^{-3} mole	6.02×10^{20}
μ, micromole	10^{-6} mole	6.02×10^{17}
n, nanomole	10^{-9} mole	6.02×10^{14}
p, picomole	10^{-12} mole	6.02×10^{11}
f, femtomole	10^{-15} mole	6.02×10^{8}
a, attomole	10^{-18} mole	6.02×10^{5}
z, zeptomole	10^{-21} mole	6.02×10^{2}

Note. Both the stock and working concentrations of the PCR competitor are usually quite low, and investigators prefer to describe the mass of the template in attomoles. For example, a PCR product mass of 1 amol corresponds to $\approx$ 600,000 molecules.

4. Heteroduplex Formation Must Be Minimized

Heteroduplexes are PCR products that form as a result of cross-hybridization (i.e., recombination) between the competitor sequence and the experimental cDNA. This is particularly common when the competitor is a deletion or insertion form of the wild-type cDNA, and these are readily identifiable as higher molecular weight products, compared to the competitor or the target cDNA alone. The greater the number of PCR cycles performed, the more likely is heteroduplex formation. It has been suggested that heteroduplex formation in the last PCR cycle is especially troublesome because these molecules will not be subsequently denatured (Ruano and Kidd, 1992; Jensen and Straus, 1993). Obviously, heteroduplex formation compromises one's capacity for accurate quantification of the authentic PCR products. By reducing the number of PCR cycles, one reduces the likelihood/severity of heteroduplex-associated difficulties.

5. Detection Method Must Be Optimized[3]

One of the benefits associated with competitive PCR is that radiolabeling the products is not required. Routinely, 25–30 cycles are run, thereby favoring the synthesis of an adequate mass of product by simple gel staining. In this regard, SYBR Green I (and GelStar, from Molecular Probes) work well because, compared to ethidium bromide, these dyes offer an enhancement in sensitivity with minimal background when viewing, photodocumenting, or digitizing directly from the gel. Although a reduc-

---

[3]Isotopes are *not* required for highly sensitive competitive PCR.

tion in the number of cycles is in order to minimize heteroduplex formation, the concomitant decrease in product accumulation need not be a detection problem; SYBR Green allows the visualization of bands to an extent not possible with ethidium bromide. Moreover, in the instance of adequate band mass, visualization with ethidium bromide is not recommended because the high background commonly associated with this dye can detract greatly from the quantitativeness of this assay.

In summary, competitive PCR is an extremely sensitive method for template quantification. Being PCR-based, the sensitivity of optimized reactions is unparalleled; data generated by Northern analysis, and even nuclease protection assay, by comparison, are lacking. The inclusion of an internal competitor, essentially mimicking the experimental template, precludes the requirement of having to determine in which cycles the exponential phase occurs. Among the disadvantages associated with the method, which are really minor inconveniences more than anything else, are the requirement for serial dilutions for each sample needed to titrate the competitor template and the fact that two sets of primers must be synthesized: a composite set to synthesize the competitor, and the gene-specific set for the actual competitive PCR reactions.

Competitive PCR: Major Steps Involved

1. Prepare PCR Competitor

A. Nonhomologous competitor

i.   Design gene-specific and composite primers (two sets of primers required). Perform primary PCR amplification with composite primers; perform secondary PCR amplification with gene-specific primers. Try 25- to 30-mers for use as gene-specific primers, with T_m of at least 60°C. This will favor higher yields and greater fidelity of cDNA amplification. As always, use standard guidelines for PCR primer design:

- long 5' tails are acceptable and may be used to add restriction enzyme sites
- 3' terminus must base pair perfectly with the template
- average to slightly above average G+C content is desirable (50–60%)
- keep track of mismatches, which may or may not be meaningful
- check for primer self-complementarity
- check for 3' primer complementarity (can promote primer–dimer formation)

  ii. Competitors are usually 200–700 bp.

  iii. Competitors should be a maximum of 200 bp larger (or smaller) than the experimental target

  iv Purify competitor product; determine concentration [spectrophotometrically or by comparison with known standards (e.g., ϕX174/*Hae*III digest)].

  v. Prepare stock solutions, usually 50–100 amol/ml.

B. Homologous competitor prepared by *in vitro* transcription

  i. Ligate cDNA adjacent to a bacteriophage RNA polymerase promoter (e.g., T3, SP6, or T7)

  ii. Perform *in vitro* transcription, purify transcript, determine concentration

  iii. Store at –80°C until ready to use

2. Isolate High-Quality RNA

- Good RNase-free technique.
- Expedient recovery from cells or tissues.
- Guanidinium–acid–phenol technique preferred.
- Sample treatment with RNase-free DNase.
- Wash RNA pellets with 70% ethanol prepared in DEPC-treated H_2O (cationic salts influence PCR reactions by quantitatively binding dNTPs).

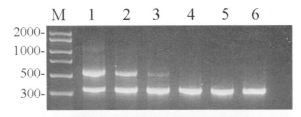

FIGURE 6

Primary 10-fold dilution series for mass determination. Coamplification of mouse β-actin cDNA and a non-homologous competitor. Competitor template (540 bp) was generated using composite primers, as described. β-actin (365 bp) cDNA synthesis was supported by oligo(dT)- and AMV-mediated reverse transcription. PCR was performed for 25 cycles, after which the products of all six reactions were electrophoresed on a 2.5% agarose gel prepared in 1X TAE, along with a molecular weight standard (Sigma). At the conclusion of the run, the gel was stained in 1X SYBR Green, also prepared in 1X TAE, and then photographed with Polaroid Type 667 film. The ten-fold dilution series represents a primary characterization of the sample, in order to estimate the approximate mass of actin cDNA in the sample. Mass of competitor in each sample is: lane 1, 10 amol; lane 2, 1.0 amol; lane 3, 0.1 amol; lane 4, 0.01 amol; lane 5, 0.001 amol; lane 6, 0.0001 amol. Best match ($IOD_{sample}/OD_{competitor} = 1.0$) is in lane 2.

- Proper long-term storage of purified RNA (ethanol precipitate at –20°C or –80°C).
- If stored in aqueous buffer, aliquot and maintain at –80°C.

3. Perform Reverse Transcription into cDNA

- Select reverse transcriptase (e.g., AMV, MMLV, etc.).
- Synthesize cDNA, purify products.
- Store at –20° until ready to use.
- Avoid Mn^{2+}-dependent reverse transcription.

4. Perform *Primary* Competitive PCR Amplification (Figure 6)

- Prepare serial 10-fold dilutions of the competitor.
- Use an aliquot of each PCR competitor dilution with first-strand cDNA samples.
- Perform PCR amplification; analyze products by electrophoresis.
- Determine competitor concentration that most closely approximates target cDNA concentration. Visual inspection of the gel is usually adequate, though image analysis software may be helpful.

5. Perform *Secondary* Competitive PCR Amplification (Figure 7)

- Prepare serial 2-fold dilutions of the competitor, beginning with a competitor stock solution that is 10-fold more concentrated than the dilution that generated competitor and experimental bands of near-equal intensity in the primary amplification.
- Use an aliquot of the first-strand cDNA with each 2-fold competitor dilution strand
- Perform PCR amplification; analyze products by electrophoresis

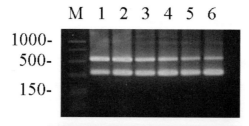

FIGURE 7

Secondary two-fold dilution series for mass determination. Reaction series and analysis was identical to that described for Fig. 6, with the exception of the competitor mass added to each tube. Primary amplification showed the best match between competitor and experimental PCR products at 1.0 amol; thus a 2-fold dilution series was utilized in the secondary amplification beginning with the next higher dilution, thus: lane 1, 10 amol; lane 2, 5 amol; lane 3, 2.5 amol; lane 4, 1.25 amol; lane 5, 0.625 amol; lane 6, 0.3125 amol. Per chance, the best match is again in lane 2, meaning that the mass of actin cDNA in the sample under investigation is about 5 amol, or approximately 120,000 molecules!

- Determine competitor concentration that most closely approximates target cDNA concentration. Visual inspection of the gel is usually adequate, though image analysis software may be helpful.
- Can expect detection/resolution of as little as a 2-fold change in gene expression (transcript abundance)

Protocol: Competitive PCR

The protocol that follows pertains to the synthesis of a nonhomologous competitor through the use of composite primers. The product of this reaction is then used to spike cDNA samples, thereby establishing competitive PCR amplification. If using an artificial control transcript generated by *in vitro* transcription, both the experimental RNA and the control RNA are pipetted together in the same tube and then reverse transcribed.

I. Synthesis of Nonhomologous Competitor

Each PCR competitor is synthesized in a separate reaction with a unique set of composite primers. Once synthesized, an aliquot of the product is used as a template for reamplification of the competitor. Note that the PCR cycling profile will vary depending on the type of thermal cycler. The following protocol works well in this lab using a thermal cycler that does not have a heated lid (e.g., Perkin Elmer 480).

1. Mix

10× PCR buffer	5 μl
25 mM MgCl$_2$	3 μl
10 mM dNTPs	1 μl
Competitor template	X μl (1 ng suggested)
10 mM 5' composite primer	2 μl
10 mM 3' composite primer	2 μl
Sterile H$_2$O	36.5–X μl
Taq polymerase	0.5 μl

2. If necessary, overlay reaction with mineral oil. With or without mineral oil, pulse centrifuge samples.
3. Perform cycling protocol:

Initial melt	94°C	3 min
Amplification	94°C	1 min
	X°C	1 min (annealing temperature is primer-dependent)

	72°C	1 min
Final extension	72°C	10 min
Soak	4°C	

Note: Because of the extensive 5' overhang structure exhibited by each of the composite primers, it may be necessary to reduce the annealing temperature to several degrees below the calculated T_m to facilitate annealing to the competitor template.

4. Prepare a 1:500 dilution of the PCR reaction from Step 3, an aliquot of which is used for reamplification. Be sure to reserve an undiluted aliquot of this same reaction as a diagnostic for electrophoresis.

5. Mix

10× PCR Buffer	5 μl
25 mM MgCl$_2$	3 μl
10 mM dNTPs	1 μl
Dilution from Step 4	2.5 μl
10 μM 5' cDNA primer	2 μl
10 μM 3' cDNA primer	2μl
Sterile H$_2$O	34 μl
Taq polymerase	0.5 μl

6. If necessary, overlay reaction with mineral oil. With or without mineral oil, pulse centrifuge samples.

7. Perform cycling protocol:

Initial melt	94°C	3 min
Amplification	94°C	1 min
	X°C	1 min (annealing temperature is primer-dependent)
	72°C	1 min
Final extension	72°C	10 min
Soak	4°C	

8. Remove an aliquot of the PCR products from Step 7 and electrophorese along with the undiluted aliquot reserved in Step 4. Stain and photograph.

9. The product generated in Step 8 is the competitor DNA, and it must be purified to remove the reaction components. The most efficient

---

[4]The PCR Mimic Construction Kit (Clontech No. K1700-1) may be used for the synthesis, purification, and subsequent use of competitor PCR templates (primers not included).

method is through the use of one of readily available spun columns for PCR product purification.[4]

10. Determine competitor concentration and yield, either by A_{260} or by image analysis, as described in Chapter 10.

11. Prepare a stock solution of competitor template by diluting a portion of the competitor to a concentration of 100 amol/μl. The dilution can be made in PCR TE buffer (10 mM Tris, pH 7.5; 0.1 mM EDTA) or in 10 μg/ml glycogen. Dilutions of the competitor and all undiluted stocks should be stored at $-20°$.

II. Synthesis of First-Strand cDNA

1. Ensure that high-quality RNA is available to support the synthesis of first-strand cDNA. This is ascertained by minigel electrophoresis and visualization of the ribosomal 28S and 18S RNA, with minimal smearing below the 18S rRNA. See Chapter 5 for isolation options.

2. Prepare the cDNA mix:

		[Final]
DEPC-H$_2$O	3 μl	—
25 mM MgCl$_2$	4 μl	5 mM
10× buffer[5]	2 μl	1×
10 mM dNTPs	2 μl	1 mM
Oligo(dT)[6]	2 μl	2.5 μM
RNasin	1 μl	2.5 U/μl

Set aside on ice.

3. Transfer 500 ng RNA (2 μl; 250 ng/μl) to a sterile microfuge tube and then add 3 μl of nuclease-free H$_2$O. Heat to 65°C for 2 min and then cool on ice for 1 min.

4. Add the entire cDNA mix (from Step 2) to the tube containing the heat-denatured RNA.

5. Incubate at room temperature for 3 min to facilitate primer annealing.

6. Add 1 μl reverse transcriptase; gently pipet up and down to mix. Do not vortex, ever.

7. Incubate at 42°C for 30–60 min (AMV) or at 37°C for 30–60 min (MMLV).

---

[5]10× first-strand buffer = 100 mM Tris, pH 8.3; 500 mM KCl. Complete systems for the synthesis of first-strand cDNA are available from every major supplier of biotechnology products.

[6]Alternatively, random primers (μg) can be substituted to support cDNA synthesis.

8. Heat to 95°C for 5 min to destroy the reverse transcriptase, and then store on ice until ready to use.

III. Competitive PCR (Primary Run)

The purpose of the primary reactions is to determine the approximate mass of the cDNA under investigation. Although the mass is expected to vary drastically among different cDNAs, reflecting the abundance of the original transcript *in vivo,* it is likely that the primary reactions will be required only once. After identifying the competitor dilution that most closely approximates the mass of the cDNA, secondary twofold dilutions of the competitor are run in order to zero in on the mass of the experimental template. Of course, profound changes in mRNA abundance, in response to an experimental challenge, will mandate repetition of the 10-fold competitor dilution series.

1. For each sample to be tested, prepare PCR master mix for 13 samples (12 tubes + 1 extra).

PCR master mix	Per 50 μl R	[Final]	6 Samples (7)	12 samples (13)
10× PCR buffer	5.0 μl	1 mM	35.0 μl	65.0 μl
MgCl₂ stock	3.0 μl	1.5 mM	21.0 μl	39.0 μl
Sterile H₂O	34.5 μl	—	241.5 μl	448.5 μl
10 mM dNTP mix	1.0 μl	0.2 mM	7.0 μl	13.0 μl
5′ primer (20 μM)	1.0 ml	0.4 μM	7.0 μl	13.0 μl
3′ primer (20 μM)	1.0 μl	0.4 μM	7.0 μl	13.0 μl
Taq polymerase (5U/μl)	0.5 μl	2 U/reaction	3.5 μl	6.5 μl
Total volume	46.0 μl	—	322.0 μl	598.0 μl

Store master mix on ice.

2. Label 6 PCR tubes P_1–P_6.

3. To prepare the primary 10-fold serial dilution series, pipet 9 μl of TE buffer (10 mM Tris, pH 7.5; 0.1 mM EDTA) into each of the 6 tubes.

4. Designate an aliquot of the competitor stock (100 amol/μl) as Tube P. Tenfold dilution of the competitor is performed as follows:

Tube	Dilution method	[Final]
P_1	add 1 μl P, mix	10 amol/μl
P_2	add 1 μl P_1, mix	1 amol/μl
P_3	add 1 μl P_2, mix	10^{-1} amol/μl
P_4	add 1 μl P_3, mix	10^{-2} amol/μl
P_5	add 1 μl P_4, mix	10^{-3} amol/μl
P_6	add 1 μl P_5, mix	10^{-4} amol/μl

5. Label six new PCR tubes.
6. To each tube add

2 μl	cDNA synthesis products
2 μl	one of the dilutions just prepared (P_1–P_6).
46 μl	PCR master (Step 1)
50 μl	total volume

If necessary, overlay with 1 drop of mineral oil.

7. Gently pulse centrifuge to collect all reaction volumes at the bottom of the tube.
8. Perform cycling protocol:

Initial melt	94°C	3 min
Amplification	94°C	1 min
(25 cycles)	X°C	1 min (annealing temperature is primer dependent)
	72°C	2 min.
Final extension	72°C	5 min
Soak	4°C	

9. Prepare a 2.5% agarose gel, made up in 1× TBE (1× TAE may be substituted).

10. Remove a 10-μl aliquot from each tube. If reactions were overlaid with mineral oil, be sure to wipe the outside of the tip with a Kimwipe before transferring the aliquot into a new tube. Add 1.5 μl 10× loading buffer to each sample.

11. Load gels and perform electrophoresis. Be sure to include a molecular weight standard.

Note 1: The slower the gel is run, the better will be the resolution of the products. Further, the concentration of the agarose can be increased or decreased, as needed.

Note 2: In this lab, Promega's PCR molecular weight marker (No.G3161) is commonly used (sizes in bp): 1000, 750, 500, 300, 150, 50.

12. At the conclusion of the run, stain gel with 1× SYBR Green I diluted in 1× TBE or 1× TAE buffer. Examine gels on transilluminator (protect eyes and skin from UV light). Photodocument.

13. Determine the dilution of competitor that generated a PCR product most similar to the mass of the experimental product. While visual inspection alone is usually quite adequate for this measurement, the inte-

grated optical density (IOD) of each band and ratio for each pair of bands can be determined using image analysis software.

IV. Competitive PCR (Secondary Run)

Having identified the competitor dilution in which the PCR products of the competitor and the cDNA are nearly identical, the competitor dilution that is 10-fold *more* concentrated will be used as the starting point for making the secondary set of dilutions. The purpose for making two-fold dilutions for use in a second set of reactions is to fine-tune the determination of gene-specific cDNA in the sample(s) under investigation. For the sake of consistency and reproducibility, use aliquots of the same PCR master mix prepared to support the primary reactions. The master mix should be maintained on ice; if prepared well in advance it may be frozen, as long as the *Taq* polymerase has not been added.

1. Label 6 PCR tubes S_1–S_6.
2. To prepare the 2-fold serial dilution series, place 5 μl of TE buffer (10 mM Tris, pH 7.5; 0.1 mM EDTA) in each of the 6 tubes.
3. Beginning with the competitor concentration that is 10-fold greater (Tube P_x) than that which matched the mass of the cDNA in the primary amplification, prepare the 2-fold dilution series as follows:

Tube	Dilution method	[Final]
S_1	add 5 μl P_x, mix	$5 \times 10^{x-1}$ amol/μl
S_2	add 5 μl S_1, mix	$2.5 \times 10^{x-1}$ amol/μl
S_3	add 5 μl S_2, mix	$1.25 \times 10^{x-1}$ amol/μl
S_4	add 5 μl S3, mix	$6.25 \times 10^{x-2}$ amol/μl
S_5	add 5 μl S_4, mix	$3.125 \times 10^{x-2}$ amol/μl
S_6	add 5 μl S_5, mix	$1.56 \times 10^{x-2}$ amol/μl

4. Label 6 new PCR tubes.
5. To each tube add

2 μl	cDNA synthesis products (this protocol, Section II)
2 μl	one of the dilutions just prepared (S_1–S_6).
46 μl	PCR master mix (this protocol, Section III)
50 μl	total volume

If necessary, overlay with 1 drop of mineral oil.
6. Gently pulse centrifuge to collect all reaction volumes at the bottom of the tube.
7. Perform cycling protocol:

Initial melt	94°C	3 min
Amplification	94°C	1 min
(25 cycles)	X°C	1 min. (annealing temperature is primer dependent)
	72°C	2 min
Final extension	72°C	5 min
Soak	4°C	

Note: Be sure to use precisely the same cycling parameters as with the primary 10-fold dilution series.

8. Prepare a 2.5% agarose gel, made up in 1× TBE (1× TAE may be substituted). Use the same type of gel and electrophoresis parameters as for analysis of the primary reactions.

9. Remove a 10 μl aliquot from each tube. If reactions were overlaid with mineral oil, be sure to wipe the outside of the tip with a Kimwipe before transferring the aliquot into a new tube. Add 1.5 μl 10× loading buffer to each sample.

10. Load gels and perform electrophoresis. Be sure to include a molecular weight standard.

11. At the conclusion of the run, stain gel with 1× SYBR Green I diluted in 1× TBE or 1× TAE buffer. Examine gels on transilluminator (protect eyes and skin from UV light). Photodocument.

12. Determine the dilution of competitor that generated a PCR product most similar to the mass of the experimental product. Because the samples represent only 2-fold dilutions of the competitor, it may be necessary to take advantage of image analysis software to determine the closest match between competitor and cDNA, that is, the sample in which $IOD_{sample}/IOD_{competitor} = 1$.

Image Analysis Considerations

The ability to make sense from the gels used to analyze competitive PCR reactions is highly dependent on the quality of the appearance and image of the gel from the start. As suggested in Chapter 10, many investigators are under the sadly mistaken notion that an investment in costly image analysis systems will compensate for poorly run gels and low-quality images. The truth is that the degree to which gel analysis is accurate and reproducible is a direct function of what the gel and/or photograph look like from the onset; to a very great extent, image analysis optimization is as important as optimizing PCR or other reactions. The reader is encouraged to review the "Digital Image Analysis" section in

Chapter 10 for suggestions pertaining to the optimization of image analysis.

With respect to competitive PCR, ethidium bromide-stained gels are alright, but gels stained with SYBR Green are clearly superior. Better quality gels facilitate greater accuracy both in visual inspection and in digital image analysis. If desired, labeled dNTPs may be included in the PCR reaction, thereby supporting detection by autoradiography and mass determinations via scintillation counting. Radiolabeling for this type of assay is not required and is often counterproductive; as such, it is not recommended.

Trouble-Shooting Competitive PCR

The exquisite sensitivity of PCR to the precise concentrations of all reaction components is unquestioned. For assays designed to generate quantitative data, accurate pipetting and the use of master mixes for both cDNA synthesis and PCR are of critical importance. The items below outline what may be done to minimize aggravation and maximize reproducibility and productivity.

1. **Nonnegotiables**

    A. High-quality RNA:

    - good RNase-free technique
    - guanidinium thiocyanate- or guanidinium HCl-based lysis buffers
    - acid–phenol extraction
    - double precipitation with guanidinium lysis buffer
    - DNase treatment; must be RNase free
    - wash RNA pellets thoroughly/extensively with 70% ethanol prepared in DEPC-treated H_2O[7]
    - proper storage of purified RNA

    B. Reverse transcription into cDNA and subsequent PCR

    - equal mass of RNA per cDNA reaction (200–500 ng)
    - enzymes lacking endogenous RNase H activity may be helpful
    - avoid Mn^{2+}-dependent reverse transcriptase activities
    - always work with a cDNA master mix
    - always work with a PCR master mix

2. **Common problems**

    A. Heavy smearing in one or more lanes:

[7]Recall that cationic salts bind dNTPs quantitatively and can cause havoc in the cDNA synthesis and PCR reactions.

- optimize Mg^{2+} concentration (usually a reduction; decrease incrementally by 0.1 mM)
- reduce the number of cycles
- decrease the amount of enzyme in the reaction
- slightly increase annealing temperature

B. No product:

- check the RNA!
- add an RNase inhibitor (e.g., RNasin)
- test the components of the RT and PCR reactions, especially the enzymes
- check off each component as the reactions are pipetted together
- check PCR cycle parameters

C. Low yield:

- incomplete homogenization or lysis of samples
- final RNA pellet incompletely redissolved
- $A_{260}/A_{280} > 1.65$ (suggests protein or other contaminants)
- RNA degradation:
  - tissue was not immediately processed/frozen after removing from the animal.
  - purified RNA was stored at –20°C, instead of –70°C
  - aqueous solutions or tubes were not RNase-free.

D. DNA contamination:

- sample was not DNase-treated

E. Nonreproducibility of data:

- use thin-walled tubes (200 μl work well); for thick walled tubes, extend cycle parameters
- consider hot start PCR (described in Chapter 15)
- prepare single, large volume master mix per cell type and aliquot for each reaction
- match T_m of upstream and downstream primer pairs more closely
- consider enzyme mixtures (e.g., *Taq* + proofreading enzyme) designed for long PCR (e.g., AccuTaq L.A., from Sigma). Working together, these enzymes generate more product and do so with high fidelity.
- consider elimination of mineral oil: use a thermal cycler with a heated lid
- RNA not isolated by the same method
- RNA not isolated on the same day
- cells/tissue not in the same state (different parts of the cell cycle or the tissues are isolated from biochemically different individuals).

- different master mix used
- pipets recalibrated?
- optimize and run reactions in the same thermal cycler (not model number)
- consider the purchase of precast gels; both agarose and polyacrylamide are available

3. General Suggestions

- Keep denaturation steps as short as possible
- Keep denaturation temperature as low as possible (90–92°C)
- Elongation temperature can be reduced to 68°C
- Extend elongation time for each successive cycle (5–20 s per cycle)
- Accurate pipeting is critical for reproducibility. Have micropipettors recalibrated before you begin, and at regular intervals thereafter.
- Competitive PCR is exquisitely sensitive to the exact input mass of RNA (for cDNA synthesis) and cDNA (for PCR amplification). As with most RT PCR reactions "less is more," meaning that it is generally counterproductive to overload the reaction with too much template. When putting the PCR reactions together, it may be helpful to actually wipe the outside of the micropipet tip before dispensing the aliquot into the destination microfuge tube so that unintentional droplets on the outside of the tip do not contribute to the outcome of the reaction.
- Observe standard methods for the prevention of carryover contamination, including the use of positive displacement micropipettors.
- Silanized microcentrifuge tubes may be helpful in minimizing losses during the various manipulations because cDNA/PCR products are present in picogram quantities (or less) prior to amplification (see Appendix F for protocol).

Real-time PCR

A major breakthrough in gene- and transcript quantification was the development of what is known as "real-time" PCR detection (Higuchi *et al.,* 1992; 1993). In this approach, PCR products are assayed as they accumulate, as opposed to performing a fixed number of cycles and assessing the amount and variety of products at the end of the reaction. Real-time PCR requires some type of fluorescent labeling reagent in the reaction, a thermal cycler equipped with a UV light source to induce the fluorescence, and a sensitive cooled CCD camera, all of which are controlled by a computer.

Real-time PCR, when first published, described the use of ethidium bromide in the reaction. This intercalating agent would associate itself with double-stranded PCR products as they accumulate; the resulting increase in fluorescence would be proportional to the product mass. While somewhat more reliable than many of the classical methods for PCR-based assay of transcript abundance, the inclusion of intercalating agents reflects the presence of both specific and nonspecific products generated in the reaction.

To address this difficulty, intrinsic to the methodology, a modification in the mechanics of detection known as probe-based detection has replaced the intercalator-based detection approach. This modified method has come to be known generically as the 5' nuclease assay[8] (Holland *et al.*, 1991) because it exploits the 5' exonuclease activity of the *Taq* polymerase to displace and then cleave a hybridized oligonucleotide probe from the target strand. In so doing, the $5' \rightarrow 3'$ polymerase activity of the *Taq* can continue through the region previously blocked by the now-displaced probe. Previously, the liberation of a $5'$ ^{32}P end label was used as indicator, with the results of probe displacement assayed at the end of reaction, most commonly using thin-layer chromatography. Currently, the method of real-time detection is mediated by the use of fluorogenic probes (Lee *et al.*, 1993), which, when destabilized from the template by *Taq* during PCR, can be measured in terms of fluorescence emitted (Figure 8). By measuring the fluorescence signal generated during thermal cycling (e.g., ABI PRISM 7700 Sequence Detector), no other handling of the sample at the end of the reaction is required.

Real-time systems are attractive for a number of obvious reasons. By linking quantitation to computer-based automation, data can be collected with each successive cycle. This approach expands the linearity of the assay to as many as six logs. Because the reaction tube(s) in which the assay is performed need not be opened at the conclusion of the run, there is, theoretically, a reduced likelihood of amplicon contamination in and about the lab.

Compared to competitive PCR, real time systems do not require spiking each reaction with an artificial template for the purpose of coamplification. However, internal controls of some type must be established, primarily for the purpose of compensating for the variations in template

---

[8]Do not confuse this nomenclature with the nuclease protection assay, described in Chapter 18.

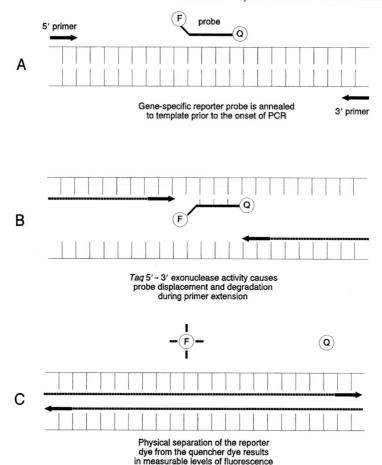

FIGURE   5

Real-time PCR. A. Fluorogenic probes, consisting of both a reporter (F = fluorescent) dye and quencher (Q), are annealed to the template. As long as probe integrity is maintained, reporter dye fluorescence is quenched due to its proximity to the quencher dye. B. During PCR, the 5′ exonuclease activity of the *Taq* polymerase causes the displacement of the probe and its subsequent cleavage. C. The fluorescence of the reporter dye, once separated from the quencher, is measured using a very sensitive cooled CCD camera. Fluorescence emission is directly related to amount of PCR product being produced. PCR primer extension then proceeds through the now-exposed region, completing the product.

mass, both DNA and RNA. These controls can also be valuable for detecting amplification inhibitors. To do this, distinct primers that recognize distinct templates are used, and many protocols suggest running the controls in different tubes. Thus, the PCR paradox: If two or more discrete primers are used in the same reaction tube, then the amplification of one template may exert a negative influence on the amplification of the other. If these "reference" reactions are run in different tubes than the experimental sample, then who is to say that the amplification efficiency is the same in both? If the same set of primers is used in the same tube to amplify the experimental template and the control, then you have competitive PCR!

REFERENCES

Becker-André, M. (1991). Quantitative evaluation of mRNA levels. *Methods Mol. Cell. Biol.* **2,** 189–201.

Becker-André, M., and Hahlbrock, K. (1989). Absolute mRNA quantification using the polymerase chain reaction (PCR). A novel approach by a PCR aided transcription titration assay (PATTY). *Nucleic Acids Res.* **17,** 9437–9446.

Blok, H. J., Gohlke, A. M., and Akkermans, A. D. L. (1997). Quantitative analysis of 16S rDNA using competitive PCR and the QPCR System 5000. *BioTechniques* **22,** 700–704.

Brenner, C. A. Tam, A., W. Nelson, P. A., Engleman, E. G., Suzuki, N., Fry, K. E., and Larrick, J. W. (1989). Message amplification phenotyping (MAPPing): A technique to simultaneously measure multiple mRNAs from small numbers of cells. *BioTechniques* **7**(10), 1096.

Chelly, J., Kaplan, J. C., Gautron, S., and Kahn, A. (1988). Nature **333,** 858–860.

Clemeti, M., Menzo, S., Bagnarelli, P., Manzin, A., Valenza A., and Varaldo, P. E (1993). Quantitative PCR and RT-PCR in virology. PCR Methods Appl. **2,** 191–196.

DiCesare, J., Grossman, B., Katz, E., Picozza, E., Ragusa R., and Woudenberg, T. (1993). A high-sensitivity electrochemiluminescence-based detection system for automated PCR product quantitation. *BioTechniques* **15,** 152–156.

Diviacco, S., Norio, P., Sentilin, L., Menzo, S., Clementi, M., Biamonti, G., Riva, S., Falaschi, A., and Giacca, M. (1992) A novel procedure for quantitative polymerase chain reaction by coamplification of competitive templates. *Gene* **122,** 313–320.

Farrell, R. E., Jr. (1993). "RNA Methodologies: A Laboratory Guide for Isolation and Characterization." Academic Press, San Diego, CA.

Gaudette, M. F., and Crain, W. R. (1991). *Nucleic Acids Res.* **19,** 1879.

Gilliand, G. S., Perrin, S., Blanchard, K., and Bunn, H. F. (1990). Analysis of cytokine mRNA and DNA: detection and quantitation by competitive polymerase chain reaction. *Proc. Natl. Acad. Sci.* **87,** 2725–2729.

Higuchi, R., Dollinger, G., Walsh, P. S., and Griffith, R. (1992). Simultaneous amplification and detection of specific DNA sequences. *Biotechnology* **10,** 413–417.

Higuchi, R., Fockler, G. Dollinger, G., and Watson, R. (1993). Kinetic PCR: Real time monitoring of DNA amplification reactions. *Biotechnology* **11,** 1026–1030.

Ikonen, E., Manninen, T., Peltonen, L., and Syvänen, A. (1992). Quantitative determination of rare mRNA species by PCR and solid-phase minisequencing. *PCR Methods and Applications* **1**, 234–240.

Jensen, M. A., and Straus. N. (1993). Effect of PCR conditions on the formation of heteroduplex and single-stranded DNA products in the amplification of bacterial ribosomal DNA spacer regions. *PCR Methods Appl.* **3**, 186–194.

Kawasaki, E. S. (1991). In "PCR Protocols: A Guide to Methods and Applications" (Innis *et al.,* Eds.), pp. 21–27. Academic Press, San Diego, CA.

Lee, L. G., Connell, C. R., and Bloch, W. (1993). Allelic discrimination by nick-translation PCR with fluorogenic probes. *Nucleic Acids Res.* **21**, 3761–3766.

Li, B., *et al.* (1991). *J. Exp. Med.* **174**, 1259.

Liang, P., and Pardee, A. B. (1992). Differential display of eukaryotic messenger RNA by means of the polymerase chain reaction. *Science* **257**, 967–971.

Murphy L. D., Herzog, C. E., Rudick, J. B., Fojo, A. T., and Bates, S. E. 1990. *Biochemistry* **29**, 10351.

Piatak, M., Luk, K., Williams, B., and Lifson, J. D. (1993). Quantitative competitive polymerase chain reaction for accurate quantitation of HIV DNA and RNA species. *BioTechniques* **14**, 70–80.

Rappolee, D. A., Mark, D., Banda, M. J. and Werb, Z. (1991). *Science* **241**, 708–712.

Ruano, G. and Kidd, K. K. (1992). Modeling of heteroduplex formation during PCR from mixtures of DNA templates. *PCR Methods Appl* **2**, 112–116.

Schneeberger, C., Speiser, Kury, F., and Zeillinger, R. (1995). Quantitative detection of reverse transcriptase-PCR products by means of a novel and sensitive DNA stain. *PCR Methods Appl.* **4**, 234–238.

Siebert, P. D. and Larrick, J. W. 1992. Competitive PCR (product review). *Nature* **359**, 557–558.

Siebert, P. D., and Larrick, J. W. (1993). PCR MIMICS: Competitive DNA fragments for use as internal standards in quantitative PCR. *BioTechniques* **14**(2), 244.

Vanden Heuvel, J. P., Tyson, F. L., and Bell, D. A. (1993). Construction of recombinant RNA templates for use as internal standards in quantitative RT-PCR. *BioTechniques* **14**(3), 395–398.

Wages, J. M., Jr., Dolenga, L., and Fowler, A. K. (1993). Eletrochemiluminescent detection and quantitation of PCR-amplified DNA. *Amplifications (PE)* **10**, 1–6.

Wang, A. M., Doyle, M. V., and Mark, D. F. (1989). Quantitation of mRNA by the polymerase chain reaction. *Proc. Natl. Acad. Sci.* **86**, 9717–9721.

Wilkinson, T. E., Cheifetz , S., and DeGrandis, S. A. (1995). Development of competitive PCR and the QPCR System 5000 as a transcription-based screen. *PCR Methods Appl.* **4**, 363–367.

Quantification of Specific Messenger RNAs by Nuclease Protection

Rationale

The study of rare or low abundance transcripts is a frequently reiterated theme in the molecular biology laboratory. While Northern analysis generally provides detailed information about the natural size and abundance of particular transcripts, both the sensitivity and resolution of this technique are often compromised by the inability to load large amounts of RNA on a gel, by the inefficient transfer from gel to blotting membrane, by nonspecific cross-hybridization between probe and transcripts exhibiting limited homology, and by nonspecific hybridization between target RNA and vector sequences, if present. Moreover, one of the fundamental drawbacks of filter-based hybridizations is that complete hybridization cannot be guaranteed, because some of the membrane-bound target sequences may not be fully accessible to probe sequences. In both theory and practice, complete hybridization is difficult, if not impossible.

One basic approach that can alleviate some of these Northern blot-associated symptoms is generically referred to as a nuclease protection assay, which is an extremely sensitive and precise method for quantifying specific transcripts. In this approach, the formation of hybrid duplexes between probe and target RNA is promoted under highly stringent hybridization, in solution. The investigator has the choice of using total cellular, total cytoplasmic, poly(A)$^+$ messenger RNA (mRNA), or any other RNA subset as starting target material, though the enhanced sensitivity of this approach usually does not mandate poly(A)$^+$ selection. Hybridization is followed immediately by exposure to a single-strand-specific nuclease that ordinarily has no affinity for DNA:RNA or RNA:RNA hybrids. Thus, any probe sequences or target RNA molecules that do not participate in duplex formation are digested, and the resultant protected fragments recovered by ethanol precipitation. The size and mass of these hybrid molecules are then deduced, usually by denaturing polyacrylamide gel electrophoresis, and the signal intensity is quantified by autoradiography (Figure 1). In some variations of this technique, the abundance of the undigested molecules is assessed by liquid scintillation counting instead.

By virtue of the solution hybridization of probe and target, followed by digestion of all other sequences that do not participate in duplex formation, the signal-to-noise ratio is boosted, often dramatically. The investigator can expect at least a 10-fold increase in sensitivity over conventional blot analysis and the ability to reproducibly detect as little as 100 femtograms (Fg) of target RNA[1] after a 3-day X-ray exposure using high spe-

---

[1]Although autoradiography is considered to be the preferred method for maximum sensitivity, detection of as little as 0.03 pg of homologous RNA using chemiluminescence coupled with the RNase protection assay has been reported.

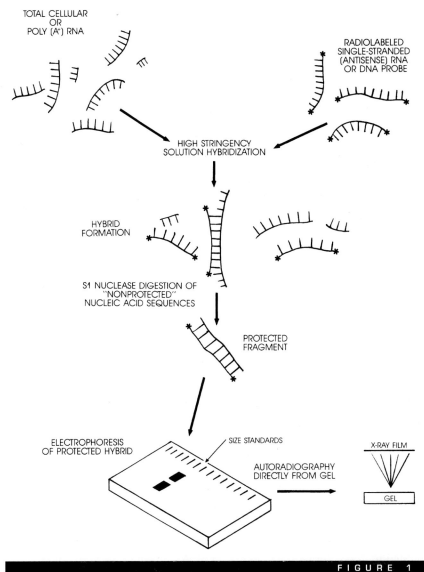

TOTAL CELLULAR
OR
POLY (A⁺) RNA

RADIOLABELED
SINGLE-STRANDED
(ANTISENSE) RNA
OR DNA PROBE

HIGH STRINGENCY
SOLUTION HYBRIDIZATION

HYBRID
FORMATION

S1 NUCLEASE DIGESTION OF
"NONPROTECTED"
NUCLEIC ACID SEQUENCES

PROTECTED
FRAGMENT

ELECTROPHORESIS
OF PROTECTED HYBRID

SIZE STANDARDS

X-RAY FILM

AUTORADIOGRAPHY
DIRECTLY FROM GEL

GEL

F I G U R E   1

The S1 nuclease assay for the quantification of specific RNA species. Purified RNA is hybridized in solution with a labeled, complementary sequence to form thermodynamically stable hybrid molecules. Any RNA and probe molecules that do not participate in the formation of hybrid molecules are digested away by the single-strand-specific nuclease S1, followed by electrophoresis of the intact hybrid molecules. The size and abundance of protected RNAs are then deduced by autoradiography, directly from the gel. The general approach is identical for the RNase protection assay.

cific activity probes. Moreover, with the incorporation of a 5′ label via the kinasing reaction, probes of high specific activity (>10^9 cpm/μg) can be generated for the precise mapping of the 5′ end of the transcript (Boorstein and Craig, 1989), as well as for the mapping of exon–intron boundaries (Weaver and Weissman, 1979). Any transcript that can be studied quantitatively by standard Northern analysis is easily detectable by a more sensitive nuclease protection assay. This is especially convenient when working with small amounts of RNA from several different samples.

Basic Approach

A variety of nucleases and approaches are used to quantify and characterize specific messages in greater detail than can be derived from Northern analysis. The selection of alternative approaches is expressly contingent upon the questions being asked in a particular investigation. These enzymes include nuclease S1 (Ando, 1966; Vogt, 1973), mung bean nuclease (Kowalski et al., 1976), and nuclease VII (Chase and Richardson, 1974a,b). Alternative approaches include the ribonuclease (RNase) protection assay, featuring a combination of RNase A and RNase T1 (Zinn et al., 1983; Melton et al., 1984; Lee and Costlow, 1987); primer extension (McKnight and Kingsbury, 1982; Jones et al., 1985; Kingston, 1988), and the nuclear runoff assay. To an extent, the polymerase chain reaction (PCR) has replaced some of these techniques, though the S1 protection assay and RNase protection assay (RPA) remain popular and reliable techniques; the key advantage they offer is that they do not require a reverse transcriptase step (see Chapter 15 for details).

The basic methodology of nuclease protection, when first introduced, described double-stranded DNA probe sequences directed against target cellular RNA, accompanied by S1 attack of nonprotected sequences (Berk and Sharp, 1977; 1978). The methodology has long since been extended to accommodate the inclusion of RNA probes as well. An issue of paramount importance when double-stranded probes are used is suppression of probe renaturation during the assay, thereby mandating empirical deduction of the optimal hybridization temperature. In general, hybridization is performed in buffer containing high concentrations of formamide, an environment in which the rate of hybridization is approximately 12-fold lower (Casey and Davidson, 1977) than in simple aqueous solution. More importantly it is an environment that tends to give very low background by promoting the formation of RNA:DNA duplexes rather than the renaturation of complementary strands of DNA.

The use of single-stranded probes, therefore, is clearly preferred, because of the potential for double-stranded probe renaturation. In addition, the hybridization temperature must be slightly below the melting temperature (T_m) of the hybrid molecules that are expected to form during hybridization. Because of the selective removal of single-stranded regions of DNA:DNA, RNA:DNA, and/or RNA:RNA structures, failure to identify this "window of hybridization" will very likely result in suboptimal probe:target duplex formation and thus yield misleading data.

Nuclease S1 is a natural product of the beast *Aspergillus oryzae*. It is the least expensive nuclease for this type of analysis and has been used interchangeably with mung bean nuclease in RNA mapping assays. The S1 nuclease is a zinc-dependent enzyme[2] (Ando, 1966; Vogt, 1973) that has a high specificity for single-stranded or imperfectly base-paired polynucleotide structures, both DNA and RNA; single-stranded nucleic acids are hydrolyzed endo- and exonucleolytically, yielding 5'-mononucleotides and 5'-oligonucleotides. The hydrolysis of DNA is about five times faster than is observed for RNA. Optimal nuclease activity is observed in the pH 4.0–4.5 range, under conditions of relatively high ionic strength,[3] and is essentially inactive at pH 7.2 (Linn and Roberts, 1982). S1 nuclease exhibits thermostable qualities (Ando, 1966), although in many applications, incubations are conducted at relatively low temperature (37°C). Moreover, the enzyme is resistant to denaturants (Vogt, 1973; Hofstetter *et al.*, 1976), including formamide, sodium dodecyl sulfate (SDS), and urea, thereby enhancing its usefulness.

While there are many advantages of an enzyme possessing these characteristics, there are potential difficulties associated with the use of the S1 nuclease enzyme as well. S1 nuclease is an extremely aggressive enzyme[4] whose activity often requires calibration. Excessive amounts of nuclease activity will result in the destabilization and rapid digestion of hybrid molecules and naturally unstable A+T-rich regions are particularly susceptible. Because of the low pH requirement for optimal activity of

---

[2]Co^{2+} is also effective as a cofactor (Vogt, 1973).

[3]At high concentrations of enzyme, as required in some applications, a concomitant increase in the ionic strength of the assay environment is in order, to minimize the nicking of double-stranded molecules that is often observed at higher enzyme concentrations. Nicking of double-stranded DNA is suppressed at 200 mM NaCl (Vogt, 1980) and at a slightly higher than optimal pH (Weigand *et al.*, 1975).

[4]The aggressive nature of the S1 nuclease enzyme was partially responsible for poor conversion of mRNA into double-stranded cDNA and failure to make full-length cDNA in the old days of molecular biology [i.e., before the Gubler and Hoffman (1983) technique became popular]. For the young scientists reading this, that is how we used to make cDNA in the days before PCR.

this enzyme, acid depurination of double-stranded DNA is an expected consequence of prolonged incubation in such an environment.

The basic approach in the RPA is precisely the same: A probe that is complementary to RNA transcripts under investigation is used in solution hybridization, followed by digestion of excess probe and all transcripts that are not locked up in a protected fragment. From a mechanical point of view, one should note that in the S1 assay, either a DNA or an RNA probe can be used, because S1 nuclease is single-strand specific, with little regard for whether its substrate is RNA or DNA. In the RPA, however, digestion is usually performed with a combination of RNase A and RNase T1, mandating the use of an antisense RNA probe. RNase A attacks pyrimidine nucleotides (Py/pN) by breaking the phosphodiester bond, thereby generating a free 3'-OH. RNase T1 attacks the phosphodiester bond between GpN, also generating a free 3'-OH. Table 1 presents a comparison of the closely related S1 nuclease and RNase protection techniques.

From a strategy perspective, it has become stylish of late to multiplex these reactions, that is, to use more than one probe in a single reaction tube. The notion is that as long as the probes manifest similar thermodynamic behavior and the sizes of the protected fragments are different, they can be assayed simultaneously, generating a different size band for each protected fragment, all of which appear in a single lane of gel.

Probe Selection

Many of potential problems associated with spurious hybridization and probe renaturation may be avoided by the use of single-stranded nucleic acid probes. With the ready availability of highly efficient, cost-effective, complete systems for *in vitro* transcription, large quantities of single-stranded probes of well-characterized sequence and high specific activity

TABLE 1

Comparison of S1 and RNase Protection Assays

	S1 nuclease assay	RNase protection assay
Target	RNA	RNA
Probe	DNA or RNA	RNA
Nuclease	S1	RNase A + RNase T1
Background potential	Low to moderate	Moderate to high
Average limit of sensitivity	0.5–1 pg	0.1 pg

can be produced in short order. This is currently the most popular method for generating single-stranded probes for nuclease protection analysis. Alternative, though less frequently used, methods for single-strand probe synthesis for this application include oligonucleotide synthesis, cloning into M13, and, of course, asymmetric PCR. Thus, when single-stranded probes are generated, the potential for probe renaturation becomes a non-issue, although intramolecular base pairing can still occur under nonstringent and semi-stringent conditions.

Single-stranded probes generated by *in vitro* transcription almost always contain at least a small quantity of the DNA template from whence the probe was derived. When present, both the cDNA template and contiguous vector sequences can potentially hybridize to and protect full-length probe sequences from S1 nuclease degradation, and can also promote artifactual hybridization, both of which become evident upon autoradiography. Further, these occurrences can radically reduce the concentration of free probe available to participate in the formation of RNA:RNA or DNA:RNA hybrids. Although the high stringency of the hybridization solution can alleviate these phenomena to some extent, they will not be completely eliminated unless vector sequences have been separated from the probe either by electrophoresis or by DNase I digestion of the vector, followed by probe precipitation to remove the degraded template. Last, failure to maintain a great molar excess of probe over target RNA will certainly put the sensitivity of the assay in jeopardy.

Another major difference between the Northern analysis and the S1 nuclease assay involves interpretation of the banding distribution observed after electrophoresis and autoradiography. In the classical Northern analysis, the observed signal is interpreted as corresponding to the native size of the transcript. For example, in humans, the major *c-myc* oncogene mRNA is 2.4 kb and is thus expected to lie between the predominant ribosomal RNAs (rRNAs), just above the 18S species. In contrast, the resulting hybrid molecule generated by nuclease protection analysis can only be as long as the probe, and is usually shorter than that: Nucleotides from both the probe and the target that do not directly participate in duplex formation are digested (Figure 2). As a happy consequence of this, nuclease protection assays are also more tolerant of partially degraded RNA samples as long as the integrity of the area of the target molecules recognized by the probe has not been compromised; recall that a single break in a transcript results in that molecule not being detected by standard Northern analysis. In a well-planned assay, the length of the hybrid (the protected fragment) after nuclease digestion should be detectably shorter than nonhybridized or renatured (probe with

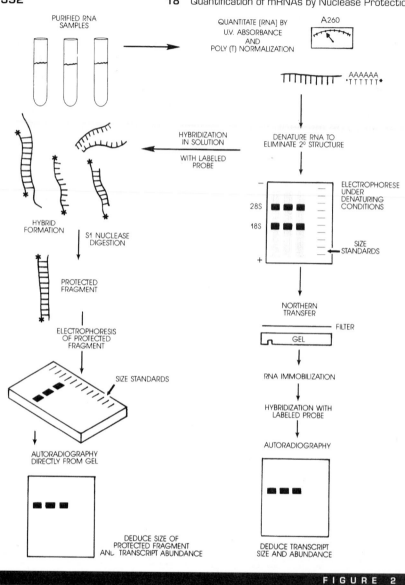

FIGURE 2

Quantification of specific RNA transcripts by S1 nuclease assay compared with conventional Northern analysis.

template) molecules. Electrophoresis of S1 nuclease-treated samples, for example, can easily resolve molecules differing in molecular weight by as little as 10% (Figure 2).

The manner in which probes are labeled is a critical concern, especially when one is attempting to characterize a transcript for the first time. End-labeled probes (5′ or 3′) are *not* preferred and should be avoided, especially in the initial characterization of probe. If the nucleotide containing the label does not directly participate in duplex formation, the label will be digested away, rendering the hybrid undetectable. This concern can be quickly alleviated with the use of continuously labeled probes (probes that are labeled along the length of the molecule); probes that are 200–500 bases tend to work best.

As indicated previously, single-stranded probes can be easily generated by *in vitro* transcription. Keep in mind that RNA probes must be handled using excellent RNase-free technique, and that the extremely high specific activity associated with these (and other) probes may result in radiolysis[5] of the probe itself. Fragmented probes translate into increased background and nonspecific hybridization. The investigator is well advised to monitor the integrity of the probe (labeled or not) on a gel prior to each use.

Optimization Suggestions

Once the type of probe and method of labeling have been established, empirical determination of the following hybridization parameters will support the most quantitative nuclease protection assay with the highest fidelity. Although it is often necessary to perform this optimization regimen for each new probe, optimization is usually necessary only once.

1. The most precise method for determining the prevalence of a given mRNA is titration in solution, using a single-stranded probe in a substantial molar excess. Assay a constant mass of probe against various quantities of a given RNA sample. If the probe is present in excess, changing the starting mass of RNA target should change the signal proportionally.

2. The stable formation of target:probe hybrids must be favored. In the absence of detailed information about the behavior of a particular probe

---

[5]Radiolysis is the phenomenon by which radiolabeled molecules are actually degraded by decay of the label they carry. This occurs most often with probes that have been labeled to extremely high activity ($>1 \times 10^9$ cpm/μg) and are then stored for more than a few days before use.

in a nuclease protection assay, begin at 45°C and run the assay at various temperatures (±5°C, ±10°C, and so forth) to determine the optimum stringency level for each probe. The temperature should be high enough to allow hybrid formation but not so high as to cause the dissociation of duplex molecules. Keep in mind that:

a. At the T_m of the probe, only 50% of all possible hybrids are formed (see Chapter 13 for discussion of T_m and hybridization temperatures). Try to adjust the hybridization temperature to about $T_m - 5°C$ (5°C below the T_m).

b. RNA:RNA duplexes are thermodynamically more stable than DNA:RNA duplexes. Thus, hybridization with a complementary RNA (cRNA; also known as antisense RNA) probe can withstand more stringent hybridization than the same assay with a DNA probe.

3. Hybridization kinetics must be driven to completion. As suggested in the protocol, it is very helpful to keep the nucleic acid components of the reaction as concentrated as possible at all times. This can be accomplished by the inclusion of carrier RNA (e.g., junk yeast RNA or transfer RNA). To demonstrate this experimentally during empirical determination, remove aliquots at specific intervals and digest each with S1 nuclease as described below. When the hybridization signal intensity no longer increases, the hybridization reaction has gone to completion.

4. The actual S1 nuclease digestion must proceed efficiently and in a very controlled fashion. S1 nuclease is a very aggressive enzyme, the activity of which can vary tremendously from lot to lot. If difficulties are encountered with this component of the assay, vary the amount of S1 nuclease so that the probe is completely degraded without loss of hybridization signal. If the activity level of a preparation of S1 nuclease is unknown, it may be useful to run an activity titration on a pilot reaction scale before attempting the S1 nuclease assay with valuable experimental material. As with most commercially available enzymes, S1 nuclease is provided in an appropriate storage buffer containing 50% glycerol, accompanied by a 10× reaction buffer. Typical 1× reaction conditions consist of 50 mM sodium acetate (pH 4.5), 280 mM NaCl, 4 mM ZnSO$_4$, 5% glycerol, accompanied by incubation at 37°C. RNase is generally provided as a lyophilized stock, with highly variable 1× reaction conditions, dependent on the nature of the assay.

5. Among the more common problems associated with transcript quantification by nuclease protection pertains to the use of antisense transcripts as probes. Probe synthesis by *in vitro* transcription will occasion-

ally produce smaller than full-length transcripts. These, in turn, often generate smaller than expected protected hybrids, manifested as more than one band upon autoradiography. This symptom can be alleviated to an extent by gel purification of the newly transcribed cRNA prior to use as a probe. Further, it is wise to electrophorese an aliquot of undigested probe alone, in a lane adjacent to the experimental samples, to convince data critics of the inclusion of full-length probe at the onset of the assay.

Potential Difficulties

Despite the aggressive nature of S1 nuclease, it tends to be more controllable than the RNase protection assay. Difficulties associated with these assays in general fall into one or more of the following categories.

1. Excessive background. This is more of a problem when working with a new probe or set of reagents. Excessive background appears as long smears in the lanes of the gel, usually due to incomplete nuclease digestion. It is normal, however, to observe fairly broad smears at the leading edge of the gel far beyond the location of discrete bands corresponding to protected fragments. One common mistake that can result in elevated background levels is waiting too long to place the samples at the hybridization temperature after the high temperature RNA denaturation step (described below). This interval affords the opportunity for some RNA molecules to regain some degree of secondary configuration, or to base pair to one another inappropriately. Efficient handling of samples is essential in this step.

Alternatively, the experimental RNA was partially degraded, thereby inviting the probe to be less discriminating. Occasionally, "noise" will be observed, resulting from the breakage (due to radiolysis or nuclease degradation) of older probes. Infrequently, smearing in the lanes of the gel may be caused by adding a huge excess of probe to the hybridization reaction mixture. More often than not, the additional radioactivity will be reflected as a large smear at the leading edge of the gel (in the form of free nucleotides).

Last, the investigator may wish to check the activity of the S1 nuclease itself prior to use, according to the recommendations of the manufacturer, though the certificate of analysis that accompanies most molecular biology enzymes is usually quite reliable. Be certain also that the integrity of the experimental RNA has not been compromised (see Chapter 7).

2. No signal. There are six common reasons for lack of signal at the end of a nuclease protection assay:

a. The probe was not labeled to sufficiently high specific activity. Further, in the case of end labeling, the nucleotide(s) containing the label were removed from the hybrid by nuclease digestion. Although continuously labeled probes (Chapter 12) are less likely to experience this phenomenon, it remains a concern unless it is known with absolute certainty that the actual labeled nucleotide(s) participate in duplex formation.

b. The experimental RNA may have been degraded. While it is true that nuclease protection assays are more tolerant of *partially* degraded RNA samples, the degradation cannot extend into the region of the RNA complementary to the probe. Always assess the integrity of experimental RNA before doing anything with it.

c. Hybridization stringency may have been too high for the type of probe. The preferred way to remedy this is to lower the hybridization temperature drastically, in order to determine if the components of the assay will permit hybridization at all. Then gradually increase the temperature until the optimal conditions have been determined (see Optimization Suggestions).

d. The precipitated pellet (sample RNA, probe, carrier RNA) may not have been thoroughly resuspended and, therefore, some or most of the sample was unable to participate in the hybridization reaction. Keeping the pellet slightly damp with ethanol will assist in dispersing and dissolving it in the hybridization buffer at the onset of the assay. In extreme cases, it is also possible that the nucleic acid pellet may have been lost during the postprecipitation ethanol washes.

e. The cells under investigation are not making the specific mRNA of interest. It may be worthwhile to perform the polymerase chain reaction if suitable primers are available. While selection of poly(A)$^+$ mRNA prior to S1 analysis may be marginally helpful, it is also potentially counterproductive. Alternatively, the investigator may be aware of a method for inducing or superinducing the specific gene(s) of interest. One may wish to pursue this assay, in such an event, with a different probe, to demonstrate that the sample was at least capable of hybridization to something (anything).

f. Inadvertent use of "sense," rather than antisense, probe.

3. Background is excellent but there are too many bands. If this is the only difficulty with this assay, only minimal adjustments should be needed to optimize it. The appearance of multiple, well-defined bands can re-

sult from the presence, during the hybridization reaction, of DNA that is homologous to the probe. One frequent source of contamination in *in vitro* transcription of cRNA probes is plasmid template sequences. A vector or template sequence generally bands closer to the origin (the wells of the gel) because of its much greater size and the fact that it was linearized just before the *in vitro* transcription reaction. Clearly, it is far more important to destroy the template by DNase I digestion when a nuclease protection assay is to be performed, than in any other application, except RNA PCR techniques. Alternatively, the hybridization stringency may have been sufficiently relaxed to permit probe hybridization to transcripts, that is, those transcribed from different members of the same gene family. Depending on the exact degree of homology, and number and distribution of mismatches, if any, fairly interesting (or frustrating!) banding patterns may be observed. To explore this possibility, conduct an identical hybridization 5°C above the hybridization temperature that results in the appearance of multiple bands, a procedural change likely to destabilize mismatched hybrids. The appearance of multiple bands may also occur in the S1 nuclease assay when denatured, double-stranded DNA is used as a probe. For this reason, many investigators simply prefer to use RNA probes directed against RNA target sequences.

Protocol: Transcript Quantification by S1 Analysis

The most important point to keep in mind is that there is no hard and fast regimen for performing the S1 nuclease assay. The procedure given here is but one of several variations and is a modification of the procedure of Quarless and Heinrich (1986). With this procedure it is possible to detect 0.5–1pg of a specific mRNA, corresponding to 2×10^{-5}% of the poly(A)$^+$ component, in 100 µg of total RNA.

1. **Wear gloves!** Use RNase-free materials and reagents. Purified RNA samples are always susceptible to RNase digestion.

Note: Be certain to check the concentration, purity, and integrity of the RNA sample before continuing with this protocol.

2. In a microfuge tube, mix 20–30 µg cellular RNA and sufficient carrier RNA to give a final mass of 100 µg. Add approximately 1×10^6 cpm of the probe. Adjust the volume of the mixture to 100 µl with DEPC-treated H_2O. Be sure to prepare a negative control tube containing carrier RNA and probe only, which should yield no protected fragment.

Note: Beginning with 10 µg total cellular or total cytoplasmic RNA is usually sufficient to quantify most mRNAs; some investigators begin *with as much as 50 μg of RNA. Although the quantity of S1 nuclease used in this protocol is more than adequate to completely digest all of the RNA in the reaction, it is critical to ensure that the probe is present in molar excess. In most cases, adding 10^6 cpm of radiolabeled probe will fulfill this requirement.*

3. Coprecipitate sample, carrier, and probe with the addition of 11 μl 3 M sodium acetate, pH 5.2, and 250 μl ice-cold 95% ethanol. Store at −20°C overnight.

Note: The experimental RNA and carrier can be coprecipitated without the probe and stored in hybridization buffer for up to 1 week at −80°C.

4. Collect precipitate by centrifugation. Decant supernatant and wash pellet with 70% ethanol (containing 30% DEPC-treated H_2O). Discard ethanol (radioactive waste) and invert tube to dry pellet for 10–15 min. An optional, second wash of the pellet with 95% ethanol will accelerate the drying process. Maintain the RNA pellet *slightly* damp with ethanol.

Note: Do not Speedvac or otherwise allow the pellet to dry to completion or it will become virtually impossible to resuspend.

5. Resuspend the pellet in 20 μl S1 hybridization buffer (80% deionized formamide; 40 mM Pipes, pH 6.4; 400 mM NaCl; 1 mM EDTA, pH 8.0). Pipet the sample up and down at least 20 times to ensure complete resuspension.

Note 1: Hybridization buffer can be prepared in advance and stored in 1-ml aliquots at −70°C.

Note 2: If experimental RNA and carrier were coprecipitated ahead of time without the probe, add the probe to the pellet first and then add hybridization buffer to a final volume of 20 μl.

Note 3: To detect rare (low abundance) mRNAs, it is necessary to keep the RNA concentration near the limit of solubility (5 mg/ml = 100 μg/20 μl) to maximize hybridization kinetics.

Note 4: Hybridization buffer consisting of 80% freshly deionized formamide is a standard formulation. If using a DNA probe, in which case the thermodynamic stability of the hybrid will not be as high as in the case of a cRNA probe, a hybridization buffer consisting of as little as 40–50% formamide may be substituted. Recall that stringency is modulated by temperature (salt and pH, too) and the presence of organic solvents.

6. Heat the sample in a water bath at 85°C for 10 min to fully denature the RNA sample.

Note: It is imperative that the sample be fully denatured, as secondary structures will have an adverse effect on the proper formation of hybrids.

7. Pulse centrifuge the tube(s) for 5 s to collect the hybridization mixture at the bottom of the tube. *Immediately* place the tube in a prewarmed water bath and incubate the reaction mixture at 45°C, or other predetermined temperature, for 3 h.

Note: Hybridization temperature is a direct function of the type of probe (RNA or DNA) used, probe length, and G+C content, and must be empirically determined. The typical range for the parameters defined in this protocol is 30–60°C. In the absence of experience with or information about a particular probe, begin with 45°C.

8. Dilute the reaction mixture 10-fold with the addition of 160 µl DEPC-treated H_2O. Add 20 µl of 10× S1 nuclease buffer (500 mM sodium acetate, pH 4.5; 2.8M NaCl; 40 mM $ZnSO_4$;[6] 50% glycerol).

9. Add 20 units of S1 nuclease per microgram of nucleic acid unless the required volume of enzyme would be greater than 5% of the total volume (recall that high concentrations of glycerol inhibit enzyme activity). In such an event, the volume of enzyme stock solution to be added must be no greater than 5% of the total volume; incubation time in the presence of fewer units of S1 per microgram of nucleic acid is extended.

Note: The addition of large amounts of S1 nuclease and extended incubation, as in this protocol, favors degradation of nonspecifically hybridized molecules when either RNA or DNA probes are used.

10. Pulse centrifuge briefly to collect the entire volume in the bottom of the microfuge tube. Incubate at 37°C for 1 h.

11. Extract the reaction once with an equal volume of phenol:chloroform:isoamyl alcohol (25:24:1) or phenol chloroform (1:1) to terminate S1 digestion. Microfuge for 2 min at top speed to separate the phases.

12. Carefully transfer the aqueous (upper) phase to a fresh microfuge tube. Add 50 µg of carrier RNA to the aqueous material. Add 2.2 volumes of 95% ethanol to precipitate the hybrids that were protected from nuclease digestion.

13. Incubate tubes on dry ice for 15 min or at −20°C for 2 h.

---

[6]Zinc chloride may be substituted, if the enzyme is stored in ZnCl.

14. Collect the precipitate by centrifugation. Carefully decant the supernatant and allow the pellet to air dry. If desired, it is acceptable to *briefly* dry down the pellet under vacuum, though it is critical not to allow the sample to dry out.

15. Resuspend the pellet in 5–10 μl TE buffer (10 mM Tris, 1 mM EDTA, pH 8.0).

16. Add an equal volume of 2× loading buffer (125 mM Tris, pH 6.8; 10% glycerol; 0.01% bromophenol blue) to the sample. Electrophorese on a 4% or 5% native polyacrylamide gel or use a percentage appropriate for characterizing the size of the expected protected fragment (Appendix K, Table 1). Use 1× TBE (Tris–borate–EDTA)[7] as the electrophoresis buffer.

Note: Hybrids larger than 500–600 bp can be studied by electrophoresis in a 3% agarose[8] gel rather than in polyacrylamide.

17. Gels may then be dried down for autoradiography at –70°C with an intensifying screen (Chapter 14). Alternatively, when high specific activity probes are used, autoradiography can be done on the bench-top by plastic wrapping the gel and overlaying a piece of Ready-Pack film (Kodak, Rochester, NY).

Protocol: Transcript Quantification by RNase Protection

As indicated in the preceding protocol, there are no hard and fast rules for transcript quantification by nuclease protection. The guiding principle is high-stringency hybridization, and the RPA offers greater thermodynamic probe:target stability and discriminatory power than is acheivable when using DNA probes in conjunction with the S1 nuclease assay. The procedure given here is but one of several variations, virtually all of which are sold in kit form. This procedure, when optimized, offers sensitivity approaching 0.1 pg of target material, or less. While the substitution of nonisotopic labeling and detection methods is a possibility, radiolabel continues to offer maximum sensitivity and resolution.

1. **Wear gloves!** Use RNase-free materials and reagents. Purified RNA samples and RNA probes prepared by *in vitro* transcription are always susceptible to RNase digestion.

---

[7]1× TBE buffer = 50 mM Tris; 50 mM boric acid; 1 mM EDTA, pH 8.0.
[8]In this lab, both NuSeive 3:1 agarose and SeaKem GTG agarose (FMC BioProducts) work extremely well in this and related applications.

2. Prepare antisense RNA probe by cloning the probe sequence downstream from a bacteriophage promoter, according to the specifications of any *in vitro* transcription system (e.g., Boehringer Mannheim kit or Promega kit) or as described elsewhere (e.g., Gilman, 1987; Sambrook *et al.*, 1989).

3. Isolate RNA from cells or tissues under investigation, using any of the protocols described herein.

Note: Be certain to check the concentration, purity, and integrity of the sample before continuing with this protocol.

4. In a microfuge tube, mix 20–30 μg cellular RNA and sufficient carrier RNA to give a final mass of 100 μg. Add approximately 1×10^6 cpm of the probe. Adjust the volume of the mixture to 100 μl with DEPC-treated H_2O. Be sure to prepare a negative control tube containing carrier RNA and probe only, which should yield no protected fragment.

Note: Beginning with 10 μg total cellular or total cytoplasmic RNA is usually sufficient to quantify most mRNAs; some investigators begin with as much as 50 μg of RNA. Although the quantity of RNase used in this protocol is more than adequate to completely digest all of the non-hybridized RNA in the reaction, it is critical to ensure that the probe is present in molar excess. In most cases, adding 10^6 cpm of radiolabeled probe will fulfill this requirement.

5. Coprecipitate sample, carrier, and probe with the addition of 11 μl 3 *M* sodium acetate, pH 5.2, and 250 μl ice-cold 95% ethanol. Store at –20°C overnight.

Note: The experimental RNA and carrier can be coprecipitated without the probe and stored in hybridization buffer for up to 1 week at –80°C.

6. Collect precipitate by centrifugation. Decant supernatant and wash pellet once with 70% ethanol (containing 30% DEPC-treated H_2O). Discard ethanol (radioactive waste) and invert tube to dry pellet for 10–15 min. An optional, second wash of the pellet with 95% ethanol will accelerate the drying process. Maintain the RNA pellet *slightly* damp with ethanol.

Note: Do not Speedvac or otherwise allow the pellet to dry to completion or it will become virtually impossible to resuspend.

7. Resuspend the pellet in 20 μl RPA hybridization buffer (80%

deionized formamide; 40 mM Pipes, pH 6.4; 400 mM NaCl; 1 mM EDTA, pH 8.0). Pipet the sample up and down at least 20 times to ensure complete resuspension.

Note 1: If necessary, the volume of hybridization buffer in which the sample is resuspended may be increased to as much as 30 μl.

Note 2: Hybridization buffer can be prepared in advance and stored in 1-ml aliquots at −70°C.

Note 3: If experimental RNA and carrier were coprecipitated ahead of time without the probe, add the probe to the pellet first and then add hybridization buffer to a final volume of 20 μl.

Note 4: To detect rare (low abundance) mRNAs, it is necessary to keep the RNA concentration near the limit of solubility (5 mg/ml = 100 μg/20 μl) to maximize hybridization kinetics.

8. Heat the sample in a water bath at 85°C for 10 min to fully denature the RNA sample.

Note: It is imperative that the sample be fully denatured, as secondary structures will have an adverse effect on the proper formation of hybrids.

9. Pulse centrifuge the tube(s) for 5 s to collect the hybridization mixture at the bottom of the tube. *Immediately* place the tube in a prewarmed water bath and incubate the reaction mixture at 45°C, or other predetermined temperature, for 3 h.

Note 1: Hybridization temperature is a direct function of the type of probe (RNA or DNA) used, probe length, and G+C content, and must be empirically determined. The typical range for the parameters defined in this protocol is 30–60°C. In the absence of experience with or information about a particular probe, begin with 45°C.

Note 2: If carrier RNA was omitted from the reaction tube (Step 4), it will be necessary to extend greatly the hybridization interval, meaning that it may be fruitful to incubate overnight.

10. At the conclusion of the hybridization period, dilute the reaction mixture 10-fold with the addition of 240 μl sterile H_2O. Add 30 μl of 10× RNase buffer (100 mM Tris, pH 7.5; 3 M NaCl, 50 mM EDTA, pH 8.0), followed by the addition of 6 μl RNase A (2 mg/ml; [final] ≈ 40 μg/ml) and 6 μl RNase T1 (0.1 mg/ml; [final] ≈ 2 μg/ml).

Note: If more than 20 µl of RPA hybridization buffer was used in Step 7, be sure to increase the volume of H_2O and $10\times$ RNase buffer proportionately.

11. Pulse centrifuge briefly to collect the entire volume in the bottom of the microfuge tube. Incubate at 37°C for 1 h.

12. Extract the reaction once with an equal volume of phenol:chloroform:isoamyl alcohol (25:24:1) or phenol chloroform (1:1) to terminate RNase digestion. Microfuge for 2 min at top speed to separate the phases.

13. Carefully transfer the aqueous (upper) phase to a fresh microfuge tube. Add 50 µg of carrier RNA to the aqueous material. Add 2.2 volumes of 95% ethanol to precipitate the hybrids that were protected from nuclease digestion.

14. Incubate tubes on dry ice for 15 min or at –20°C for 2 h.

15. Collect the precipitate by centrifugation. Carefully decant the supernatant and allow the pellet to air dry. If desired, it is acceptable to *briefly* dry down the pellet under vacuum, though it is critical not to allow the sample to dry out.

16. Resuspend the pellet in 5–10 µl TE buffer (10 mM Tris, 1 mM EDTA, pH 8.0).

17. Add an equal volume of $2\times$ loading buffer (125 mM Tris, pH 6.8; 10% glycerol; 0.01% bromophenol blue) to the sample. Electrophorese on a 4% or 5% native polyacrylamide gel, or use a percentage appropriate for characterizing the size of the expected protected fragment (Appendix K, Table 1). Use $1\times$ TBE (Tris–borate–EDTA)[9] as the electrophoresis buffer.

Note 1: Hybrids larger than 500–600 bp can be studied by electrophoresis in a 3% agarose[10] gel rather than in polyacrylamide.

Note 2: Although denaturing polyacrylamide sequencing gels are often used, they are usually not necessary for this type of analysis. If background or smearing are chronic problems, then one should switch to the denaturing electrophoresis approach.

18. Gels may then be dried down for autoradiography at –70°C with an intensifying screen (Chapter 14). Alternatively, when high specific activity probes are used, autoradiography can be done on the bench-top by plastic wrapping the gel and overlaying a piece of Ready-Pack film (Kodak, Rochester, NY).

[9]See Footnote 7.
[10]See Footnote 8.

References

Ando, T. (1966). A nuclease specific for heat-denatured DNA isolated from a product of *Aspergillus oryzae. Biochim. Biophys. Acta* **114**, 158.

Berk, A. J., and Sharp, P. A. (1977). Sizing and mapping of early adenovirus mRNAs by gel electrophoresis of S1 endonuclease-digested hybrids. *Cell* **12**, 721.

Berk, A. J., and Sharp, P. A. (1978). Spliced early mRNAs of simian virus 40. *Proc. Natl. Acad. Sci. USA* **75**, 1274.

Boorstein, W. R., and Craig, E. A. (1989). Primer extension analysis of RNA. *Methods Enzymol.* **180**, 347.

Casey, J., and Davidson, N. (1977). Rates of formation and thermal stabilities of RNA:DNA and DNA:DNA duplexes at high concentration of formamide. *Nucleic Acids Res.* **4**, 1539.

Chase, J. W., and Richardson, C. C. (1974a). Exonuclease VII of *Escherichia coli*. Purification and properties. *J. Biol. Chem.* **249**, 4545.

Chase, J. W., and Richardson, C. C. (1974b). Exonuclease VII of *Escherichia coli*. Mechanism of action. *J. Biol. Chem.* **249**, 4553.

Gilman, M. (1987). Ribonuclease protection assay. *In* "Current Protocols in Molecular Biology" (F. M. Ausubel, R. Brent, R. E. Kingston, D. D. Moore, J. G. Seidman, J. A. Smith, and K. Struhl, Eds.), pp. 4.7.1–4.7.8 Wiley, New York.

Gubler, U., and Hoffman, B. J. (1983). A simple and very efficient method for generating cDNA libraries. *Gene* **25**, 263.

Hofstetter, H., Schambock, A., Van Den Berg, J., and Weissmann, C. (1976). Specific excision of the inserted DNA segment from hybrid plasmids constructed by the poly(dA)·poly (dT) method. *Biochim. Biophys. Acta* **454**, 587.

Jones, K. A., Yamamoto, K. R., and Tjian, R. (1985). Two distinct transcription factors bind to the HSV thymidine kinase promoter *in vitro. Cell* **42**, 559.

Kingston, R. E. (1988). Primer extension. *In* " Current Protocols in Molecular Biology" (F.M. Ausubel, R. Brent, R. E. Kingston, D. D. Moore, J. G. Seidman, J. A. Smith, and K. Struhl, Eds.), pp. 4.8.1–4.8.3. Wiley, New York.

Kowalski, D., Kroeker, W. D., and Laskowski, M. (1976). Mung bean nuclease. I. Physical, chemical, and catalytic properties. *Biochemistry* **15**, 4457.

Lee, J. J., and Costlow, N. A. (1987). A molecular titration assay to measure transcript prevalence levels. *Methods Enzymol.* **152**, 633.

Linn, S., and Roberts, R. J. (1982). "Nucleases." Cold Spring Harbor Laboratory Press, Cold Spring Harbor, NY.

McKnight, S. L., and Kingsbury, R. (1982). Transcriptional control signals of a eukaryotic protein-coding gene. *Science* **217**, 316–324.

Melton, D. A., Krieg, P. A., Rebagliati, M. R., Maniatis, T., Zinn, K., and Green, M. R. (1984). Efficient *in vitro* synthesis of biologically active RNA and RNA hybridization probes from plasmids containing a bacteriophage SP6 promoter. *Nucleic Acids Res.* **7**, 1175.

Quarless, S. A., and Heinrich, G. (1986). The use of complementary RNA and S1 nuclease for the detection of low abundance mRNA transcripts. *BioTechniques* **4**(5), 434.

Sambrook, J., Fritsch, E. F., and Maniatis, T. (Eds.). (1989). "Molecular Cloning: A Laboratory Manual." Cold Springn Harbor Laboratory Press, Cold Spring Harbor, NY.

Vogt, V. M. (1973). Purification and further properties of single-strand-specific nuclease from *Aspergillus oryzae. Eur. J. Biochem.* **33**, 192.

Vogt, V. M. (1980). Purification and properties of S1 nuclease from *Aspergillus*. *Methods Enzymol.* **65,** 248.

Weaver, R. F., and Weissman, C. (1979). Mapping of RNA by a modification of the Berk–Sharp procedure: The 5′ termini of 15S β-globin mRNA precursor and mature 10S β-globin mRNA have identical map coordinates. *Nucleic Acids Res.* **7,** 1175.

Weigand, R., Godson, G., and Radding, C. (1975). Specificity of the S1 nuclease from *Aspergillus oryzae. J. Biol. Chem.* **250,** 8848.

Zinn, K., DiMaio, D., and Maniatis, T. (1983). Identification of two distinct regulatory regions adjacentn to the human β-interferon gene. *Cell* **34,** 865.

Analysis of Nuclear RNA

Introduction

The analysis of nuclear RNA is rightly defined from two perspectives. First, knowledge of the rate at which various loci are being transcribed may provide key evidence supporting the hypothesized behavior of genes in a model system under the conditions defined by an investigation. The relative rates at which genes are transcribed may be discerned by the nuclear runoff assay, the subject of the first part of this chapter. Second, examination of steady-state nuclear RNA by Northern analysis may be used to provide a qualitative aspect to the nuclear runoff analysis; it is also an invaluable tool when investigating the processing pathways of heterogeneous nuclear RNA (hnRNA). A comparison of the traditional Northern analysis, nuclease protection assays, nuclear runoff assay, and the polymerase chain reaction (PCR) is presented in Table 1.

This chapter contains five protocols for the analysis of nuclear RNA: Part I contains three protocols for assessment of the rate of transcription by the nuclear runoff assay, and Part II contains two protocols for the isolation of steady-state nuclear RNA in a form suitable for Northern analysis.

Part I: Transcription Rate Assays

Rationale

The modulation of key regulatory molecules is an integral cellular response to both intracellular and extracellular challenge. One fundamental goal in the assessment of any biological model system is the elucidation of the level of gene modulation. Although potential levels of regulation are infinite, they are broadly categorized as transcriptional or due to some posttranscriptional event(s). The initial characterization of these systems commonly involves isolation, hybridization, and subsequent detection of specific RNA species by dot blot, Northern blot, nuclease protection, and PCR-related techniques (Chapters 8, 11, 18, and 15, respectively). While these approaches may furnish excellent qualitative and quantitative data with respect to steady-state levels of message, that is, the final accumulation of RNA in the cytoplasm, nucleus, or both, RNA prepared by total cellular lysis provides information neither about the rate of transcription nor about the intracellular compartmentalization (nuclear or cytoplasmic) of the RNA of interest. Knowledge of this aspect of the cellular biochemistry is necessary to elucidate the level of gene regulation because many messenger RNAs (mRNAs) naturally have much longer half-lives than

TABLE 1

Comparison of the Relative Merits of the Traditional Northern Analysis, Nuclease Protection Assay, Nuclear Runoff Assay, and the Polymerase Chain Reaction

	Northern analysis	Nuclease protection assay	Nuclear runoff assay	Polymerase chain reaction
Advantages	Provides a qualitative component to RNA analysis	Much higher sensitivity than Northern analysis	Characterizes relative rate of transcription	Provides unparalleled sensitivity when properly designed
	Supports several rounds of hybridization with different probes	Requires less handling of RNA than other types of analysis	Natural geometry of the chromatin is maintained	Provides unparalleled resolution
	Is compatible with total, cytoplasmic or poly(A)$^+$ RNA	Is tolerant of partially degraded RNA	Permits simultaneous study of several genes	Supersedes many of the classical techniques
	RNA is relatively stable on filter	Solution hybridization is more quantitative than filter hybridization	Can be used to discern transcriptional vs. posttranscriptional gene regulation when used in conjunction with data from Northern analysis	Minimizes the amount of handling of the RNA
	Is able to assess integrity of the sample	Can be used for steady-state or rate assays		Very rapid technique; favors laboratory productivity
Disadvantages	Is the least sensitive assay	Protected fragment is smaller than native RNA	Nuclear isolation requires a fair amount of skill	More sensitive to the precise reaction components than other assays
	Denaturants can be toxic	Nucleases, esp. S1, can be difficult to control	Probe complexity is very large	Exquisitely sensitive to contaminants, especially genomic DNA
	Requires extensive handling of RNA	Assay is more sensitive to exact hybridization parameters than other assays	Unlabeled endogenous RNA can compete with labeled RNA during hybridization	Carryover contamination must be addressed
	Is a time-consuming process	Double-stranded probes can compromise quantitativeness of the assay if reannealing occurs	Mechanics of the assay support transcript elongation and not initiation during labeling	Optimization, in many cases, can be time-consuming and costly
	Provides ample opportunity for RNase degradation			
	Characterizes steady-state RNA only			

others and because the half-life of many mRNA species can be tremendously modified in response to a particular xenobiotic regimen or environmental stimulus.

A variety of factors influence the prevalence of mature mRNA molecules in the cytoplasm as well as translatability, including, but not limited to, regulatory factors and sequences acting in trans and cis, respectively. Transcriptional efficiency, transcription rate, splicing efficiency of precursor hnRNA molecules, nucleocytoplasmic transport, and accessibility of mRNA to the protein translation machinery are but a few potential levels of gene regulation that ultimately govern phenotypic manifestation (Figure 1).

To address these questions, two basic approaches have been employed to study the mechanism of transcription of specific genes and the subsequent processing of the resulting primary transcripts in eukaryotic cells. In one approach, the rate of transcription is measured in intact nuclei by the incorporation of labeled precursor nucleotides into RNA transcripts initiated on endogenous chromatin at the time of nuclear isolation. This elongated, labeled nuclear RNA is subsequently purified for hybridization to complementary, membrane-bound target DNA sequences. This widely used technique, known as the nuclear runoff assay (Figure 2), is currently the most sensitive method for measuring transcription as a function of cell state (Marzluff, 1978; Marzluff and Huang, 1984). The other technique employs specific cloned DNA fragments as templates for transcription (Manley, 1984); it is used to dissect the factors and sequences necessary for faithful transcription *in vitro* using whole cell extracts.

The principal advantage of the nuclear runoff assay is that labeling occurs while maintaining the natural geometry of the transcription apparatus. Because the nuclear runoff assay yields unspliced precursor RNA, however, the investigator is uncertain as to the natural posttranscriptional fate of these RNA molecules. The mechanics and reaction conditions of the nuclear runoff assay promote the elongation of initiated transcripts but are not believed to support new initiation events.[1] The degree of labeling of any particular RNA species (and hence, the relative transcription rate of specific genomic sequences) may then be assessed by Cerenkov counting coupled with autoradiography. Alternatively, if labeling is performed by the incorporation of nonisotopic label, detection can

---

[1] Large amounts of unlabeled, endogenous nuclear RNA can competitively interfere with an assay designed to measure newly synthesized RNA, one aspect of nuclear biochemistry that this assay does not support. Details for quantifying transcript initiation have been described elsewhere (Marzluff and Huang, 1984, pp. 103–119).

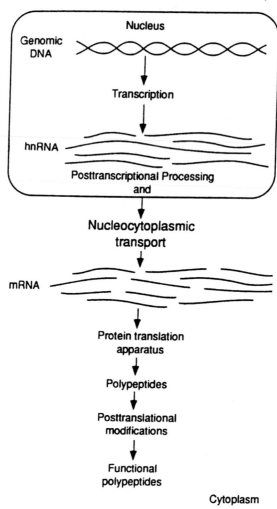

FIGURE   1

mRNA biogenesis. Transcription of genes by RNA polymerase II produces hnRNA, the direct precursor of mRNA. Events in the cellular biochemistry that affect mRNA maturation are designated as being transcriptional or posttranscriptional in nature, depending on where, precisely, they exert their influence. Nuclear or cellular lysis yields only those transcripts already accumulated; to assess the relative rates at which transcripts are synthesized, the nuclear runoff assay is utilized.

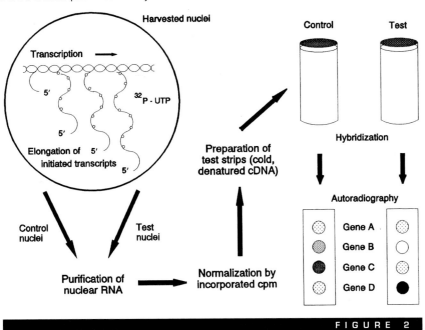

FIGURE 2

Nuclear runoff assay. Relative rate of transcription of all genes can be assessed by incubation of intact nuclei with an NTP cocktail containing labeled UTP.

be carried out by chromogenic detection or chemiluminescence. Under a defined set of experimental conditions, these data correlate directly with the number of RNA polymerase molecules engaged in transcribing a specific gene and indirectly with the transcriptional efficiency of regulatory sequences associated with the gene(s) of interest. When used in conjunction with the Northern blot analysis of cytoplasmic RNA species (see NP-40[2] lysis procedure), data from the nuclear runoff assay may be used to assess whether observed gene modulation is a result of a change in the synthesis (transcriptional control) or a change in splicing/nucleocytoplasmic transport/mRNA stability (posttranscriptional control).

The nuclear runoff assay permits the simultaneous assay of transcriptional activity at several specific genomic loci, all of which are presumably transcribed in the same relative amounts in isolated nuclei as in intact cells. This is a key advantage of this technique. Because a "typical" eukaryotic cell is actively transcribing tens of thousands of different

[2]Nonidet P-40; now known as Igepal CA-630.

genes at any given point in the cell cycle, the complexity of the labeled RNA (which acts as a heterogeneous probe in this assay) is enormous; the concentration of the transcribed RNA sequences of immediate interest in the experiment may be only a very small percentage of the mass of the probe. Despite this, the sensitivity of the assay is excellent for the following reasons:

1. The labeled RNA is single stranded (renaturation of a double-stranded probe reduces the effective concentration of the probe in hybridization solution).
2. The RNA is labeled continuously along its length, to extremely high specific activity (compare with end-labeled probes, Chapter 12).
3. The enhanced thermodynamic stability of RNA:RNA hybrids, compared to RNA:DNA and DNA:DNA duplexes, permits very stringent posthybridization washes, which increases the signal-to-noise ratio.

The nuclear runoff assay consists of several distinct phases, each of which has a distinct goal:

1. Nuclear isolation. The goal is the rapid isolation of intact nuclei capable of supporting transcript elongation.
2. Transcript elongation in the presence of labeled NTP precursor. The goal is to label nascent transcripts while maintaining the natural geometry of the transcriptional apparatus.
3. Recovery of labeled transcripts. The goal is removal of protein and DNA from the now labeled probe.
4. Hybridization of labeled RNA to specific, membrane-bound DNA target sequences. The goal is to hybridize only those labeled transcripts for which there is a corresponding complementary DNA (cDNA). There is no limit to the number of genes that can be assayed at once.
5. Quantification by autoradiography and/or Cerenkov counting. The goal is to assess the relative transcription rates for genes of interest. The hybridization signal correlates with the degree to which RNAs of interest are present in the pool of all transcribed RNAs. A typical example of how these data might be reported is shown in Table 2.

The most critical parameter by far is the preparation of nuclei prior to labeling; most protocols appearing in the literature vary with respect to the postlabeling RNA purification steps only.

The degree of success of the nuclear runoff assay is almost entirely dependent on the speed with which nuclei can be isolated from intact cells

TABLE 2

Dependence of Protooncogenes on Cell Shape

Cell line	Proliferation phenotype	Contact inhibited	Relative transcription rate			
			c-ras	c-myc	c-fos	c-fes
CUA-1	Shape dependent	+	30.0	16.0	0.92	0.93
HT-1080	Shape independent	–	0.86	0.67	0.93	0.92
HT-IFN^R	Shape dependent	+	0.38	0.51	0.64	0.66

Note. Values reflect the abundance of gene-specific nuclear transcripts in flat cells divided by the transcript abundance in rounded cells plated onto the ethanol-soluble polymer poly[HEMA]. Abundance levels are determined from image analysis performed on nuclear runoff autoradiograms (Farrell and Greene, 1992).

and radiolabeled with precursor UTP. Failure to generate high specific activity RNA probe is usually a direct result of inexperienced handling of the nuclei prior to the labeling step. Nuclear isolation must also be carried out in such a way as to preserve RNA polymerase activity and nuclear structure during the isolation from cells cultured *in vitro* (Marzluff *et al.*, 1973; 1974) or if absolutely necessary, from tissue (Ernest *et al.*, 1976; Marzluff and Huang, 1984). Methods for the isolation of chromatin that retains endogenous RNA polymerase activity have also been described (Marzluff and Huang, 1975).

In general, nuclei are isolated from whole cells by incubation in hypotonic lysis buffer featuring the inclusion of the gentle, nonionic detergent NP-40 (Price and Penman, 1972; Marzluff and Huang, 1984; Greenberg, 1988); in isoosmotic sucrose buffer containing Triton X-100 (Marzluff *et al.*, 1973, 1974), or by nonaqueous methods[3] (Gurney and Foster, 1977; Lund and Paine, 1990). Nuclei may then be labeled immediately or stored frozen in liquid nitrogen or at $-70°C$ for several months without significant loss of labeling potential.

Protocol: Nuclear Runoff Assay[4]

Harvesting of Cells and Preparation of Nuclei

In this laboratory, the best labeling has occurred when cells are maintained as cold as possible until the initiation of the labeling reaction. It is

---

[3]Nonaqueous methods of nuclear isolation consist of tissue/cell lyophilization, homogenization of the dry powder in glycerol (for example) and then nuclear sedimentation in nonaqueous buffer.

[4]The protocols presented here are improvements of the procedures of Groudine *et al.* (1981), Nevins (1987), and Greenberg (1988).

very helpful to fill two or three large trays (such as those often used for autoclaving bottles) with ice so that tissue culture vessels can be set directly on ice for the first three steps in this protocol. This precludes having to work in a cold room. Further, the recovery of cells from tissue culture is greatly accelerated when cells are grown in 100- or 150-mm tissue culture dishes rather than in tissue culture flasks.

1. Decant and discard cell growth medium from adherent cells growing in tissue culture.

Note: While as few as 5×10^6 nuclei may be utilized in a labeling reaction, it is wise to begin with $3–5 \times 10^7$ nuclei to achieve efficient labeling. Sufficient counts per minute (cpm) must be generated to titrate the RNA sample at the hybridization step. Nuclei and cell number can be quantified by hemocytometer counting.

2. Wash the monolayer once with 10–15 ml ice-cold phosphate-buffered saline (PBS). Decant most of the PBS, leaving approximately 1 ml in the tissue culture vessel.

3. Using a sterile cell scraper or a rubber policeman, harvest the cells by gentle but rapid scraping.

Note 1: Cells may be harvested by trypsinization[5] although overtrypsinization will greatly reduce the labeling efficiency in this assay.

Note 2: While the synthesis of hnRNA in "quasi-transformed" cell populations appears to be unaffected by the cell shape distortions that undoubtedly accompany trypsinization (Ben-Ze'ev et al., 1980), its synthesis in diploid cells shows a much greater shape-responsiveness to even subtle changes in morphology (Wittelsberger et al., 1981). For this reason, scraping cells from dishes resting on ice is the method least likely to cause distortion of the cellular morphology and is therefore strongly recommended.

Note 3: Collect cells grown in suspension by centrifugation at 500g for 5 min at 4°C. Decant and discard the supernatant. Then resuspend the cell pellet in 10 ml of ice-cold 1× PBS and pellet as described above. Remove the supernatant as completely as possible and then proceed to Step 5.

4. Transfer the cells to a prechilled centrifuge tube (e.g., a 15- or 30-ml nuclease-free Corex glass tube) and collect cells by centrifugation at 500g for 5 min at 4°C. Remove the supernatant as completely as possible.

[5]For trypsinization protocol see Appendix H.

Note: Transport tubes containing cells and/or nuclei to and from the centrifuge in an ice bucket.

Cellular Lysis with NP-40 Buffer

5. Prior to resuspending the cells, loosen the pellet by very gentle vortexing for 5 s. Slowly add 4 ml of NP-40 lysis buffer (10 mM Tris, pH 7.4; 10 mM NaCl; 5 mM MgCl$_2$; 0.5% NP-40) while continuing to vortex gently. Incubate on ice for 5 min.

Note: This technique will prevent the formation of clumps of cells.

6. Collect the nuclei by centrifugation at 500g for 5 min at 4°C.

Note: If desired, cytoplasmic RNA can be purified from the resulting cytoplasmic supernatant, for Northern analysis. Store supernatant on ice and continue working up nuclei first.

7. Resuspend the nuclear pellet in 4 ml of NP-40 lysis buffer as described in Step 5. Collect the nuclei by centrifugation at 500g at 4°C.

8. Carefully decant and discard the supernatant. Resuspend the nuclei in 200–300 μl of prechilled glycerol storage buffer (50 mM Tris, pH 8.0; 30% glycerol; 2 mM MgCl$_2$; 0.1 mM EDTA) by gentle vortexing as described in Step 5. The nuclei are now ready for immediate labeling or may be stored in sealed ampules in liquid nitrogen or at –70°C for several months.

Alternative Protocol for Preparation of Nuclei: Isolation of Fragile Nuclei from Tissue Culture Cells[6]

In this procedure, fragile nuclei are prepared by cell lysis, in isoosmotic sucrose buffer, and pelleted into a dense cushion of sucrose. This approach helps maintain nuclear integrity by preventing nuclei from slamming into the bottom of the centrifuge tube. Keep in mind that the density of nuclei varies from one cell type to the next and may have to be adjusted to accommodate nuclear purification from a specific cell type. As in the previous protocol, the recovery of cells from tissue culture is greatly accelerated when cells are grown in 100- or 150-mm tissue culture dishes rather than in tissue culture flasks.

1. Decant and discard cell growth medium from adherent cells growing in tissue culture.

---

[6]This is a modification of the procedure of Marzluff and Huang (1984).

Note: Begin with 3–5 × 10⁷ nuclei to generate sufficient cpm to titrate the RNA sample at the hybridization step. Nuclei and cell number may be quantified by hemocytometer counting.

2. Wash the monolayer once with 10–15 ml cold PBS. Decant most of the PBS, leaving approximately 1 ml in the tissue culture vessel.

3. Using a sterile cell scraper or a rubber policeman, harvest the cells by gentle but rapid scraping.

Note 1: Cells may be harvested by trypsinization,[7] though overtrypsinization will greatly reduce the labeling efficiency in this assay.

Note 2: While the synthesis of hnRNA in "quasi-transformed" cell populations appears to be unaffected by the cell shape distortions that undoubtedly accompany trypsinization (Ben-Ze'ev et al., 1980), its synthesis in diploid cells shows a much greater shape-responsiveness to even subtle changes in morphology (Wittelsberger et al., 1981). For this reason, scraping cells from dishes resting on ice is the method least likely to cause distortion of the natural cellular morphology and is therefore strongly recommended.

Note 3: Collect cells grown in suspension by centrifugation at 500g for 5 min at room temperature (25°C). Decant and discard the supernatant. Optional: Wash the cell pellet one time with an aliquot of cold PBS and pellet as described earlier. Remove the supernatant as completely as possible and then proceed to Step 5.

4. Transfer the cells to a suitable centrifuge tube (e.g., 15- or 30-ml nuclease-free Corex glass tube) and collect cells by centrifugation at 500g for 5 min at room temperature. Remove the supernatant as completely as possible.

5. Resuspend the cells in prechilled buffer I (0.32 *M* sucrose; 5 m*M* $CaCl_2$; 3 m*M* magnesium acetate; 0.1 m*M* EDTA; 0.1% Triton X-100; 1 m*M* DTT; 10 m*M* Tris, pH 8.0) at a density of 1 × 10⁷ cells/ml.

Note: This buffer can be prepared in advance; however, do not add DTT or Triton X-100 until just prior to use.

6. Homogenize the cells in a Dounce homogenizer with a maximum of 10 strokes of a tightly fitting pestle.

Note: One common cause for poor label incorporation is the use of a loosely fitting pestle.

---

[7]For trypsinization protocol see Appendix H.

7. Place a cushion of buffer II (2 M sucrose; 3 mM magnesium acetate; 1 mM DTT; 10 mM Tris, pH 8.0) into the bottom third of an RNase-free centrifuge tube. Be certain to use only those centrifuge tubes that can withstand ultracentrifugation. Dilute the homogenate with an appropriate aliquot of buffer II to produce a sufficient volume to occupy the remaining two-thirds of the ultracentrifuge tubes. Carefully layer the resulting mixture over the cushion.

8. Collect the nuclei by centrifugation at 30,000g for 45 min at 4°C.

9. Resuspend the nuclear pellet in storage buffer (40% glycerol; 5 mM magnesium acetate; 0.1 mM EDTA; 5 mM DTT; 50 mM Tris, pH 8.0) at a density of 10^8 nuclei/ml.

10. Immediately freeze nuclei in aliquots in liquid nitrogen or initiate labeling protocol.

Alternative Protocol for Preparation of Nuclei: Isolation of Nuclei from Whole Tissue[8]

1. Harvest tissue according to standard procedures. Rapidly mince tissue and rinse with an ice-cold solution of 0.14 mM NaCl; 10 mM Tris, pH 8.0.

2. Homogenize minced tissue in cold lysis buffer (0.32 M sucrose; 5 mM CaCl$_2$; 3 mM magnesium acetate; 0.1 mM EDTA; 0.1% Triton X-100; 1 mM DTT; 10 mM Tris, pH 8.0). Use 5 ml lysis buffer per gram of tissue.

3. *Rapidly* filter homogenate through several layers of sterile cheesecloth. This will facilitate removal of connective tissue and other debris.

4. Dounce homogenize filtrate with 10–20 strokes of a tightly fitting pestle. Remove a 5-ml aliquot of the homogenate and place on a microscope slide, overlay with a coverslip, and assess the extent of cellular lysis by light microscopy.

5. Add 1–2 volumes of prechilled solution of 2.0 M sucrose;[9] 3 mM magnesium acetate; 0.1 mM EDTA; 1 mM DTT; 10 mM Tris, pH 8.0) to the homogenate. Layer the resulting mixture over a cushion of this same buffer that occupies one-third of the volume of an RNase-free centrifuge tube. Be certain to use only centrifuge tubes that can withstand ultracentrifugation.

---

[8]This is a modification of the procedures of Derman *et al.* (1981) and Marzluff and Huang (1984).

[9]The exact sucrose concentration is a function of the cell type under investigation and should be empirically determined before using valuable starting material. The usual range is 1.8–2.2 M.

6. Collect nuclei by centrifugation at 30,000 g for 45 min at 4°C.

7. Resuspend the nuclear pellet in glycerol storage buffer (25% glycerol; 5 mM magnesium acetate; 0.1 mM EDTA; 5 mM DTT; 50 mM Tris, pH 8.0) at a density of 1×10^8 nuclei/ml.

8. Immediately freeze nuclei in aliquots in liquid nitrogen or initiate labeling protocol.

Labeling of Transcripts

Note: It is very important to have the 2× reaction buffer prepared before thawing nuclei. Permitting thawed nuclei to sit for just a few minutes on ice will greatly diminish labeling efficiency.

1. Prepare a 2× reaction buffer (10 mM Tris, pH 8.0; 5 mM MgCl$_2$; 300 mM KCl; 1 mM ATP, 1 mM CTP, 1 mM GTP; 5 mM DTT).

Note: Prepare individual NTP stock solutions by dissolving in sufficient 0.5 mM EDTA, pH 8.0, to give a final concentration of 100 mM; adjust pH to 7.0–8.0. Store nucleotide stock solutions in aliquots at –20°C, and thaw each aliquot only once. Do not add nucleotides to reaction buffer until just prior to use. Although the minimum concentration of NTPs is 50 μM, the usual working concentration is in the 0.5–1.0 mM range. Alternatively, a stock solution of NTPs may be purchased from almost any supplier of biotechnology reagents.

2. Add 200 μl of the 2× reaction buffer to each 200 μl aliquot of nuclei. If nuclei were previously stored frozen, add 200ml of the 2× reaction buffer to 200 μl of thawed nuclei.

Note 1: In this laboratory, optimum labeling efficiency is observed when 200 μl of 2× reaction buffer is added to the ampule containing the frozen nuclei, thereby allowing them to thaw on ice in the reaction buffer. Do not wash nuclei after thawing, and thaw only once.

Note 2: Do not be overly alarmed if the mixture of thawed nuclei appears viscous. Although frozen and thawed nuclei may lyse to some extent, transcriptional complexes remain intact and are competent to support the elongation of initiated transcripts and concurrent incorporation of label (Marzluff and Huang, 1975; Nevins, 1987). Of course, care should be taken in each step prior to the labeling reaction to minimize the amount of nuclear damage.

3. *Immediately* add 10 μl[α-^{32}P]UTP 800 Ci/mmol, 10 mCi/ml, and incubate for 30 min at 26°C, with gentle shaking.

Note 1: Labeling can be carried out directly in the freezing ampule. Alternatively, freshly prepared nuclei are best labeled in a conical 15-ml polypropylene tube to minimize the radioactive mess that will invariably result from the mechanics of this assay.

Note 2: In this laboratory, the most reproducible degree of label incorporation was observed when the labeling reaction is shaken at 50 rpm on a temperature-controlled orbital platform (Lab-Line Instruments, Inc.).

Note 3: Incubation temperature may range from 25°C to 37°C. Incubation at 25°C may be extended to 1 h while a 37°C incubation beyond 30 min will not increase the degree of labeling/elongation of initiated transcripts. Alternatively, some procedures recommend incubating nuclei with label for a little as 15 min at 37°C (Nevins, 1987).

Recovery of Labeled Transcripts

1. In a separate, RNase-free tube, mix 1 ml of high salt buffer (500 mM NaCl; 50 mM MgCl$_2$; 2 mM CaCl$_2$; 10 mM Tris, pH 7.4) with 50 μl of 1 mg/ml RNase-free DNase I, just prior to use. Add 750 μl of the resulting mixture to each tube containing labeled nuclei.

Note 1: The high salt buffer will terminate the labeling reaction and disrupt all of the remaining intact nuclei. The nuclear disruption greatly increases the viscosity of the solution, because of the liberation of chromatin. Use a silanized,[10] RNase-free Pasteur pipet, a micropipettor, or a 25-gauge needle attached to a 1-ml tuberculin syringe to shear the DNA several times.

Note 2: The inclusion of CaCl$_2$ in the high salt buffer enhances DNase I activity.

2. Incubate this mixture for 5 min at 37°C.

Note: The purpose of this brief treatment with DNase I is to partially fragment genomic DNA so that the solution is physically more manageable and less likely to trap labeled RNA molecules in the subsequent purification steps. Partially degraded DNA is removed in subsequent steps.

3. Add 200 μl sodium dodecyl sulfate (SDS)–Tris buffer (5% SDS;

---

[10]See Appendix F for hints on silanizing glassware.

500 m*M* Tris, pH 7.4; 125 m*M* EDTA) and 15 μl of 20 mg/ml proteinase
K. Incubate for 30 min at 42°C to ensure extensive protein hydrolysis.

4. Extract each sample with 1.25 ml of phenol:chloroform:isoamyl alcohol (25:24:1). Centrifuge for 8–10 min at 500–800 *g* to separate the phases.

5. Transfer the upper aqueous phase to a fresh tube and add, in order:

2 ml sterile (nuclease-free) H_2O

3 ml 10% trichloroacetic acid (TCA) in 60 m*M* sodium pyrophosphate

15 μl 10 mg/ml carrier RNA (e.g., junk yeast RNA)

Incubate on ice for 45 min.

Note: These steps will result in precipitation of the RNA.

6. Filter the precipitate onto Whatman GF/A glass fiber filters using a vacuum filtration bank (Millipore, Bedford, MA). Wash each filter three times with 10-ml aliquots of 5% TCA, 30 m*M* sodium pyrophosphate.

Note: Omission of the TCA precipitation step and filtration will usually result in unacceptably high levels of background even after the most stringent posthybridization washes. This is a key step contributing to the quantitativeness of this assay.

7. Transfer each filter to a glass scintillation vial and add 1.5 ml DNase I buffer (20 m*M* Hepes (free acid), pH 7.5; 5 m*M* $MgCl_2$; 1 m*M* $CaCl_2$). Add 50 μl 1 mg/ml RNase-free DNase I and incubate for 30 min at 37°C.

8. Terminate the digestion with the addition of 40 μl of 500 m*M* EDTA, pH 8.0, and 70 μl of 20% SDS.

9. Elute the purified RNA from each filter paper by heating the samples to 65°C for 10 min. *Carefully* transfer all of the supernatant to a fresh tube.

10. Repeat the elution step by adding 1.5 ml of elution buffer (1% SDS; 10 m*M* Tris, pH 7.5; 5 m*M* EDTA) to the scintillation vials containing the filters. Heat to 65°C for 10 min and then *carefully* recover the supernatant and combine it with the original supernatant from Step 9.

11. Extract the resulting RNA solution one time with an equal volume of phenol:chloroform:isoamyl alcohol (25:24:1), followed by a single extraction with an equal volume of chloroform.

Note: In both instances, the RNA will partition to the aqueous (upper) phase.

12. Transfer the aqueous material (approximately 3 ml) to a silanized, RNase-free Corex glass tube (or the equivalent). Add 650 μl of 1 N NaOH and incubate on ice for no more than 5 min.

Note: The limited hydrolysis of RNA tends to improve hybridization efficiency (Jelinek et al., 1974). Incubation for more than 5 min runs the great risk of RNA degradation by alkaline hydrolysis. Further, limited hydrolysis reduces the likelihood of high-molecular-weight RNA self-hybridization, in favor of forward, intermolecular hybridization kinetics.

13. Neutralize the reaction with the addition of 1.5 ml 1 M Hepes (free acid), pH 7.4.

14. Precipitate the RNA with the addition of 0.1 volume of 3 M sodium acetate, pH 5.2 (approximately 500 μl) and 2.5 volumes of ice-cold 95% ethanol (approximately 15 ml). Store overnight at –20°C.

Preparation of Target DNA

1. DNA sequences corresponding to the genes of interest may be prepared by PCR amplification or by insert excision from the plasmids into which they are cloned. In the latter case, use a restriction endonuclease that will cut only the vector and not the DNA insert. Retention of these target DNA sequences on nylon is a non-issue; however, the target DNA should be at least 1.0 kb to ensure quantitative retention on nitrocellulose. In this regard, and in consideration of its poor binding capacity, nitrocellulose should be avoided, and nylon used instead.

2. Denature linear target DNA with the addition of 0.1 volume of 1 N NaOH. Incubate at room temperature for 30 min or at 37°C for 10 min.

3. Neutralize the alkaline pH of the mixture by the addition of 10 volumes of ice-cold 6× SSC.

4. Store denatured DNA on ice until ready to blot, but for no longer than 15–20 min. It is best to complete blotting within 15 min after the addition of 6× SSC to preclude the reannealing of the denatured sample.

5. Assemble a Minifold I (dot blot) or Minifold II (slot blot) apparatus (Schleicher & Schuell, Inc.) and begin to apply a low vacuum. Pipet 2–5 μg (or other predetermined mass) of denatured DNA to each well in a volume of 100–200 μl. After the buffer has been pulled through the filter, wash each well with 300 μl 6× SSC.

    a. This mass of DNA is generally in excess for most genes, with the exception of extremely abundant sequences such as ribosomal RNAs and small nuclear RNAs. That the target DNA is in excess can be determined by the following:

- varying the amount of input RNA, in which case the hybridization signal should be proportional to the amount of input RNA;
- varying the amount of target DNA on the filter paper, in which case there should be no effect on the amount of RNA hybridized if the DNA is in excess;
- rehybridizing RNA that does not bind to the target DNA or the filter paper, none of which should participate in hybridization.

b. Keep in mind that the DNA binding capacity of nitrocellulose is only about 80 μg/cm^2. Particularly when target DNA cloned into a plasmid is not excised from vector, meaning that linearized vector plus insert are to be blotted, it may be advantageous to use nylon membranes, many of which have a nucleic acid binding capacity in excess of 400 μg/cm^2. This is especially important when a variety of different size vectors are involved or when variable sized DNA inserts are cloned into the same type of vector because an equivalent mass of the DNA insert (and not the combined mass of vector and insert) must be applied to each well. Normalization in this fashion is necessary to ensure that observed variations in hybridization signal intensity truly reflect differential transcription rates. The normalization of the mass of target DNA samples may be accomplished by the addition of an appropriate amount of linearized, nonrecombinant vector DNA (vector with no insert), or other carrier DNA, to each sample before applying the sample to the wells of the blotting manifold.

c. Because the nuclear runoff assay can be used to simultaneously assess the extent of transcription of a variety of genes, one worthwhile blotting strategy is to apply the target DNAs corresponding to the different genes of interest in the same vertical column of the blotting manifold. Do not forget to include DNA corresponding to a gene with a purported "housekeeping" function, that is, a gene whose level of expression is not expected to vary under the defined experimental conditions. Thus, the filter can be cut into narrow strips and hybridized to the extremely heterogeneous RNA probe in a minimum volume of hybridization buffer.

d. Remember to pre-wet the filter in nuclease-free water for 5 min and then in 6× SSC for at least a few seconds prior to assembling the dot blot apparatus.

e. The slot blot apparatus will concentrate the sample into a smaller area (6 mm^2) than the dot blot apparatus (32 mm^2); thus slot-blotting is expected to yield data that are easier to quantify than dot-blotting because the slot configuration permits easier quantitation by scanning densitometry. For image analysis software, however, this is a non-issue.

f. Remember to include a suitable negative control (such as nonrecombinant vector) and a positive control (if possible) on each filter paper strip to be hybridized.

g. To demonstrate the linearity of the assay, it is advisable to make at least triplicate filter membrane strips, onto which samples of target DNA have been immobilized. The heterogeneous labeled RNA generated by nuclear runoff will be subjected to multiple twofold dilutions based on its measured specific acticity (cpm). These dilutions are then used to hybridize to individual filter paper strips containing DNAs of interest. The signal intensity obtained upon hybridization detection should, of course, reflect the twofold dilutions of the labeled RNA. This approach is far more reliable than making dilutions of the DNA on a single filter paper and then subjecting it to hybridization with the labeled RNA at only one concentration.

6. Immobilize the target DNA onto the filter paper for subsequent hybridization according to the instructions of the manufacturer. In general, DNA may be crosslinked to nylon membranes using a calibrated UV light source while nitrocellulose is baked for 2 h at 80°C *in vacuo*. Alternatively, nylon filters may be baked instead, for as little as 1 h at temperatures as low as 65°C (vacuum not required). The filter may then be used immediately or stored in a cool, dry location for later use. Details pertaining to nucleic acid immobilization on solid supports can be found in Chapter 11.

Preparation of RNA for Hybridization

1. Recover the precipitated RNA by centrifugation at 10,000 g for 30 min at 4°C. Carefully decant as much of the ethanol as possible into rad waste. Allow the tubes to drain briefly, and then wash once with 1 ml 70% ethanol (dilute 95% ethanol with sterile H_2O). Centrifuge, if neces-

sary, and then decant as much of this ethanol wash as possible. Allow the tubes to drain briefly.

2. Resuspend the RNA pellet in 1 ml of N-tris(hydroxymethyl)methyl-2-aminoethane sulfuric acid (TES) solution (10 mM TES, pH 7.4; 10 mM EDTA; 0.2% SDS). Shake the sample at 100 rpm on an orbital platform for 30 min at room temperature to redissolve the RNA.

3. Remove a 5-μl aliquot from each sample and count for 1 min to determine cpm/ml. Normalize the samples with respect to cpm/ml by the addition of TES solution, to provide nearly equal counts per hybridization. In general, the cpm/ml averages about 10^7 cpm/ml per 5×10^7 nuclei.

Note: The normalization of samples by cpm is based on the assumption that the overall level of RNA synthesis is not changing as a function of cell state. The investigator must empirically determine the reasonableness of this assumption for each new set of cell state conditions and also ensure equal numbers of nuclei at the onset of the experiment.

4. Cut the filter containing the immobilized target DNA sequences into strips such that each strip contains one complete set of target DNAs to be hybridized to the labeled RNA probes.

5. Hybridization is most efficiently and reproducibly executed directly in a clean 5-ml scintillation vial. This may be accomplished by coiling the filters strips containing the DNA and carefully pushing each strip to the bottom of individual scintillation vials such that the side of the filter onto which the DNA is fixed is facing the center of the tube.

Note: Hybridization in this fashion will minimize the amount of probe required and will maintain the maximum probe concentration. Moreover, this approach will minimize the radioactive mess that is invariably associated with hybridization in plastic bags.

6. Mix 1 ml of RNA solution with 1 ml of TES/NaCl (10 mM TES, pH 7.4; 10 mM EDTA; 0.2% SDS; 600 mM NaCl). Add the resulting 2 ml of RNA probe solution to the scintillation vials containing the target DNA filter strips, close the tubes tightly, and hybridize for 24–36 h at 65°C. Be certain that filters in the vials are completely immersed in hybridization solution.

Note: In this lab, the most efficient method for hybridization is to insert individual scintillation vials into the disposable polystyrene racks in which conical 15-ml centrifuge tubes are often packaged. The rack is then placed on the platform of an orbital shaker incubator preheat-

ed to 65°C and agitated slowly (50–75 rpm) for the duration of the hybridization.

Posthybridization Washes and Detection

7. At the conclusion of the hybridization period, wash the filters in batch for 1 h in an excess volume of 2× SSC, preheated to 65°C.

8. Place filters in individual glass scintillation vials containing 8 ml of 2× SSC and 12 µl of 10 mg/ml RNase A. Alternatively, nonspecific RNA molecules can be digested with a combination of RNase T1 (5 U/ml) and RNase A (2.5 µg/ml) in 2× SSC. Incubate at 37°C for 30 min without shaking.

9. Wash the filters in batch for 1 h in an excess volume of 2× SSC at 37°C. *Briefly* blot the filters, ensuring that they do not dry out completely. Plastic-wrap the damp filters and then tape them to a solid support such as Whatman 3MM paper or even a piece of cardboard. Then set up autoradiography (see Chapter 14).

Note: Signal intensity on autoradiograms is most easily quantified through the use of image analysis software, after first ensuring that the observed signal is within the linear range of the film.

10. Following autoradiography, the hybridized RNA probe can be stripped from the membrane-bound target DNA by washing the filters in 0.4 *N* NaOH at 42°C for 30 min, followed by a 60-min wash in 1× SSC, 0.5% SDS at 65°C. This technique removes all hybridized RNA, leaving the efficiency of subsequent hybridization unimpaired.

11. If the filter membrane is not to be used for subsequent hybridization, the magnitude of the observed autoradiographic signal can be precisely quantified by simply cutting the appropriate samples from the filter membrane strip and counting directly in scintillation fluor.

Protocol: Nuclear Runoff Assay: Alternative Procedure

A simplified version of the traditionally cumbersome and labor-intensive nuclear runoff assay has been described (Celano *et al.*, 1989b), and a modification of it is included here. In this procedure, labeled RNA transcripts are purified by acid–phenol–guanidinium thiocyanate extraction (Chomczynski and Sacchi, 1987; see Chapter 5). The principal advantage of this technique is that RNA is rapidly purified immediately after labeling, with all of the benefits associated with the use of guanidinium buffers, and without TCA precipitation. Be sure to wear gloves throughout and observe good RNase-free technique.

1. Isolate nuclei as described above, or elsewhere (Celano *et al.*, 1989b; deBustros *et al.*, 1986).

2. Collect isolated nuclei (up to 5×10^7) in sterile 2.2-ml microfuge tubes in 160 μl of nuclear storage buffer (40% glycerol; 50 mM Tris, pH 8.3; 5 mM MgCl$_2$; 0.1 mM EDTA). Then freeze nuclei in an ethanol–dry ice bath and store at –70°C for up to 1 year. *Do not store this type of tube in liquid nitrogen.*

3. Thaw aliquots of frozen nuclei on ice. For each 200 μl of nuclei, add 25 μl of 10× runoff buffer (40 mM ATP, GTP, CTP; 5 mM DTT; 50 mM MgCl$_2$; 800 mM KCl) and 20 μl[α-^{32}P]UTP (200 uCi, 3000 Ci/mmol). Incubate for 15 min at 26°C.

4. Lyse the nuclei and initiate DNA digestion by adding 12 μl 20 mM CaCl$_2$ and 12 μl 10 mg/ml RNase-free DNase I. Incubate for 5 min at 26°C.

5. Add 30 μl of 10× SET (5% SDS; 50 mM EDTA; 10 mM Tris, pH 7.4) and 7 μl 10 mg/ml carrier RNA (e.g., junk yeast RNA). Initiate peptide hydrolysis with the addition of 3 μl 10 mg/ml proteinase K. Incubate the samples at 37°C for 30 min.

6. Shear genomic DNA by drawing the lysate repeatedly through a 25-gauge needle on a 1-ml tuberculin syringe.

7. Extract the RNA from the sample by adding the following to each tube containing labeled nuclear RNA (mix after the addition of each):

> 500 μl GTC (4M guanidinium thiocyanate; 25 mM sodium citrate, pH 7.0; 0.5% sarcosyl; 0.1 M 2-mercaptoethanol)
>
> 80 μl sodium acetate (2 M, pH 4.8)
>
> 800 μl water-saturated phenol
>
> 160 μl chloroform:isoamyl alcohol (49:1)

Incubate samples on ice for at least 15 min.

8. Microcentrifuge samples at 12,000 g for 10 min, at 4°C if possible.

9. Transfer the upper aqueous phase to a fresh microfuge tube, taking care to avoid the interface. Add an equal volume of isopropanol to this aqueous material to precipitate the RNA. Store for at least 1 h at –70°C or overnight at –20°C.

10. Collect the precipitate RNA by microcentrifugation at 12,000 g for 10 min, at 4°C if possible.

Recommended: Redissolve the RNA pellet in 300 μl of guanidinium thiocyanate solution (see Step 7), transfer to a 1.7-ml microfuge tube, and then reprecipitate the RNA by the addition of 300 μl of isopropanol. Collect the reprecipitated RNA by centrifugation as described above.

11. Wash the pellet two or three times with 70% ethanol (95% ethanol diluted in sterile H_2O) and briefly air dry. If desired, a final wash with 95% ethanol will accelerate the drying process. In either case, do not allow the RNA to dry completely. Dissolve the RNA pellet in 250–500 μl in a solution of 10 mM Tris, pH 7.2; 1 mM EDTA; 0.1% SDS.

12. Determine the activity of the RNA probe by counting a 2-μl aliquot. The total expected activity is approximately 10^7 cpm for the assay.

13. Hybridize labeled RNA to DNA target sequences as described above, or elsewhere (Celano et al., 1989b; deBustros et al., 1986).

Protocol: Nuclease Protection-Pulse Label Transcription Assay

The mechanics of the nuclear runoff assay (label incorporation into nascent transcripts in isolated nuclei) differ markedly from pulse labeling techniques. Subjecting cell cultures to a 5-min pulse with [^{3}H]uridine (Nevins and Darnell, 1978) would require a sufficiently high rate of transcription in order to derive meaningful incorporation of label. This certainly is not the case with single-copy genes or other sequences that are transcribed with relatively low efficiency. This issue and other concerns pertaining to the inherent difficulties of preparing nuclei capable of meaningful label incorporation have been partially circumvented by the nuclease protection of pulse-labeled nuclear RNA (Greene and Pearson, 1994).

This technique involves pulse labeling tissue culture cells for 10 min, followed by RNA isolation, solution hybridization with cold (unlabeled) gene-specific sequences, and finally incubation with S1 nuclease; with this approach, cDNA probes produce significantly higher fidelity data than oligonucleotide probes. The nuclease-resistant hybrids are precipitated and the relative rate of transcription inferred from the incorporated label, as determined by Cerenkov counting. As with most transcription assays, the data derived from this protocol have the advantage of labeling transcripts while maintaining natural nuclear geometry and, in this protocol, maintaining cellular geometry during labeling. At the conclusion of the labeling period, RNA is isolated from the cells for hybridization analysis. In short, this is a rapid, highly quantitative assay. Moreover, without the requirement for X-ray film, the inherent limitations associated with autoradiography (extended exposure time, nonlinearity of the film) are superseded by scintillation counting; this assay has the ability to detect subtle changes in transcriptional activity that might otherwise by missed by the more traditional nuclear runoff approach.

1. Cells growing in tissue culture are labeled with the addition of 25 μCi/ml[^{3}H]uridine (specific activity 27.1 Ci/mmol) added directly to the growth medium tissue culture vessel.

Note 1: RNA from 10^7 cells is generally required for one assay in duplicate, including all controls (see Step 11).

CAUTION   At the conclusion of the labeling period, be sure to dispose of radioactive tissue culture media and RNA isolation by-products in a safe manner, according to departmental guidelines.

2. Harvest cells and wash cell pellets twice with ice-cold PBS. All subsequent centrifugations are carried out at 4°C.

3. Resuspend cells (up to 10^7) in 1.8 ml NP-40 lysis buffer (10 mM Tris, pH 7.4; 10 mM NaCl; 3 mM MgCl$_2$; 0.5% NP-40). Transfer the cell suspension to an autoclaved 2.2-ml microfuge tube. Incubate on ice for 3 min.

4. Centrifuge at 4000 g for 5 min at 4°C to collect nuclei.

5. Carefully remove and discard supernatant. Resuspend nuclei in 800 μl SDS buffer (0.5% SDS; 1 mM CaCl$_2$; 5 mM MgCl$_2$; 20 mM Hepes, pH 7.5).

6. Add 50 U RNase-free DNase I and incubate for 15 min at 37°C.

7. Split the nuclear lysate into two microfuge tubes, and then extract each with an equal volume of phenol:chloroform:isoamyl alcohol (25:24:1).

8. Centrifuge to separate the phases. Carefully transfer the upper aqueous phase to a fresh tube.

9. Precipitate RNA with the addition of 0.1 volume 3 M sodium acetate, pH 5.2, and 2.5 volumes of 95% ethanol. Store at –20°C for several hours to overnight.

10. Carefully wash sample once with 70% ethanol (95% ethanol diluted with sterile H$_2$O), and centrifuge again if necessary. Briefly air dry the sample but do not allow it to dry out.

11. Resuspend the RNA pellet in 125 μl S1 hybridization buffer (40% deionized formamide; 40 mM Pipes, pH 6.4; 400 mM NaCl; 1 mM EDTA, pH 8.0). Be sure to redissolve the RNA by repeated pipetting. This is a critical step. Do not vortex.

Note 1: Dissolve the RNA pellet in enough hybridization buffer to allow 20 μl for each hybridization reaction to be performed.

Note 2: Always run each sample in duplicate; do not forget about control reactions. Appropriate controls include (1) a tube with no probe and no S1 nuclease, (2) a tube with no probe and 250 U S1 nuclease, and (3) a tube with probe, but no S1 nuclease.

12. Heat each resuspended RNA sample to 50°C for 2 min.

13. Add denatured cold probe (unlabeled) to the samples to a final concentration of 25 ng/μl. This corresponds to 0.5 μg probe per 20 μl hybridization reaction.

Note: Be sure to use linearized, denatured DNA. Double-stranded probes can be easily denatured by boiling them for 10 min and then plunging them into ice until further use. Alternatively, the RNA/probe mix can be heated to 65°C for 10 min to ensure all-inclusive denaturation at the onset of the hybridization.

14. Hybridize overnight at an experimentally determined temperature. For optimization suggestions, see Chapter 18. If uncertain where to begin, hybridize at 37°C.

15. At the conclusion of the hybridization period, add to each sample:

160 μl 2× S1 nuclease buffer (0.5 M NaCl; 0.1 M sodium acetate, pH 5.5; 9 mM ZnSO$_4$)

142 μl H$_2$O

6 μl boiled salmon DNA (10 mg/ml; [Final] ≈ 0.02 μg/ml)

12 μl S1 Nuclease (25 U/μl; [Final] ≈ 1 U/μl)

Incubate at 32°C for 1 h.

Note: Be sure to prepare one control tube without the addition of any S1 nuclease. This will provide the total counts in the system.

16. Add 80 μl stop buffer (3 M sodium acetate, pH 5.2; 20 mM EDTA, pH 8.0; 40 μg/ml carrier RNA) to each sample.

17. Add 1.0 ml ice-cold 95% ethanol to each tube. Invert to mix thoroughly and then incubate on ice for 20 min.

18. Collect precipitate on GF/F glass fiber filters using a vacuum filtration bank apparatus (Millipore, Bedford, MA).

19. Wash each filter twice with 500 μl cold 70% ethanol.

20. Transfer filters to scintillation vials and add an appropriate volume of fluor. Count samples in triplicate for 5 min. The activity remaining on each filter is directly proportional to the extent of hybridization

between probe and target RNA, and hence proportional to the relative rate of synthesis.

Distinguishing among the Activities of RNA Polymerases I, II, and III

The relative rate of transcription of several genomic sequences can be evaluated simultaneously by the nuclear runoff assay. In eukaryotic cells, the products of transcription are, collectively, the result of three different RNA polymerases. The activities of these enzymes may be distinguished from each other by using the fungal cyclic octapeptide α-amanitin, which binds to RNA polymerase molecules, and to which the three eukaryotic RNA polymerases exhibit differential sensitivity (Roeder, 1976). For example, at a concentration of 0.01 μg/ml, RNA polymerase II activity is reduced by 50% (Roeder, 1976; Marzluff and Huang, 1984). RNA polymerase III is also sensitive to the inhibitory activity of α-amanitin, although a concentration of 25 μg/ml is required to achieve an equivalent 50% inhibition of activity. In contrast, RNA polymerase I exhibits no sensitivity to this peptide. While α-amanitin is extremely toxic, it is a very useful tool for determining the amount of RNA synthesis attributable to each polymerase. The inclusion of α-amanitin in one or more labeling reactions is an important and an excellent negative control by which blot hybridization signals are validated as authentic and not artifactual. This type of negative control is especially valuable when quantifying hybridization in dot blot format.

In nuclei isolated from cultured cells, the contribution to the total observed RNA synthesis due to RNA polymerase I, RNA polymerase II, and RNA polymerase III is about 45%, 50%, and 5%, respectively. Clearer definition may be given to any model system by performing three separate labeling reactions in which α-amanitin is added as a reaction buffer component *before* the addition of the nuclei:

1. no α-amanitin added to labeling reaction;
2. 1 μg/ml α-amanitin added to the labeling reaction;
3. 100 μg/ml α-amanitin added to the labeling reaction.

Following addition of nuclei, the reaction mixtures are briefly incubated on ice and then transferred to the temperature at which the labeling will take place.

The difference in the mass of transcription products in Reactions 1 and 2 represents RNA polymerase II activity. The difference between Reactions 2 and 3 represents RNA polymerase III activity. The activity in Reaction 3 is due to RNA polymerase I activity. These differences associated with α-amanitin poisoning may be assessed by dot blot analysis (Chapter 8), by the separation of transcription products on a denaturing agarose gel, or electrophoresis on a denaturing 10% polyacrylamide gel; electrophoresis will permit visualization of the size distribution of all transcription products by autoradiographic exposure directly from the gel.

Alternatively, the size distribution of transcribed species can be determined by sucrose density gradient centrifugation (Marzluff and Huang, 1984). Clearly, this is a much more time consuming procedure than simple gel electrophoresis and requires specialized equipment. Briefly,

1. Dissolve RNA samples in 0.1% SDS; 1 mM EDTA, pH 7.5, load 0.5 ml of this onto a 17-ml 10–40% (w/v) sucrose gradient made up in 0.1 M NaCl, 1 mM EDTA, 0.1% SDS, and 10 mM Tris, pH 7.5. The gradient must be prepared in a tube suitable for ultracentrifugation.
2. Centrifuge at 90,000g for 16 h in a suitable rotor. Under the parameters defined here, examination of all RNA species from 4S to 45S will be possible, with the 28S RNA sedimenting 60% of the way down the gradient.
3. Gradients may then be fractionated by UV absorbance at 254 nm, for example, with an ISCO UA-6 monitor/fraction collector, and the radioactivity in each fraction may be determined by assaying TCA-precipitable counts in a 10-μl aliquot from each fraction.

Part II: Extraction of Nuclear RNA for Steady-State Analysis

Rationale

The mechanics of the nuclear runoff assay facilitate the assessment of the transcription rate of individual genes by the elongation of nascent polyribonucleotides in the presence of labeled NTP precursor. In addition, the study of nuclear RNA by Northern analysis is essential for understanding the processing of the primary transcript. In both types of assay, exciting quantitative data can be derived from hybridization signal intensity, assayed in any number of ways. To assess the size distribution of labeled nuclear transcripts, an aliquot of the reaction mixture can be studied by minigel electrophoresis coupled with in-gel autoradiography.

Alternatively, target DNA may be digested, electrophoresed, and Southern blotted onto a filter membrane, followed by attempted hybridization with the heterogeneous population of labeled nuclear RNA. A major drawback of this technique, however, is the lack of information about the size of the nuclear RNA molecules themselves, since the restriction digestion of the target DNA yields fragments of defined length. In extreme cases, extensive degradation of full-length hnRNA molecules might well result in nonspecific hybridization of the resulting oligoribonucleotides to target DNA molecules, with which full-length, undegraded hnRNAs would not normally hybridize. High-stringency washes, in conjunction with RNase digestion of nonspecific RNA molecules, will alleviate some, but not all, of the background due to this phenomenon.

Northern analysis of nuclear RNA is an integral part of an investigation involving the turnover of RNA as one parameter of gene expression, especially when attempting to elucidate gene regulation at a posttranscriptional level. The successful Northern analysis of nuclear RNA yields quantitative data based on signal intensity of discrete bands, but also provides a qualitative component that cannot be discerned in dot blot format alone. The appearance of a reproducible banding pattern strongly supports the existence of a primary RNA transcript and, in so doing, attests to the reliability of the nuclear runoff assay. A series of bands is associated with the systematic splicing together of coding (exon-associated) sequences and concomitant removal of noncoding (intron-associated) sequences.

The isolation of nuclear RNA for Northern blot analysis begins with the preparation of nuclei from whole cells or tissue samples as described above. Invariably the highest quality RNA always results from the most expedient extraction procedures—those which minimize the interval between cellular disruption and immobilization of RNA on a filter membrane, or other experimental use. Inhibition of RNase activity in problematic cells can be accomplished by incorporating guanidinium-based techniques (Chirgwin et al., 1979; Chomczynski & Sacchi, 1987). Once the nuclei have been isolated from the cytosolic contents, one may elect to purify the RNA from the nuclei as if starting with intact cells. Although one should realize the imperative to control nucleases, very satisfactory results are attainable by extraction with organic solvents alone. It is worth noting that virtually all protocols describing the isolation of nuclear RNA recommend heating the sample in the presence of phenol, accompanied by vigorous shaking to shear the genomic material.

Protocol: Direct Isolation of Nuclear RNA

The following protocol is a modification of the procedure of Soeiro and Darnell (1969) in which the use of hot phenol for RNA purification was originally described.

1. Thaw previously frozen nuclei on ice and pellet at 750 g for 3 min at 4°C. If starting with freshly prepared nuclei, proceed to Step 2.

2. Wash nuclear pellet with 1 ml ice-cold modified RSB/K buffer (10 mM Tris, pH 7.9; 10 mM NaCl; 10 mM MgCl$_2$; 100 mM KCl). Transfer nuclei to an RNase-free polypropylene tube.

Note: If this extraction is scaled down, it is possible to isolate the RNA in a 2.2-ml microcentrifuge tube. As always, it is best to keep nucleic acids as concentrated as possible and work with microliter volumes, if possible.

3. Resuspend the nuclear pellet to a concentration of 1–5 × 10^7 nuclei/ml in HSB buffer (10 mM Tris, pH 7.4; 500 mM NaCl; 50 mM MgCl$_2$; 2 mM CaCl$_2$), to which RNase-free DNase I has been added to a final concentration of 50 U/ml.

Note 1: HSB buffer can be prepared ahead of time; however, add RNase-free DNase I just prior to use.

Note 2: CaCl$_2$, at a concentration of 1–5 mM is a stabilizer of DNase I.

4. Pipet this mixture up and down for 30 s at room temperature. This is necessary for lysate homogeneity.

Note: The objective here is not the complete digestion of genomic DNA, but rather a partial degradation, in order to decrease the viscosity and increase the manageability of the prep. The DNA will be selectively removed later in this procedure.

5. Add an equal volume of SDS extraction buffer (10 mM Tris, pH 7.4; 20 mM EDTA; 1% SDS).

6. Add an equal volume of phenol saturated with NETS buffer (10 mM Tris, pH 7.4; 100 mM NaCl; 10 mM EDTA; 0.2% SDS).

Note: See Appendix A for tips on phenol saturation.

7. Add an equal volume of chloroform *or* a mixture of chloroform: isoamyl alcohol (49:1). Mix thoroughly and *carefully.*

8. Heat to 55°C for 10 min, with periodic shaking.

9. Cool on ice or in an ice-water bath for 5 min. Centrifuge at 2500 g for 3 min to separate the phases.

10. Remove the lower organic phase by carefully sliding a sterile Pasteur pipet or microcentrifuge tip along the side of the tube, through the aqueous phase and interphase, and aspirate from the bottom of the tube. Leave the aqueous phase and interphase in the tube and repeat Steps 6 and 7.

Note: This approach is especially important when working with small quantities of nuclear RNA, where there is a potential for loss of RNA by entrapment in chromatin.

11. Centrifuge at 2500 g for 3 min to separate phases.

12. Carefully recover the aqueous phase, taking care not to disrupt the lower organic phase or interphase, if present. Add an equal volume of chloroform to the aqueous material, mixing thoroughly and carefully. Pulse centrifuge briefly to separate the phases.

13. Transfer the aqueous phase to a new tube. Precipitate the nuclear RNA at –20°C with the addition of 2.5 volumes of ice-cold 95% ethanol.

14. Collect the precipitate by centrifugation at 4°C. Wash at least once with 70% ethanol (95% ethanol diluted with sterile H_2O) to remove excess salt. Quantify yield and perform quality control as described in Chapters 5 and 7.

Protocol: Preparation of Nuclear RNA from Cells Enriched in Ribonuclease[11]

1. Prepare nuclei as described above; nuclei can be prepared ahead of time and stored frozen. Allow frozen nuclei to thaw on ice for 3–5 min.

2. Lyse intact nuclei with the addition of GTC Solution D (4 *M* guanidinium thiocyanate; 25 m*M* sodium citrate, pH 7.0; 0.5% sarcosyl; 100 m*M* 2-mercaptoethanol). Use 100 μl per 2×10^6 nuclei.

Note: The volumes indicated for the remaining steps in this procedure assume a starting volume of 1 ml of Solution D. For scaled-up extractions, remember to increase the volumes of all other reagents appropriately. If less than 800 μl of Solution D is used, the entire extraction can be carried out in a 2.2-ml microfuge tube.

3. Transfer the homogenate to a 4- or 15-ml disposable polypropylene tube and add, in order:

0.1 ml 2 *M* sodium acetate, pH 5.2

---

[11]Modification of the procedure of Chomczynski and Sacchi (1987).

1.0 ml water-saturated phenol (molecular biology quality)

0.2 ml chloroform:isoamyl alcohol (49:1).

Cap tube and mix carefully and thoroughly by inversion following the addition of reagent. Shake tube vigorously for 10 s after all reagents have been added.

4. Cool sample on ice for at least 15 min and then centrifuge at 4°C to separate the phases.

5. Transfer aqueous phase (contains RNA) to a fresh tube and mix with 0.75 volume of isopropanol (approximately 1ml). Store at –20°C for at least 1 h to precipitate RNA.

6. Collect precipitate by centrifugation at 10,000 g for 20 min at 4°C. Carefully decant and discard supernatant.

7. Completely dissolve RNA pellet in 300 μl of Solution D (see Step 1) and then transfer to an RNase-free 1.5-ml microcentrifuge tube.

8. Reprecipitate the RNA by adding 0.75 volume of ice-cold isopropanol at –20°C for 1 h.

9. Collect precipitate at top speed in a microcentrifuge for 10 min at 4°C. Carefully decant and discard supernatant.

10. Wash pellet with 70% ethanol and recentrifuge. The pellet may be washed once more with 95% ethanol to accelerate air drying. Do not allow the sample to dry out completely.

11. Redissolve damp RNA pellet in 50 μl of TE buffer, pH 7.5, or 50 μl of sterile H_2O. Incubation at 65°C for 10 min may facilitate solubilization. Determine sample concentration (Chapter 5); store RNA in suitable aliquots at –70°C. As always, avoid repeated freezing and thawing.

References

Benecke, B. J., Ben-Ze'ev, A., and Penman, S. (1978). The control of mRNA production, translation, and turnover in suspended and reattached anchorage-dependent fibroblasts. *Cell* **14,** 931.

Ben-Ze'ev, A., Farmer, S. R., and Penman, S. (1980). Protein synthesis requires cell-surface contact while nuclear events respond to cell shape in anchorage-dependent fibroblasts. *Cell* **21,** 365.

Blumberg, D. D. (1987). Creating a ribonuclease-free environment. *Methods Enzymol.* **152,** 20.

Celano, P., Baylin, S. B., and Casero, R. A. (1989a). Polyamines differentially modulate the transcription of growth associated genes in human colon carcinoma cells. *J. Biol. Chem.* **264,** 8922.

Celano, P., Berchtold, C., and Casero, R. A., Jr. (1989b). A simplification of the nuclear runoff assay. *BioTechniques* **7**(9), 942.

Chirgwin, J. M., Przybyla, A. E., MacDonald, R. J., and Rutter, W. J. (1979). Isolation of biologically active ribonucleic acid from sources enriched in ribonuclease. *Biochemistry* **18**, 5294.

Chomczynski, P., and Sacchi, N. (1987). Single-step method of RNA isolation by acid guanidinium thiocyanate–phenol–chloroform extraction. *Anal. Biochem.* **162**, 156.

deBustros, A., Baylin, S. B., Levin, M. A., and Nelkin, B. D. (1986). Cyclic AMP and phorbol esters separately induced growth inhibition, calcitonin secretion, and calcitonin gene transcription in cultured human medullary thyroid carcinoma. *J. Biol. Chem.* **261**, 8036.

Derman, E., Krauter, K., Walling, L., Weinberger, C., Ray, M., and Darnell, J. E. (1981). Transcriptional control in the production of liver-specific mRNAs. *Cell* **23**, 731.

Ernest, M. J., Schultz, G., and Feigelsen, P. (1976). RNA synthesis in isolated hen oviduct nuclei. *Biochemistry* **15**, 824.

Farrell, R. E., Jr. (1989). Methodologies for RNA characterization. I. The isolation and characterization of mammalian RNA. *Clin. Biotechnol.* **1**, 50.

Farrell, R. E., Jr. (1990). Methodologies for RNA characterization. II. Quantitation by northern blot analysis and the S1 nuclease assay. *Clin. Biotechnol.* **2**(2), 107.

Farrell, R. E., Jr., and Greene, J.J. (1992). Regulation of *c-myc* and *c-Ha-ras* oncogene expression by cell shape. *J. Cell. Physiol.* **153**, 429.

Greenberg, M. E. (1988). Identification of newly transcribed RNA. In "Current Protocols in Molecular Biology" (F. M. Ausubel, R. Brent, R. E. Kingston, D. D. Moore, J. G. Seidman, J. A. Smith, and K. Struhl, Eds.), p. 4.10.1. Wiley, New York.

Greenberg, M. E., and Ziff, E. B. (1984). Stimulation of 3T3 cells induces transcription of the *c-fos* proto-oncogene. *Nature* **311**, 433.

Greene, J. J., and Pearson, S. L. (1994). Measurement of gene-specific transcription by nuclease protection of pulse-labeled nuclear RNA. *J. Biochem. Biophys. Methods* **29**, 179.

Groudine, M., Peretz, M., and Weintraub, H. (1981). Transcriptional regulation of hemoglobin switching on chicken embryos. *Mol. Cell. Biol.* **1**, 281.

Gurney, T., and Foster, D. N. (1977). Nonaqueous isolation of nuclei from cultured cells. *Methods Cell Biol.* **16**, 45.

Jelinek, W., Molloy, G., Fernandez-Munoz, R., Salditt, M., and Darnell, J. E. (1974). Secondary structure in heterogeneous nuclear RNA: Involvement of regions from repeated DNA sites. *Mol. Biol.* **82**, 361.

Lund, E., and Paine, P. L. (1990). Nonaqueous isolation of transcriptionally active nuclei from *Xenopus* oocytes. *Methods Enzymol.* **181**, 36.

Manley, J. L. (1984). Transcription of eukaryotic genes in a whole-cell extract. In "Transcription and Translation: A Practical Approach" (B. D. Hames and S. J. Higgins, Eds.), p. 71. IRL Press, Oxford, England.

Marzluff, W. F. (1978). Transcription of RNA in isolated nuclei. *Methods Cell Biol.* **19**, 317.

Marzluff, W. F., and Huang, R. C. C. (1975). Chromatin directed transcription of 5S and tRNA genes. *Proc. Natl. Acad. Sci. USA* **72**, 1082.

Marzluff, W. F., and Huang, R. C. C. (1984). Transcription of RNA in isolated nuclei. In "Transcription and Translation: A Practical Approach" (B.D. Hames and S.J. Higgins, Eds.), p. 89, IRL Press, Oxford, England.

Marzluff, W. F., Murphy, E. C., and Huang, R. C. C. (1973). Transcription of ribonucleic acid in isolated mouse myeloma nuclei. *Biochemistry* **12**, 3440.

Marzluff, W. F., Murphy, E. C., and Huang, R. C. C. (1974). Transcription of the genes for 5S ribosomal RNA and transfer RNA in isolated mouse myeloma cell nuclei. *Biochemistry* **13**, 3689.

Melton, P. A., Drieg, P. A., Rebagliati, M. R., Maniatis, J., Zinn, K., and Green, M. R. (1984). Efficient *in vitro* synthesis of biologically active RNA and RNA hybridization probes from plasmids containing bacteriophage SP6 momoter. *Nucleic Acids Res.* **12**, 7035.

Nevins, J. R. (1987). Isolation and analysis of nuclear RNA. *Methods Enzymol.* **152**, 234.

Nevins, J. R., and Darnell, J. E. (1978). Steps in the processing of Ad2 mRNA: Poly(A)$^+$ nuclear sequences are conserved and poly(A)$^+$ addition precedes splicing. *Cell* **15**, 1477.

Price, R., and Penman, S. (1972). A distinct RNA polymerase activity, synthesizing 5.5S, 5S, and 4S RNA in nuclei from adenovirus 2-infected HeLa cells. *J. Mol. Biol.* **70**, 435.

Roeder, R. G. (1976). Nuclear RNA polymerase. *In* "RNA Polymerase" (R. Losick and M. Chamberlain, Eds.), pp. 283–329. Cold Spring Harbor Laboratory Press, Cold Spring Harbor, NY.

Sambrook, J., Fritsch, E. F., and Maniatis, T. (Eds.). (1989). "Molecular Cloning: A Laboratory Manual." Cold Spring Harbor Laboratory Press, Cold Spring Harbor, NY.

Soeiro, R., and Darnell, J. E. (1969). Competition hybridization by "pre-saturation" of HeLa cell DNA. *J. Mol. Biol.* **44**, 551.

Tilghman, S. M., and Belayew, A. (1982). Transcriptional control of the murine albumin/alpha-fetoprotein locus during development.

Wittelsberger, S. C., Kleene, K., and Penman, S. (1981). progressive loss of shape-responsive metabolic controls in cells with increasingly transformed phenotype. *Cell* **24**, 85.

An RNA Paradigm

A Typical(?) Experiment
What to Do Next

The protocols contained in the preceding chapters of this text illustrate various methods for the expedient isolation of RNA from eukaryotic systems. The judicious application of various combinations of these techniques constitutes a systematic method for understanding the role of transcriptional products in the regulation of gene expression. An experimental model can be extremely well characterized by assaying specific messenger RNAs (mRNAs) as one parameter of gene expression. A typical experimental design includes an assessment of steady-state levels of RNA and, perhaps, the rate of transcription in control and experimental cell populations. In so doing, it is likely that changes in the prevalence of certain mRNAs may be ascribed to transcriptional control or to some posttranscriptional event(s).

Northern analysis of cytoplasmic RNA affords the investigator an opportunity to examine directly the extranuclear transcript population. In subsequent experiments involving nuclear RNA, very definitive conclusions can be drawn about the transcriptional and posttranscriptional regulation of gene expression. Side-by-side analysis of poly(A)$^+$ and poly(A)$^-$ has, in the past, been an excellent means of enriching mRNAs under investigation, though with the advent of the polymerase chain reaction (PCR), enrichment of this nature is usually no longer necessary and may even be counterproductive.

Classical Northern analysis is limited, in part, by the very mechanics of the assay. It is well established that the physical immobilization of target RNA on a solid support (e.g., a nylon membrane) in some way prevents complete accessibility of those molecules to a nucleic acid probe. A more complete method of transcript quantification is known as solution hybridization, epitomized in the form of the S1 nuclease- and RNase protection assays and, of course, the polymerase chain reaction. The enhanced sensitivity of these techniques commonly precludes the requirement for poly(A)$^+$ selection. The results of these experiments are often quantified by autoradiography or chemiluminescence, and/or Cerenkov counting; these may be embellished further by coupling them with digital image analysis.

The induction or repression of transcription is commonly reported in the literature in terms of relative abundance, a term used to compare the observed prevalence of specific mRNA species in control and experimental cell populations. For example, it might be reported that contact inhibition results in a 50-fold decrease in the abundance of *c-myc* oncogene mRNA in a newly derived cell line, compared to the abundance of *c-myc* mRNAs in exponentially proliferating, subconfluent cultures of the same. Thus, statements can be made about the increase or decrease in the prevalence of transcripts without having knowledge of the absolute mass of these transcripts in the cell.

RNA characterization is most often employed for some form of assessment of gene expression, and the greater the sensitivity, the better. PCR has facilitated probing the depths of the cellular biochemistry to levels inconceivable 10 years ago, and continues to offer greater potential in the form of the many permutations of this technique, such as differential display PCR and competitive PCR, to name but two.

Molecular scrutiny of a model system does not necessarily stop at transcriptional characterization. Having assayed the prevalence of transcriptional products and, perhaps, the relative transcription rates of different genes by the nuclear runoff assay, an investigator may wish to begin to characterize the molecular basis of observed variations from the norm (control systems). Gene expression is, after all, only partly about the biogenesis of RNA. At the level of genome organization, the modulation of mRNA may, in part, be the result of a gene rearrangement. It is entirely possible that an aberration within the coding portion of a locus, or the flanking sequences that influence its expression, constitute at least one underlying basis for the up- or down-regulation observed upon transcript quantification. Subsequent investigations may involve Southern analysis to probe for such structural changes.

By the same token, gene expression at the cellular level culminates in the form of a functional polypeptide. To give more complete definition to a system, characterization of translational and posttranslational regulation of gene expression may also be helpful. While the presence of one mRNA species or another suggests that they are translated, experimentally induced changes in the cellular biochemistry may compromise access of these same mRNAs to the translational apparatus. Moreover, the posttranslational modifications commonly associated with eukaryotic gene expression may also be compromised under the parameters defined in a particular experimental situation. Thus, appraisal of the cellular biochemistry from several perspectives permits a more complete understanding of naturally occurring gene regulation.

A Typical(?) Experiment

Before countless hours and thousands of dollars are invested investigating a particular problem, an initial characterization of a new system is always in order (Figure 1). The type of "quick and dirty" assessment required is a direct function of the model being proposed. For example, an abundant transcript might easily be assayed by something as unsophisti-

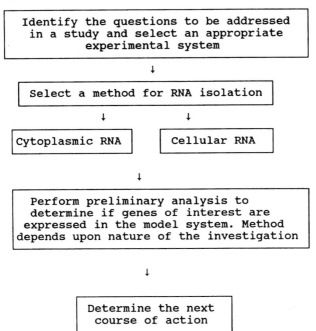

FIGURE 1

Preliminary analysis is used to determine whether a selected experimental system is appropriate. In-house equipment, expertise, and the nature of the questions being asked determine the most reasonable "quick-and-dirty" approach.

cated as dot blot analysis; likewise, amplification of specific genomic sequences might be suggested by DNA dot blot analysis. If expression of sequences known to be a part of the genomic complement is of interest and transcription is not detected, it may be necessary to modify the experimental design. Experimental modifications could include (a) selecting an alternative biological source, (b) selecting a more sensitive assay, or (c) induction of the genes of interest by chemical or physiological stimulation.

In an initial characterization, analysis of total cellular *and* total cytoplasmic RNA may be very useful because many transcribed RNAs are degraded in the nucleus and hence do not appear in the cytoplasm in the form of mRNA; further, probe hybridization to cellular RNA samples and not to cytoplasmic samples suggests some type of posttranscriptional regulatory mechanism in action. Assuming that initial assessment yields encouraging data, a more detailed examination of the system is in order.

The Northern analysis is a standard tool with which to quantify transcripts and at the same time examine their respective sizes (Figure 2). This is necessary because dot blot analysis lacks the qualitative aspect associated with the electrophoresis of RNA. The purpose of the Northern analysis is to gain salient information pertaining to the expression of genes, particularly with respect to the fate of their transcriptional products. Knowledge of the size of the transcript as well as its abundance demonstrates the specificity of hybridization and at the same time permits direct comparison to previously published data (if any) on the size and abundance of the messages under investigation. The investigator must be fully aware, however, that a great deal of contemporary research involves the assay of genes that are undetectable by standard blot analysis; often, low and very low abundance transcripts remain below the level of detection by this method.

The more sensitive quantitative assays of transcription are those in-

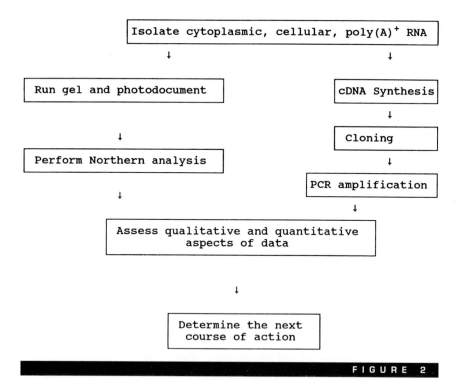

FIGURE 2

Northern analysis can be used following preliminary evaluation to further assess the system. This level of investigation does not preclude concomitant synthesis of cDNA and PCR-based analyses.

volving solution hybridization, such as nuclease protection (Figure 3). It may also be salient to the experimental design to include one or more studies of the rate at which various transcripts are produced in control cell populations compared to experimentally manipulated cells. This is readily accomplished by performing the nuclear runoff assay. Transcription rate data are then supported by data from the Northern analysis of nuclear RNA, as well as previous data pertaining to the cytoplasmic abundance (above) of the RNAs of interest. If the underlying molecular basis of observed differences in experimental cells is genomic in nature, then this possibility might be suggested by Southern analysis. For example, biochemical differences in normal versus malignant cells at the transcriptional level might be the result of a change or rearrangement in

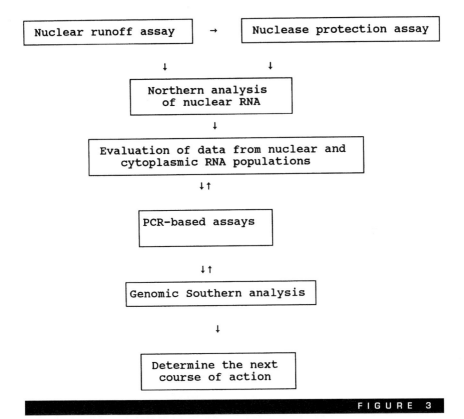

FIGURE 3

Nuclease protection assays and the nuclear runoff assay offer greater sensitivity and insight than filter-based assays (e.g., Northern analysis). The outcome of these experiments often suggests further experiments pertaining to the investigation of gene expression and sequence organization at the genomic level.

the locus itself or in the flanking regulatory elements that influence it expression.

Well above all other techniques with respect to resolution and sensitivity is the polymerase chain reaction. This revolutionary approach to the amplification of gene sequences, when applied to RNA templates, is among the surest of methods to assay transcription in virtually any experimental context. Of course, PCR requires transcript- or gene-specific sequencing data to generate useful primer sequences. Although RT PCR can be a stand-alone technique in the assay of gene expression, it is wise to seek corroborating data from some of the previously described, more classical methods.

What to Do Next

The observation that one or more RNAs of interest have been induced or repressed may be the basis for a decision to expend further laboratory resources to continue to characterize a system under investigation. At this point, it is often in the best interests of the laboratory to retrieve the corresponding complementary DNAs (cDNAs) from a newly synthesized or previously existing library, an imperative mandated by the extremely labile nature of RNA. Isolation of specific cDNAs will support the continued use of the sequence as a probe and many of the other experiments suggested below.

If a suitable library already exists, the cDNA(s) of interest may already be present in the library, a circumstance that will save considerable time and expense. The investigator is cautioned that older libraries, that is, those constructed in the late 1980s, may be severely lacking in longer cloned sequences and are not likely to be as representative as libraries constructed more recently. There are two principal reasons for this. First, the synthesis of older libraries formerly was mediated mostly by oligo(dT)-priming from the poly(A) tail of poly(A)$^+$ mRNA. Inefficient synthesis of first-strand cDNA, a commonplace occurrence, results in a marked reduction in the size of the cDNA molecules; a lack of sequence in the region corresponding to the probe obviously will not produce a hybridization signal. In more extreme cases, many mRNAs might not be represented in the library at all. Second, oligo(dT)-primed libraries may be underrepresented because nonpolyadenylated mRNAs (poly(A$^-$) mRNA) would not be in the library. More rationally, the synthesis of first-strand cDNA is random-primed for completeness, and currently, entire libraries can be made by PCR.

Libraries can be screened by traditional approaches, such as plating out an aliquot of the library followed by a plaque screening assay using nucleic acid probes or, if applicable, with antibodies. If some sequence information is available, either from studies of a closely related member of the same gene family or from the same gene in another species, it may be possible to synthesize primers that will support screening the library by PCR. Failure to recover specific members of the library by traditional screening approaches or by PCR either mandates the design of a new probe or primers or may well necessitate construction of a new library. The utility and significance of a cDNA library are derived from the fact that it represents a permanent biochemical record of the biological material under investigation at the moment of cell lysis.

Once isolated, cDNA can be propagated for use as a probe for Northern and Southern analysis, nuclease protection, DNA sequencing, and any of a number of other applications. About 90% of the eukaryotic genome is believed to be transcriptionally silent; thus, cDNA sequencing is a rapid and inexpensive means of identifying discrete loci and measuring genetic expression. If previously unknown, novel DNA sequence information can be used to generate primers for PCR, greatly facilitating detection of the gene sequence under investigation, or expression thereof. Beyond this level of sophistication, subsequent experiments are directed toward study of the particular questions being asked in an investigation. Gene expression is a multifaceted phenomenon; subsequent research may involve, but certainly is not limited to, some of the following areas:

- Development of PCR primers for diagnostic applications.
- Recovery of DNA fragments from a genomic library, to eventually reconstruct the entire gene.
- Determination of the exon–intron structure of the loci under investigation.
- Assessment of the splicing pattern of heterogeneous nuclear RNA.
- Analysis of flanking sequences of the loci for possible regulatory roles and protein binding sites, by gel shift assay and DNA footprinting.
- Site-directed mutagenesis of suspected regulatory sequences.
- *In situ* hybridization, including *is situ* PCR, to determine the histological distribution of cells expressing genes of interest.
- Determination of the cellular distribution of corresponding protein products.

- High-level protein expression in prokaryotic system and eukaryotic systems for clinical or biophysical studies.

- Eukaryotic transfection studies to assess the effects of inappropriate gene expression. Transfection is also used to study putative regulatory elements by ligation to so-called reporter genes.

- Attempted homologous recombination of the gene under investigation with stem cells, to examine the role of the locus in differentiation and development.

- *Responsible* production of transgenic animals to facilitate *in vivo* production of life-saving pharmaceuticals.

Epilogue:
A Few Pearls of Wisdom

The discipline that has come to be known as molecular biology encompasses little more than the innovative application of fewer than two dozen standard reactions in the context of a variety of research problems. Upon mastery of these techniques, many of which have been presented here, the research potential is enormous. Molecular biology *is not* about kits; rather, it is about planning and determining how best to use the tools at one's disposal in order to get the job done. The thoughtful, well-considered planning of strategy and the implementation of the correct techniques, in the correct order, will ultimately govern the outcome of an investigation. In this regard, never forget this author's definition of the 5 R's of molecular biology:

Rapid
Representative
Reproducible
RNase-free
Reliable
* * *

1. Running a gel is the single best diagnostic that can be used to assess the integrity of a nucleic acid preparation. A small aliquot of sample electrophoresed on a minigel (e.g., 7 × 8 cm) will reveal the probable usefulness of the sample in downstream applications. This can be done at several points in a series of experiments to check for degradation of RNA, improper denaturation, efficient conversion into complementary DNA (cDNA), and so forth.

2. When pipetting microliter volumes, it may be very helpful to pipet individual reagents onto the inside wall of the microfuge tube, close to

but not at the bottom. The individual reagents can then be mixed by pulse centrifugation. This avoids the problem of inaccurate volume delivery to the tube due to capillary action in the tip of the micropipet, and shows convincingly that small aliquots have actually been delivered into the tube.

3. RNA pellets do not adhere to the side of microfuge tubes as firmly as DNA pellets. When decanting an alcohol supernatant, keep the pellet in sight at all times. If you are afraid of losing a pellet when decanting, pour off into another microfuge tube so that the pellet will not be lost if it slides out of the tube.

4. Take logistical measures to keep nucleic acids, especially RNA, as concentrated as possible at all times. This will facilitate UV spectrophotometric measurements, and subsequent dilution, as needed. Further, it is easier to recover small RNA yields in minimal volumes from 1.7-ml microfuge tubes than from 2.2-ml tubes or from larger Corex tubes, because of the shape of the bottom of these tubes.

5. Wash nucleic acid pellets with 70% ethanol to remove excess salt. This is necessary because nucleic acid molecules and salt form aggregates in solution that demonstrate dramatically reduced solubility in alcohol. Thus, the precipitate at the bottom of the tube is both salt and nucleic acid. Washing the pellet with 70% ethanol will remove much of the salt, yet it is not sufficiently aqueous to redissolve nucleic acids. After two or three washes, do a final wash with 95% ethanol to facilitate rapid drying of the pellet.

6. Do not allow nucleic acid pellets to dry out completely unless explicitly specified in the procedure. RNA and DNA pellets that are slightly damp with ethanol will dissolve more readily than those that are completely dry.

7. Buy phenol already redistilled (molecular biology grade). Phenol distillation is an extremely dangerous process. Excellent quality redistilled phenol can be purchased from any of a number of suppliers. In this lab, the standard procedure is to melt the phenol upon arrival and then store in 25-ml aliquots at $-20°C$ until use (Appendix A). This approach precludes the extremely detrimental process of repeated thawing and refreezing phenol.

8. In the course of extraction with organic solvents to deproteinate nucleic acid samples, quantitative recovery of the aqueous phase without disturbing the protein interface is often difficult. To maximize the recovery of aqueous material after most of it has been removed, it may be useful to tilt the tube to a 45°C angle, causing a "bubble" of the remaining aqueous material to form up against the side of the tube (Figure 1). At

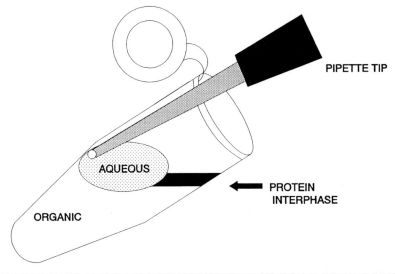

PIPETTE TIP

AQUEOUS

PROTEIN
INTERPHASE

ORGANIC

F I G U R E   1

Technique for quantitative recovery of aqueous material during organic extraction. Be sure to wear appropriate eye and skin protection.

this point, one may slide the micropipet tip along the inside wall of the tube, penetrate the aqueous bubble, and remove a great majority of the remaining aqueous volume without disturbing the protein interphase or organic material.

C A U T I O N   Use extreme care when handling tubes containing organic solvents in this fashion. As always, wear appropriate eye protection and skin protection should be worn and use common sense.

9. When purifying nucleic acids with phenol, always perform a final extraction with chloroform or a mixture of chloroform:isoamyl alcohol (24:1). Because of the great solubility of phenol in chloroform, even trace amounts of carryover phenol will be removed in the final chloroform extraction. This is necessary because quinones, the oxidation product of phenol, can compromise the integrity of the nucleic acid sample.

10. Do not use polystyrene plasticware (e.g., centrifuge tubes, serological pipets) with organic solvents. The plastic will dissolve in phenol or chloroform and will be difficult to remove from the sample.

11. When handling RNA preps, always keep tubes containing the samples on ice, unless the protocol specifically dictates otherwise. If reading through the protocol, let the samples rest on ice until the next

course of action has been discerned. It is also useful to prechill centrifuge tubes on ice when carrying out a sequential extraction regimen. Remember that cool temperatures partially inhibit nuclease activity.

12. Sodium dodecyl sulfate (SDS) is often added to buffers to control nuclease activity. In many applications it may be desirable to eliminate it because it precipitates out of solution at cold temperatures (the sodium salt of SDS is especially insoluble). Moreover, even traces may be incompatible with further applications. Other types of RNase inhibitors, such as RNasin, may be used instead.

13. When preparing buffers and reagents that include EDTA, be sure to use the disodium salt (Na_2-EDTA) rather than the tetrasodium salt (Na_4-EDTA). This is important because of the great sensitivity of many molecular biology enzymes to salt concentration. Buffers prepared with Na_4-EDTA will initially be more alkaline than those made with Na_2-EDTA. The subsequent addition of HCl to establish pH 7–8 will greatly increase the NaCl concentration in the final buffer and, more than likely, have a negative influence on the outcome of an experiment.

14. In addition to the RNase associated with the purification of RNA, it is important to keep in mind that heavy metals can cause RNA degradation when they are present for extended periods. If this is a suspected problem, filtering buffers through Chelex (Bio-Rad) may be helpful. In addition, the inclusion of 8-hydroxyquinoline, which chelates heavy metals and is itself a partial RNase inhibitor, may optimize RNA purification efficiency.

15. When collecting nucleic acid precipitates by centrifugation, be certain not to exceed the recommended g-force for a particular type of tube. Attention to this detail is especially important when scaling up or scaling down RNA isolation procedures, as significant g-forces may be required. For smaller scale procedures, polypropylene microfuge tubes are used almost exclusively. Properly cushioned, nuclease-free Corex glass tubes are handy for larger scale preps because of their ability to withstand the g-force needed to pellet precipitated RNA.

16. The ubiquitous instruction to store labile materials in "suitable aliquots" is quite vague at best. Different applications require different quantities of RNA or DNA. If the precise definition of a suitable aliquot is unknown, purified RNA is best stored in precipitated form at –80°C until such a determination can be made. It is unwise to store the sample dissolved in buffer until UV analysis has been performed and the fate of the sample determined. Repeated thawing and refreezing of RNA samples dramatically reduces the utility of the material. If a redissolved sample must be stored, aliquot it and then store at –80°C.

17. A_{260}/A_{280} less than 1.65 is usually indicative of contaminating proteins, which absorb heavily at 280 nm. A low ratio may also be indicative of contamination with phenol, which can remedied by final chloroform extraction; it may also be indicative of incomplete solubilization of the RNA pellet.

18. Do not mouth-pipet anything, ever.

19. When RNA is to be recovered from a lysate that also contains DNA, organic extraction at pH below 6.0 will drive DNA to the aqueous/organic interface. Keep in mind, however, that samples isolated in this manner are almost certainly tainted with genomic DNA. DNase I treatment is a must, especially for techniques involving the polymerase chain reaction (PCR).

20. To avoid many of the problems associated with staining RNA in agarose gels, prepare duplicate samples that are electrophoresed in one or two end lanes, into at least one of which a molecular weight marker is loaded. At the conclusion of the electrophoresis period, these lanes can be physically separated from the rest of the gel and stained for molecular weight determinations and to assess the integrity of the sample. The bulk of the gel containing "true" experimental samples can then be used directly for Northern analysis.

21. If denaturing RNA with dimethyl sulfoxide (DMSO; and glyoxal), especially for dot blot analysis, be aware that DMSO dissolves nitrocellulose.

22. If denaturing RNA with formaldehyde for any application, be aware that the sample may become degraded if the formaldehyde pH is below 4.0

23. Be sure to check for ß-mercaptoethanol incompatibility with ultracentrifuge tubes.

24. Contrary to popular belief, it is not detrimental to hybridize more than one filter in the same hybridization chamber. As long as there is sufficient buffer to keep the filters from sticking together and the filters are free to move in the chamber with the motion of the incubator (orbital shaker incubators work well), hybridization of two or more filters should proceed unencumbered.

25. Define the linear range of X-ray films when preparing to record experimental data by autoradiography or chemiluminescence. If performing dot blot analysis, preparing dilutions of each sample will facilitate accurate interpretation of data. Remember that changing X-ray film exposure time and exposure time/aperture settings with Polaroid film will minimize problems associated with the linear response of film in general and highlight the differences among the samples.

26. To permanently record the location of size standards on X-ray film, the investigator may wish to label a very small amount of probe that will be able to hybridize to the size markers, making detection of these nucleic acid species possible as well. This approach will eliminate guess-work regarding the precise location of molecular weight standards and comparing them to experimental samples.

27. All chemicals should be considered potentially hazardous. Indi-viduals should be trained in good laboratory technique and safety prac-tices, by their respective departments. In a molecular biology setting, there are many inherent dangers associated with what is considered "the daily routine." Always follow the safe handling, containment, and person-al safety recommendations of the manufacturers of chemicals and equip-ment.

28. Do not store microfuge tubes in liquid nitrogen.

29. Prepare recipe cards that delineate the exact formulation of com-monly used buffers and solutions. Having a card at hand will save a great deal of time when reagents are in short supply, or when they must be pre-pared quickly in the middle of an experiment.

30. Always test enzyme and reaction components when opening a new package. Nonfunctional enzyme translates into one thing . . . no bands!

31. Get in the habit of making up master mixes for cDNA- and PCR-related applications. This is *especially* important for differential display PCR and quantitative analyses. Even the most subtle variations in micro-liter reaction volumes or nanomolar concentrations can have a strong negative impact on the accuracy of the resulting data.

32. Watch the final enzyme concentration. Recall that molecular biol-ogy enzymes are often stored in 60% glycerol. The volume of a reaction occupied by the enzyme should not exceed 5% of the total volume. Be-yond this amount, the enzymes will not function optimally.

33. PCR optimization may be easier if a thermal cycler with a heated lid is used. This may also be helpful because of the small volumes in-volved, the elimination of mineral oil, and the potentially large number of samples involved. Moreover, many investigators have reported a marked decrease in amount of product generated when a PCR reaction is overlaid with mineral oil.

Phenol Preparation

CAUTION Phenol and chloroform are both volatile, caustic agents. Disposable gloves, protective eyewear, and the use of a chemical fume hood are necessary safety precautions when handling these and other organic solvents.

Phenol must be purified or redistilled prior to use in any molecular biology application. Redistilled phenol is commercially available, at a very reasonable cost, from all major suppliers of molecular biology reagents. The distillation of phenol in the laboratory is a very dangerous process that should be carried out by experience personnel only, and even then it should be aggressively discouraged. Details of in-lab phenol distillation can be found elsewhere (Wallace, 1987).

In this lab, the standing procedure is to purchase molecular-grade (i.e., redistilled) phenol and store it frozen in suitable aliquots, wrapped in aluminum foil; an aliquot should not be saturated until just before use for the first time.

Phenol is very light sensitive and it oxidizes rapidly. The oxidation products, known as quinones, form free radicals that break phosphodiester bonds and crosslink nucleic acids. Quinones can usually be detected by the pinkish tint they impart to phenol reagents. To reduce the rate of phenol oxidation, 8-hydroxyquinoline may be added (Chapter 5) after the phenol has been saturated with aqueous buffer. Moreover, adding an equal volume of chloroform, in the preparation of organic extraction buffer, stabilizes the phenol, imparts a greater density to the mixture, improves the efficiency of protein removal from the sample, and facilitates removal of lipids from the RNA prep. Very often, isoamyl alcohol is used with chloroform or with mixtures of phenol and chloroform; isoamyl alcohol reduces the foaming of proteins that would normally accompany the mechanics of most extraction procedures. When prepared in this way, a saturated phenol reagent is known as an extraction buffer. Common organic extraction buffer formulations consist of phenol:chloroform: isoamyl alcohol in a 25:24:1 ratio.

Example: Redistilled phenol is typically supplied in 500-g bottles. Upon arrival in the lab, the phenol is melted in a 65°C water bath and aliquoted in a fume hood, using a sterile glass pipet, into conical 50-ml polypropylene tubes (*never* polystyrene) or 100-ml glass bottles. Then, the individual aliquots are sealed, wrapped in aluminum foil, and stored at −20°C. This precludes the repeated thawing and refreezing of the unused phenol. When preparing to use the phenol, an aliquot is thawed only once and can then be saturated as described here, or according to the specifications of a particular procedure. If 50 ml is a convenient volume of extracting buffer to prepare, about 13 ml of melted phenol is aliquoted into a 50-ml tube in advance and stored frozen.

To prepare the extraction buffer, melt the aliquot and then saturate it with the addition of an equal volume of a Tris-containing buffer (described below), thereby doubling the volume in the tube to about 25ml. This is usually followed by the addition of an equal volume (25 ml) of chloroform:isoamyl alcohol (24:1). This will produce 50 ml of saturated extraction buffer, which is compatible with a variety of applications. At this point, 8-hydroxyquinoline can be added to a final concentration of 0.1% (w/v); otherwise, the extraction buffer can be used as is. This basic strategy can be applied to the preparation of any volume of extraction buffer. For convenience, saturated phenol and ready-to-go extraction buffer may be purchased from Sigma Chemical Company (Cat. Nos. P-4557 and P-3803, respectively).

Phenol Saturation

There are two standard types of phenol saturation and several variations under each heading. One of these methods, Tris saturation, is used when buffered phenol is to be prepared and maintained as a stock of known pH in the lab, or when a particular protocol specifies that a phenol-containing extraction buffer be prepared at a specific pH prior to use. It is important to note here that the rate of phenol oxidation increases as pH increases. Therefore, phenol buffers greater than pH 8.0 should be prepared fresh each time they are required. Saturated phenol buffers equilibrated to more acidic pH are considered stable for 3–4 weeks if wrapped in foil and stored at 4°C. The other method, water saturation, is used when the *starting* pH of the phenol is not of critical importance. Phenol should never be water saturated unless specifically required of a particular protocol. When water saturation is used, the final pH of the organic material is usually achieved by mixing the phenol with a suitable volume of an appropriate buffer [for example, sodium acetate, pH 4.5, as in the method of Chomczynski and Sacchi (1987); see Chapter 5]. It is judicious to prepare only the amount of extraction buffer that will be required for a given experiment; any residual material is handled according to lab policy regarding the disposal of organic waste.

The choice of pH at which to saturate phenol is strongly dependent on whether DNA or RNA is to be isolated. A principle of paramount importance is that, in a phenolic lysate, the partitioning of nucleic acids between the aqueous and organic phases is heavily pH dependent. Only RNA remains in the aqueous phase if the pH is acidic, and both RNA and DNA are in the aqueous phase if the pH is alkaline (Brawerman *et al.,* 1972). For example, at pH < 5.0, DNA is selectively retained in the organic phase and interphase, while RNA can be efficiently recovered from the aqueous phase. One key advantage of this approach is the reduced nuclease activity that is observed at acidic pH. One the other hand, if DNA is the target molecule in an extraction, adjusting the pH of the extraction buffer to pH > 7.5 is the industry standard. Further, near-neutral pH equilibration very effectively safequards against excessive depurination of the sample.

The proper pH at which to equilibrate organic solvents is also directly related to whether phenol is being used alone or in combination with chloroform and/or isoamyl alcohol. For example, extraction of aqueous lysates with phenol of acidic pH will cause the depurination of RNA (and DNA if present) and poly(A)$^+$ mRNA would be lost to the organic phase. This phenomenon can be avoided by including sodium dodecyl sulfate (SDS), chloroform, or both in the lysis and/or extraction buffers.

Procedure 1: Tris Saturation

CAUTION   **Be sure to wear gloves and eye protection throughout.**

1. Melt a suitable aliquot of redistilled phenol completely in a 65°C water bath.

2. Add an equal volume of 1 M Tris, pH 8.0 (or other pH, according to the particular application). Alternatively, phenol can be equilibrated us-

ing TE buffer (10 mM Tris, pH 8.0; 1 mM EDTA), or Tris–SDS buffer (100 mM NaCl; 1 mM EDTA; 10 mM Tris–Cl, pH 8.5; 0.5% SDS). In any event, always saturate with an equal volume of aqueous buffer.

3. Carefully mix; allow the phases to separate completely. Phase separation can be accelerated by centrifuging the tubes briefly in a desktop (clinical) centrifuge.

4. Observe the lower phenol phase and ensure that it is colorless. Phenol with any coloration, especially a pink tint, should be discarded because of the presence of quinones. The volume of the phenol phase should be greater than that of the upper aqueous phase, due to the preferential absorption of some of the aqueous component into the organic phase.

5. Remove the upper (aqueous) phase and discard. To the remaining phenol phase, add an equal volume of 100 mM Tris, or other Tris buffer, adjusted to the same pH as in Step 2.

6. Carefully mix; allow the phases to separate completely. Phase separation can be accelerated by centrifuging the tubes briefly in a desktop (clinical) centrifuge.

7. Remove most of the upper aqueous phase and measure its pH. If the aqueous material is pH > 7.5, the phenol is now considered Tris-saturated and is ready for use. If pH < 7.5, repeat Steps 5 and 6.

Note: Be aware that repetition of Steps 5 and 6 is likely to cause a decrease in the lower organic phase as the phenol will gradually be drawn into the aqueous phase. As such, excessive equilibration is to be avoided.

8. Store saturated phenol wrapped in foil at 4°C. If desired, mix with an equal volume of chloroform:isoamyl alcohol (24:1) and/or 0.1% (w/v) 8-hydroxyquinoline.

Procedure 2: Water Saturation

CAUTION   **Be sure to wear gloves and eye protection throughout.**

1. Melt a suitable aliquot of redistilled phenol completely in a 65°C water bath.

2. Add an equal volume of dH$_2$O.

3. Allow the phases to separate completely before use. Phase separation can be accelerated by centrifuging the tubes briefly in a desktop (clinical) centrifuge.

4. The upper (aqueous) material can be drawn off and discarded or allowed to remain in the tube with the phenol as evidence of saturation.

5. Store saturated phenol wrapped in foil at 4°C. If desired, mix with an equal volume of chloroform:isoamyl alcohol (24:1) and/or 0.1% (w/v) 8-hydroxyquinoline.

References

Brawerman, G., Mendecki, J., and Lee, S. Y. (1972). A procedure for the isolation of mammalian messenger ribonucleic acid. *Biochemistry* **11**, 637.

Chomczynski, P., and Sacchi, N. (1987). Single-step method of RNA isolation by acid guanidinium thiocyanate–phenol–chloroform extraction. *Anal. Biochem.* **162**, 156.

Wallace, D. M. (1987). Large- and small-scale phenol extractions. *Methods Enzymol.* **152**, 33.

Disposal of Ethidium Bromide and SYBR Green Solutions

Ethidium bromide (EtBr), SYBR Green, and similar dyes used to stain nucleic acids in gels are very powerful mutagens (MacGregor and Johnson, 1977). As intercalating agents and nucleic acid-binding dyes, they should be treated as toxic waste; the proper handling of these materials includes preventing environmental contamination after use. Only after proper treatment can ethidium bromide- and SYBR Green-tainted wastes be disposed of with minimal apprehension.

The salient issues surrounding the treatment of ethidium bromide waste have been described at length (Lunn and Sansone, 1987, 1990; Bensaude, 1988). The fairly widespread practice of adding bleach to ethidium waste is not recommended, because such treatment, while reducing the mutagenicity of ethidium bromide, converts the dye into another mutagenic compound (Quillardet and Hofnung, 1988). Instead, one of the following protocols should be adopted as standard lab policy, and applied to the handling of SYBR Green as well. The following protocols suggest the handling of working concentrations of ethidium bromide (0.5–1.0 μg/ml) and SYBR Green (1–4×); at higher concentrations, these dyes can be diluted down and processed as described in protocol 1, 2, or 3.

Protocol 1

The Extractor (Scheicher & Schuell; Cat. No. 448031) is a one-step filtration method for the removal of ethidium bromide and SYBR Green from gel-staining solutions (Figure 1). The device is capable of filtering 10 liters of electrophoresis buffer containing ethidium bromide, with

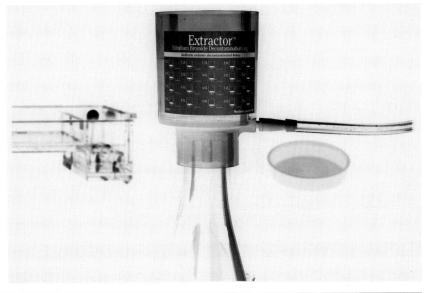

FIGURE 1

Extractor filtration device for the removal of ethidium bromide from electrophoresis and staining buffers. Photograph courtesy of Scheicher & Schuell.

greater than 99% removal (Figure 2). The filter itself, an activated carbon matrix, is disposed of according to departmental guidelines, while the filtrate can be discarded safely down the drain. This is an extremely cost-effective method for dealing with this common waste product in the molecular biology laboratory and is used extensively in this laboratory.

Ancillary Protocol: Quantitative Assay for Residual EtBr in Filtrate

1. Prepare a solution of 100 μg/ml salmon sperm DNA in running buffer or staining buffer. Prepare dilutions of ethidium bromide in the range of 0 to 500 ng/ml to prepare a standard curve.

2. Add salmon sperm DNA to 1-ml aliquots from each volume of filtrate to a final concentration of 100 μg/ml.

3. Read standards and unknowns in a fluorimeter (excitation 526 nm, emission 586 nm). Plot the standard curve, and read the EtBr concentration of the unknown from the standard curve. This assay is usually linear ($r = 0.999$) for EtBr concentrations as low as 4 ng/ml.

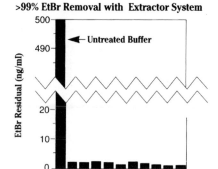

10 successive liters of buffer containing 500 ng/ml of EtBr were filtered through the Extractor device. Residual EtBr in the filtrate was measured by fluorescence spectroscopy.

FIGURE 2

Data courtesy of Schleicher & Schuell.

Protocol 2

Solutions containing working concentrations of ethidium bromide, SYBR Green, and related dyes can be decontaminated by adding about 3 g of Amberlite[1] XAD-16 (Sigma Cat. No. XAD-16) for each 100 ml of solution (Joshua, 1986; Lunn and Sansone, 1987). This resin is a nonionic, polymeric absorbent. The solution may be shaken intermittently for 12–16 h at room temperature, after which it is filtered with Whatman No. 1 filter paper, or the equivalent. The filter and Amberlite should be treated as toxic waste and disposed of according to in-house laboratory policy. The filtrate can then be discarded.

Protocol 3

Add 100 mg of activated charcoal (Sigma Cat. No. C-2889) to each 100 ml of ethidium bromide or SYBR Green at the working concentration (Bensaude, 1988). The resultant mixture may be shaken intermittently for 1 h at room temperature, after which it is filtered with Whatman No. 1 filter paper, or the equivalent. The filter and charcoal should be treated as toxic waste and disposed of according to in-house lab policy. The filtrate can be then be discarded.

[1]Amberlite resins are manufactured by Rohm and Hass, Inc. (Philadelphia, PA).

References

Bensaude, O. (1988). Ethidium bromide and safety—Readers suggest alternative solutions. Letter to editor. *Trends Genet.* **4,** 89.

Chromatographia (1990). *29,* 167.

Joshua, H. (1986). Quantitative absorption of ethidium bromide solutions from aqueous solutions by macroreticular resins. *BioTechniques* **4**(3), 207.

Lunn, G., and Sansone, E. B. (1987). Ethidium bromide: Destruction and decontamination of solutions. *Anal. Biochem.* **162,** 453.

Lunn, G., and Sansone, E. B. (1990). Degradation of ethidium bromide in alcohols. *BioTechniques* **8**(4), 372

MacGregor, J. T., and Johnson, I. J. (1977). *Mutation Res.* **48,** 103–108.

Quillardet, P., and Hofnung, M. (1988). Ethidium bromide and safety-readers suggest alternative solutions. Letter to editor. *Trends Genet.* **4,** 89.

DNase I Removal of DNA from an RNA Sample

It often becomes necessary to purge RNA samples of contaminating DNA for any of several reasons, especially when preparing for RNA-based polymerase chain reaction (PCR) applications. Although isopycnic centrifugation has been used in the past to partition grossly contaminated samples, it is time-consuming, requires expensive and highly specialized equipment, does not favor high productivity, and is not the preferred approach when the mass of both DNA and RNA in the sample totals no more than a few micrograms. Moreover, the very popular acid–phenol method for separating RNA from DNA using a very old pH trick, while very efficient, is not efficient enough to ensure that only RNA remains in the aqueous environment (Chapter 5), and the closer to the organic phase one pipets, the greater is the likelihood that DNA is carried over. This presents a significant compromise to the quantitative nature of RNA-based assays; probes and primers often have great difficulty distinguishing between RNA targets and DNA targets, when both are present. In any of these circumstances, a brief incubation of the sample(s) with ribonuclease-free deoxyribonuclease I (RNase-free DNase I) will eliminate DNA by nuclease digestion. Although the resulting RNA solution can then be used directly, it should be extracted with phenol:chloroform, and/or concentrated by standard salt and alcohol precipitation (Chapter 5), or otherwise cleaned-up. Applications for DNase include, but are not limited to, preparation of RNA free of contaminating DNA, degradation of DNA template molecules from *in vitro* transcription reactions, nick translation of DNA probes, and studying DNA protein interactions by DNase I footprinting.

Commercial preparations of RNase-free DNase I, such as RQ 1

(Promega), are typically purified by any of a number of methods, to remove all traces of detectable RNase, and are certified by the manufacturer to harbor no intrinsic RNase activity. These preparations are very handy in the lab, and well worth the purchase cost. There is no easy, cost-effective way to prepare RNase-free DNase in the lab because of the labile nature of DNase I.

Protocol

1. To a sample of RNA suspected to be tainted with DNA, add RNase-free DNase I to a final concentration of about 3 U/μg nucleic acid.

Note 1: DNase I typically will show optimal activity in a buffer consisting of about 40 mM Tris–Cl, pH 7.9; 10 mM NaCl, 10 mM CaCl$_2$; 5 mM MgCl$_2$.

Note 2: The activity of RNase-free DNase I is typically described in units, where 1 unit of DNase I will degrade 1 μg of DNA in 10 minutes at 37°C, most often performed in a 20–50 μl reaction volume. Keep in mind that DNase I is strongly inhibited in buffers containing EDTA, sodium dodecyl sulfate, and other denaturants, or if heated above 37°C.

2. Incubate the mixture for 15 min at 37°C.

3. Terminate the reaction by heating the sample to 60°C for 5 min. Alternatively, the activity of DNase I can be arrested by the addition of EDTA to a final concentration of 25 m*M*.

Note: The addition of EDTA is not recommended for RNA PCR.

4. Assess the extent of DNA digestion as well as the integrity of the RNA in the sample by electrophoresing an aliquot of the sample under denaturing conditions.

5. If necessary, concentrate the RNA by precipitation with salt and alcohol.

RNase Incubation to Remove RNA from a DNA Sample

In some applications, such as the isolation of high-molecular-weight genomic DNA (Appendix I), the experimental design may require the removal of all contaminating RNA without compromising the integrity of the DNA. In these cases, purging of all intrinsic DNase activity from the RNase stock solution must be carried out before incubation. As indicated in Appendix C, isopycnic centrifugation could be employed to partition grossly contaminated samples, although it is time-consuming, requires expensive and highly specialized equipment, and is not the preferred method when the combined mass of DNA and RNA in the sample totals no more than a few micrograms. In these circumstances, a brief incubation of the mixture with DNase-free RNases will do the job. Commercially available DNase-free RNase is certified by the manufacturer to harbor no intrinsic DNase activity. Crude preparations of RNases, if not specially treated, can harbor significant levels of DNase activity as well.

Unlike the removal of RNase activity from DNase, RNase preparations can be made DNase-free in the most basic molecular biology laboratories. The most commonly used RNases are the enzymes RNase A and RNase T1.

Protocol: Removal of DNase Activity from RNase

1. Prepare a 10 mg/ml stock solution of RNase A, RNase T1, or both in sterile H_2O or TE buffer (10 mM Tris–Cl, 1 mM EDTA, pH 8.0).

2. Heat RNase stock solutions to near boiling (90°C) for 10 min. Cool on ice and store frozen in aliquots.

Note: Although some RNase preparations maintain considerable activity after boiling, this practice is not encouraged for this particular application; at 90°C DNase activity will be quickly eliminated without compromising the RNase activity of the reagent.

Protocol: Digestion of RNA

1. Add DNase-free RNase to sample to a final concentration of 10 μg/ml.

2. Incubate for 10–30 min at 37°C.

3. Terminate RNase digestion by extraction with an equal volume of phenol:chloroform (1:1) or phenol:chloroform:isoamyl alcohol (25:24:1).

4. Concentrate DNA by salt and alcohol precipitation, or use directly for restriction endonuclease digestion, polymerase chain reaction, etc.

5. Store genomic DNA at 4°C. Plasmid DNA can be stored for extended periods at –20°C.

Reference

Gillespie, D., and Spiegelman, S. (1965). A quantitative assay for DNA–RNA hybrids with DNA immobilized on a membrane. *J. Mol. Biol.* **12,** 829.

Deionization of Formamide, Formaldehyde, and Glyoxal

Formamide, formaldehyde, and glyoxal are routinely utilized for a wide variety of molecular biology applications. The rapid oxidation of these reagents mandates deionization prior to use for optimal efficacy in hybridization and nucleic acid denaturation applications. Two common approaches for deionization are presented here.

Deionization of commonly used molecular biology reagents can be achieved by the addition of the reagent to a small mass of mixed bed resin (Sigma; Catalog No. M-8032). This resin is strongly acidic and strongly basic, facilitating rapid deionization. Briefly, mix 50 ml of the reagent to be deionized with 5 g of mixed bed resin. Swirl the mixture every 2–3 min. As the capacity of the resin is reached, its color will change from blue to amber. Deionization should be complete in 30 min and may require the use of more than one aliquot of the resin. Remove the resin by filtration or by centrifugation. Store deionized glyoxal in tightly sealed, convenient aliquots at –20°C; deionized formamide and formaldehyde should be used shortly after deionization. Once opened, the unused portion of a previously deionized aliquot should be discarded.

Alternatively, formamide, formaldehyde, and glyoxal can be deionized with AG 501-X8 mixed bed resin (Bio-Rad). AG 501-X8 resin is a 1:1 equivalent mixture of AG 1-X8 resin (OH⁻) and AG 50W-X8 resin (H⁺). It is recommended that 1 g of AG 501-X8 resin be added to each gram of material to be deionized. Deionization requires about 1 h, although the mixture can be left overnight. Then the resin can be removed by filtration, and the deionized reagent stored as described above. The Bio-Rad AG 501-X8 resin is available with or without a conjugated dye that will manifest a color change as the deionization process proceeds.

Silanizing Centrifuge Tubes and Glassware

The intrinsically "sticky" nature of nucleic acids frequently mandates the silanizing of glassware (and polypropylene tubes, in some applications) for maximum sample recovery. Silanizing Corex glass tubes and microfuge tubes, for example, minimizes loss of RNA during larger scale isolation and complementary DNA in many downstream applications, though some loss is inevitable. Silanizing is especially helpful when small quantities of sample are being manipulated.

Protocol 1

Sigmacote (Sigma; Cat. No. SL-2) is a very useful reagent for the rapid silanizing of glassware and certain types of plasticware. Sigmacote is a special silicone solution in heptane. It forms a tight, microscopically thin, water-repellent film of silicone on surfaces to which it is exposed.

1. Be sure to follow the manufacturers instructions for the safe handling of Sigmacote.
2. Fill articles to be coated with Sigmacote. Avoid dipping, as tubes will become very slippery if the outside becomes silanized, and it will be very difficult to write on them.
3. Drain and allow articles to air dry, which is usually complete in 5–10 min. Dry surfaces are essentially neutral.

Note: Sigmacote is reusable if kept free of moisture.

Protocol 2[1]

Instead of using Sigmacote, glassware and microfuge tubes can be silanized with chlorotrimethylsilane or dichloromethylsilane, though this is a far more hazardous and cumbersome approach.

CAUTION Chlorotrimethylsilane and dichloromethylsilane are toxic, volatile, and highly flammable. Exercise caution when handling these chemicals, and do so in a fume hood.

1. In a fume hood, place items (glassware, microfuge tubes, etc) to be silanized into a desiccator, along with a beaker containing 1–3 ml of chlorotrimethylsilane or dichloromethylsilane.

Note: The main drawback of this approach is that outer surfaces are also silanized, making them rather slippery and impossible to write on.

2. Connect the desiccator to a vacuum pump; apply vacuum until silane begins to boil. Close connection to the pump, maintaining vacuum in the desiccator. Leave the desiccator evacuated and closed until silane is gone (1–3 h).

3. Open the desiccator in a fume hood and leave open for several minutes to clear silane vapors.

4. Bake or autoclave silanized materials.

Note: Do not bake or autoclave anything that would not ordinarily be baked or autoclaved, and be sure that any flammable solvent is completely evaporated first.

[1] Adapted from B. Seed (1989). Silanizing glassware. *In "Short Protocols in Molecular Biology"* (F. M. Ausubel, R. Brent, R. E. Kingston, D. D. Moore, J. G. Siedman, J. A. Smith, and K. Struhl, Eds.), p. 365. Wiley, New York.

Centrifugation as a Mainstream Tool for the Molecular Biologist

Centrifugation is a separation technique based on the fact that objects moving in a circular path are subjected to an outward-directed force. The magnitude of this force, commonly expressed in terms of the earth's gravitational force [relative centrifugal force (RCF) or the "number times g"], is a direct function of the radius of rotation and the angular velocity. Centrifugation is an indispensable procedure for the separation of whole cells, organelles, and macromolecules (also referred to as particles) from a solution, based on size and density.

Types of Centrifuges

Desk-top clinical centrifuges

- usually operate below 3000 rpm and at ambient temperature

High-speed centrifuges

- operate between 20,000 and 25,000 rpm and are usually refrigerated

Desk-top microcentrifuges (microfuges)

- operate up to 14,000 rpm (12,000 g)
- are sometimes equipped with refrigeration
- indispensable for nucleic acid samples in small volumes

Ultracentrifuges

- operate up to 500,000 g (75,000 rpm with $r = 8$ cm)
- permit fractionation of subcellular organelles
- permit fractionation of molecules based strictly on density
- have a wide temperature range available
- are vacuum-operated (air friction severe above 40,000 rpm).

Rotors

Centrifuges are designed to accommodate the specific requirements for a given separation. Parameters include temperature, RCF, volume of sample, duration of run, shape of gradient (e.g., linear or step), choice of differential or density gradient separation, and type of rotor. A fixed angle rotor (Figure 1) is one in which the sample is maintained at a defined angle during the centrifugation period. In contrast, swinging bucket rotors (Figure 2) allow the holder or bucket into which the sample is placed to swing outward to a position 90°C with respect to the axis of rotation. Vertical rotors maintain sample tubes upright throughout the run, parallel to the axis of rotation. The choice of fixed angle, swinging bucket, or vertical rotor is completely dependent on the intended application. Fixed an-

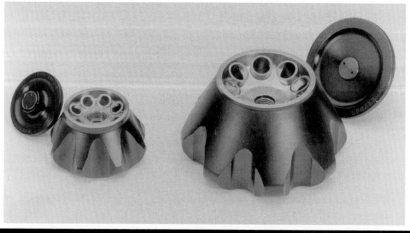

FIGURE 1

Fixed angle rotor. Samples are held in a fixed position during centrifugation. Courtesy of Sorvall, Inc. (Newtown, CT).

FIGURE 2

Swinging bucket rotor. Sample buckets holding the samples swing outward 90° with respect to the axis of rotation. Courtesy of Sorvall, Inc. (Newtown, CT).

gle rotors are most compatible with differential centrifugation techniques, while efficient gradient-based separations are supported by swinging bucket and vertical rotors.

A useful method by which to correlate performance among rotors is comparison of their respective k or k' factors, values used to compare efficiency among rotors in a particular application. The k factor estimates the time required to pellet a particle, when the sedimentation coefficient of the particle is known; the k' factor is indicative of the time required to move a zone of particles to the bottom of a centrifuge tube. Succinctly, the lower the k factor, the more efficient the rotor. The characteristics of several commonly used rotors are presented in Chapter 5, Tables 4 and 5.

Applications[1]

In the molecular biology laboratory, centrifugation fractionation techniques are utilized at several levels during the isolation of RNA, including the pelleting of whole cells prior to cell lysis, phase separation during nucleic acid extraction, gradient formation during isopycnic separation, and concentration of salt and alcohol-precipitated material.

[1] Adapted, in part, from Griffith (1986).

Differential Centrifugation

Differential centrifugation is the simplest, most straightforward centrifugation technique for sample fractionation. In this method, centrifuge tubes containing a homogeneous sample mixture are subjected to a brief centrifugation (usually less than 30 min). At the conclusion of the run, the pellet at the bottom of the tube includes all material sedimented during the run. It should be clear that the pellet is also contaminated with anything that was at the bottom of the tube from the onset. The supernatant, containing nonsedimented material, is usually removed by decanting; in the case of a very firm pellet, the supernatant can be removed carefully by aspiration. Should further fractionation be required, centrifugation of the supernatant in a fresh tube for a longer period than the initial centrifugation, or at higher speeds, will produce a new pellet and supernatant.

Density Gradient Centrifugation—Sedimentation Velocity

A more sophisticated preparative technique is density gradient centrifugation, in which sample particles move through a density gradient to achieve separation. The density of the gradient is lowest at the top of the tube and increases toward the bottom of the tube. In one type of density gradient centrifugation, known as sedimentation velocity, or rate zonal centrifugation, a sample is layered on top of a preformed gradient. When subjected to a centrifugal force, particles in the sample begin to move downward through the gradient in discrete zones; the rate of zone movement is governed by the sedimentation rate of individual particles within a zone. The hallmark of this type of separation is centrifugation through a relatively shallow gradient at low speeds for a short time (compared to buoyant density centrifugation). Successful sedimentation velocity centrifugation requires the following conditions:

1. The density of particles in the sample must be greater than the density of the gradient at every point throughout the gradient.
2. Centrifugation must be terminated before any of the separated material (specifically the zone of greatest density) reaches the bottom of the tube.

Example: In one application, a heterogeneous sample of RNA can be size-fractionated by centrifugation through a sucrose gradient (Benecke *et al.,* 1978; Nevins and Darnell, 1978). A typical sucrose gradient is as shallow as 5–20% to as steep as 5–40%; in the case of the higher viscosity 5–40% gradient, resolution can be improved by increasing the centrifugation speed. According to the parameters defined here, RNA molecules would move through the gradient according to their sedimentation rate. At the conclusion of the run, the RNA will be distributed throughout the gradient, based on size. In one type of enrichment, fractions of the chromatographed RNA are removed from the gradient for further characterization.

Density Gradient Centrifugation—Isopycnic Technique

Another type of separation through a density gradient is known as isopycnic, buoyant density, or density equilibrium centrifugation. In this application, particles in a sample move through a gradient only to the point at which the density of the gradient is equal to the density of the particle and at which particles of identical density float or band. Unlike sedimentation velocity centrifugation, extending the period of centrifugation will not result in the continued downward migration of the sample through the gradient. The gradient itself must be very steep, and at its greatest point, it must exceed the density of the particles of interest. Note that very dense components of a sample may become pelleted while other components migrate only to their isopycnic positions. Commonly used materials for this type of separation include CsCl, Cs_2SO_4, and CsTFA. Cesium chloride, historically the salt used for the classical studies of semi-conservative replication (Meselson *et al.,* 1957; Meselson and Stahl, 1958), is routinely used to establish gradients with densities ranging up to about 1.8 g/cm^3. Cs_2SO_4 can be used to form a gradient twice as steep as that which is achievable with the chloride salt and is preferred for separating DNAs with widely different buoyant densities.

A gradient for buoyant density centrifugation need not be preformed. For example, in several applications, solid CsCl is added to a nucleic acid mixture (chromosomal DNA, RNA, protein, plasmid DNA). Under the centrifugal force experienced during ultracentrifugation, the gradient material redistributes, due in part to the intrinsic density of cesium salts. These self-forming gradients require several hours at ultracentrifugation speeds before macromolecules become isopycnically banded, though the required centrifugation time can be significantly shortened by using a microultracentrifuge. In the resulting linear gradient generated during the run, components of the sample will either sediment or float to their isopycnic locations, based only on their respective densities.

References

Benecke, B. J., Ben-Ze'ev, A., and Penman, S. (1978). The control of mRNA production, translation, and turnover in suspended and reattached anchorage-dependent fibroblasts. *Cell* **14,** 931.

Griffith, O. M. (1986). "Techniques of Preparative, Zonal, and Continuous Flow Ultracentrifugation." Beckman Instruments. Palo Alto, CA

Meselson, M. and Stahl, F. W. (1958). The replication of DNA in *E. coli. Proc. Natl. Acad. Sci. USA* **44,** 671.

Meselson, M., Stahl, F. W., and Vinograd, J. (1957). Equilibrium sedimentation of macromolecules in density gradients. *Proc. Natl. Acad. Sci. USA* **43,** 581.

Nevins, J. R., and Darnell, J. E. (1978). Steps in the processing of Ad2 mRNA: Poly(A)$^+$ nuclear sequences are conserved and poly(A)$^+$ addition precedes splicing. *Cell* **15,** 1477.

Trypsinization Protocol for Anchorage-Dependent Cells

The preparation of nucleic acids from biological material begins with cellular lysis or tissue disruption. In the case of the former, it is possible to add a lysis buffer directly to a flask or dish containing cells grown in culture. This approach, however, often necessitates a volume of lysis buffer larger than would be required had the cells been harvested from tissue culture and lysed in a separate tube. In the case of suspension cultures, cells are collected by simple centrifugation of the culture medium. For anchorage-dependent cells, the following is the protocol used routinely in this lab when preparing for RNA or DNA isolation.

Protocol

1. Decant and save growth medium from the tissue culture vessel. It is usually convenient to decant directly into a conical 15- or 50-ml tube.

2. Wash the cell monolayer twice with 1× phosphate-buffered saline, calcium magnesium free (CMF-PBS). Decant and discard PBS.

Note: 1× PBS (per liter) = 8 g NaCl, 0.2 g KCl, 1.15 g Na_2HPO_4, 0.2 g KH_2PO_4. If desired, concentrated stock solutions of PBS can be prepared, and diluted with autoclaved H_2O just prior to use.

3. Add 2 ml 1× trypsin–EDTA to each T-75 tissue culture flask, or the equivalent. Incubate for 30–60 s with gentle rocking. The exact time will depend on the cell type, degree of confluency, and temperature of the trypsin. The trypsin must be removed from the tissue culture vessel *before* the cells begin to detach.

Note: 10× trypsin–EDTA = 0.5% trypsin, 0.2% EDTA, and is usually purchased, rather than formulated in the lab. Dilute 10× trypsin-EDTA stock with 1× PBS. Alternatively, 1× trypsin–EDTA can be purchased and used directly.

4. Decant and discard trypsin. Observe the rounding of the cells.

Note: In the absence of a tissue culture microscope, the rounding of cells can be assessed by holding the flask up to a window or toward the lights in the laboratory. Because the refractive index[1] changes as the cells round, the side of the flask on which the cells are attached will appear cloudy or fogged, the extent of which is dependent on the degree of cell rounding and the density of cells on the tissue culture plastic. It is extremely important that the culture not be overtrypsinized, as this may result in the premature lysis of cells and significant loss of RNA or DNA.

5. Strike the flask sharply against the palm of the hand to dislodge the cells from the substratum. If cells fail to detach completely, warm the flask in the palm of the hand or place the flask in a 37°C incubator for an additional 60 s. Strike the flask again, to ensure detachment.

Note: It should not be necessary to strike the flask more than three or four times to completely dislodge the cells. If some cells appear to remain attached, longer contact with trypsin was required. A cell culture can be trypsinized again, but only after washing it with PBS. Moreover, if retrypsinization is necessary, be sure to save the growth medium supernatant that was added back to the flask after the first attempt at trypsinization because it is likely that large numbers of cells have become dislodged from the substratum. Cells in suspension should be stored on ice until the entire cell population is trypsinized. The cells may then be pooled into a common tube.

6. Immediately after the cells have detached from the substratum, add 5 ml of medium saved from Step 1 back into the flask in order to inactivate residual trypsin.

Note: Some cell cultures exhibit exquisite sensitivity to trypsin. In these cases, the cells would benefit from trypsin inactivation by the addition of a specific inhibitor of trypsin after the cells have detached. This is also true if the detached cells are to be resuspended in serum-free medium. Consult the literature for information about specific cell lines and medium formulations.

---

[1]The refractive index is the angle at which light bends when it moves from air through another medium, in this case, cytoplasm.

7. Transfer the suspension to a suitable centrifuge tube and pellet cells by centrifugation at 150–200 g for 3–5 min.

8. Following centrifugation, be sure to remove the supernatant as completely as possible. This is necessary to ensure that the key components of the lysis buffer are not diluted or rendered useless by interaction with residual culture media.

9. To facilitate cell resuspension in lysis buffer, or any other reagent, it is very helpful to tap the tube containing the cells *after* the supernatant from Step 8 has been removed, and before the addition of the lysis buffer. In this lab, tubes containing cell pellets are dragged along the top of 15-ml tube racks two or three times to help break up the cell pellet, immediately after which the lysis buffer is added.

Isolation of High-Molecular-Weight DNA by the Salting-Out Procedure

The following is a modification of the procedure of Miller *et al.*, (1988) for the isolation of high-molecular-weight (HMW) DNA without the use of toxic organic solvents. Although other procedures have been described for the recovery of HMW DNA, including dialysis (Longmire *et al.*, 1987) and the use of filters (Leadon and Cerutti, 1982), they are time-consuming and can be cumbersome in inexperienced hands.

The procedure described here is relatively brief and involves the salting-out of cellular proteins; it begins with peptide hydrolysis by proteinase K, followed by precipitation with a saturated NaCl solution. Protein-free genomic DNA is subsequently recovered by standard salt and ethanol precipitation. This approach has been used successfully for DNA isolation from nucleated blood cells (Miller *et al.*, 1988) and, in this laboratory, from a variety of trypsinized cell types of mesenchymal origin (Farrell, 1991–1998, unpublished data). In this laboratory, the resulting DNA is routinely 10–15% larger, on average, than genomic DNA purified with phenol:chloroform. Interestingly, many of the commercially available kits for genomic DNA isolation are based on the salting-out approach. DNA purity, as assessed by scanning UV spectrophotometry and examination of A_{260}/A_{280} (Chapter 5), is consistently comparable to that routinely achievable with phenol:chloroform protein extraction techniques.

Protocol

1. Harvest cells and pellet them in a 15-ml centrifuge tube by centrifugation at 200 g for 5 min. Use 1–2×10^7 cells per tube.

2. Decant supernatant. Completely resuspend cell pellet (up to 2×10^7 cells) in a total volume of 4.5 ml TE-9 lysis buffer (500 mM Tris–Cl, pH 9.0; 20 mM EDTA; 10 mM NaCl).

Optional: Wash cell pellet with PBS[1] and centrifuge again before addition of lysis buffer.

Note 1: The alkaline pH of this lysis buffer is desirable because it will induce partial hydrolysis of RNA in the sample. It is not sufficiently alkaline, however, to cause severe denaturation of double-stranded DNA (Chapter 1).

Note 2: For fewer starting numbers of cells, the volumes of all reagents should be scaled down proportionally.

3. Add 500 μl 10% sodium dodecyl sulfate (SDS) to the lysate. Invert tube sharply to mix.

4. Add 125 μl 20 mg/ml proteinase K (0.5 mg/ml final concentration) to the lysate and invert tube sharply to mix. Incubate at 48°C for 4–20 h.

Note: The inclusion of proteinase K is necessary to at least partially hydrolyze the nucleosome proteins. If not removed, histones and other nuclear proteins will copurify with the DNA and will interfere with restriction enzyme digestion of the DNA, and other downstream applications.

5. At the end of the incubation period, add 1.5 ml saturated NaCl solution (approximately 6 M). Shake vigorously for 15 s. Allow the tube to sit on the bench for 1–2 min in order to observe the appearance of the lysate.

Note: Additional small aliquots of saturated NaCl may be needed to cause the salting-out of peptides, which is essentially a dehydration technique. This will become evident if the lysate clears within 1–2 min after shaking the tube. Precipitated polypeptides will cause the lysate to appear very cloudy and remain as such. If additional salt is required, add a 200-μl aliquot, shake, and observe the appearance of the lysate. If required, repeat with additional aliquots of saturated NaCl solution until the lysate remains cloudy. Do not be overly zealous with the use of the saturated NaCl solution, however, as a point will be quickly reached where solid NaCl begins to come out of solution, resulting in a significant decrease in DNA yield.

---

[1] $1 \times$ PBS (per liter) = 8 g NaCl, 0.2 g KCl, 1.15 g Na_2HPO_4, 0.2 g KH_2PO_4. If desired, concentrated stock solutions of PBS can be prepared, and diluted with autoclaved H_2O just prior to use.

6. Centrifuge the tube containing the lysate at 1000–1500 × g for 10 min. Be sure not to exceed the maximum recommended g-force for the centrifuge tubes in use.

7. Decant the supernatant into a fresh 15-ml tube to dislodge any residual protein that was trapped by the foaming of the SDS. Centrifuge for an additional 10 min.

8. Decant the supernatant (which contains the DNA) into a fresh conical 50-ml tube. Avoid disturbing the protein pellet, if any. Discard the protein pellet.

9. Add 2 volumes of room temperature 95% ethanol to the tubes containing the supernatant from Step 8. Swirl the tube or invert *gently* until the precipitated DNA is visible.

Note: Room temperature ethanol is recommended here because only DNA will precipitate out immediately; ice-cold ethanol will drive the precipitation of some of the residual RNA as well.

10. Recover DNA using a silanized Pasteur pipet (Sigmacote is very helpful; see Appendix F). Make sure the DNA is not aspirated into the Pasteur pipet beyond the level that is silanized. Alternatively, precipitated DNA can be collected by centrifugation at 500 g for 1 min. In either case, transfer the DNA to a microfuge tube.

Note: In some cases, the tips of Pasteur pipets can heated and bent (in a Bunsen burner flame) to resemble a hook and used to fish the DNA out of solution. Because DNA sticks tenaciously to untreated glass surfaces, silanization is essential.

11. Wash the DNA three times with 500 μl aliquots of 70% ethanol to rinse away residual salt. A final wash with 95% ethanol will accelerate the required drying of the DNA pellet. Do not allow the DNA to dry out completely.

12. Dissolve the DNA completely in 100–200 μl TE buffer (10 mM Tris–Cl, pH 7.5, 1 mM EDTA) before using the DNA for any purpose, including concentration and purity determinations. Dissolved samples of genomic DNA can be stored for extended periods at 4°C.

Note 1: Warming the sample to 45–50°C will help redissolve the DNA. Incubation at 37°C may promote DNase activity, if present. In this lab, purified samples of genomic DNA are allowed to redissolve overnight at 4°C.

Note 2: In addition to UV analysis, the integrity of the sample should be assessed by electrophoresing 250–500 ng of the sample on a minigel along with λ-HindIII size standards.

13. If necessary, the sample can be treated with DNase-free RNase, and the DNA reprecipitated, to remove any contaminating RNA.

14. Store purified genomic DNA at 4°C. Freezing may compromise the integrity of the sample.

References

Leadon, S. A. and Cerutti, P. A. (1982). A rapid and mild procedure for the isolation of DNA from mammalian cells. *Anal. Biochem.* **120,** 282.

Longmire, J., Albright K., Lewis, A., Meincke, K., and Hildebrand, C. (1987). A rapid and simple method for the isolation of high molecular weight cellular and chromosome-specific DNA in solution with the use of organic solvents. *Nucleic Acids Res.* **15,** 859.

Miller, S. A., Dykes, D. D., and Polesky, H. F. (1988). A simple salting out procedure for extracting DNA from nucleated cells. *Nucleic Acids Res.* **16**(3), 1215.

RNA Isolation from Plant Tissue

Messenger RNA prepared from flora share many features with their counterparts in the animal kingdom and, in many cases, demonstrate remarkable stability under a variety of conditions. Although details pertaining to the study of gene expression in plants are related to many of the main themes of this text, the art of RNA isolation from plants is beyond the scope of this volume. As such, the reader is directed to the following.

References

Ainsworth, C. (1994). Isolation of RNA from flora tissue of *Rumex acetosa. Plant Mol. Biol. Rep.* **12,** 198–203.

Chang, S., Puryear, J., Cairney, J. (1993). A simple and efficient method for isolating RNA from pine trees. *Plant Mol. Biol. Rep.* **11,** 113–116.

Graham, G. C. (1993). A method for extractions of total RNA from *Pinus radiata* and other conifers. *Plant Mol. Biol. Rep.* **11,** 32–37.

Lewinsohn, E., Steele, C. L., Croteau, R. (1994). Simple isolation of functional RNA from woody stems of gymnosperms. *Plant Mol. Biol. Rep.* **12,** 20–25.

Logemann, J., Schell J., and Lothar, W. (1987). Improved method for the isolation of RNA from plant tissues. *Anal. Biochem.* **163,** 16–20.

López-Gómez, R., and Gómez-Lim, M. A. (1992). A method for extracting intact RNA from fruits rich in polysaccharides using ripe mango mesocarp. *Hort. Sci.* **27,** 440–442.

Mitra, D., and Kootstra, A. (1993). Isolation of RNA from apple skin. *Plant Mol. Biol. Rep.* **11,** 326–332.

Palmiter, R. D. (1974). Magnesium precipitation of ribonucleoprotein complexes: Expedient techniques for the isolation of undegraded polysomes and messenger ribonucleic acid. *Biochemistry* **13,** 3606.

Soni, R., and Murray, J. A. H. (1994). Isolation of intact DNA and RNA from plant tissues. *Anal. Biochem.* **218,** 474–476.

Strommer, J., Gregerson, R., Vayada, M. (1993). Isolation and characterization of plant mRNA. *In* "Methods in Plant Molecular Biology and Biotechnology." B. R. Glick and J. E. Thompson (eds.). CRC Press. Boca Raton, FL.

Wallace, D. M. (1987). Precipitation of nucleic acids. *Methods Enzymol.* **152,** 41–48.

Wan, C. Y. and Wilkens, T. A. (1994). A modified hot borate method significantly enhances the yield of high quality RNA from cotton. *Anal. Biochem.* **223,** 7–12.

Electrophoresis: Principles, Parameters, and Safety

Electrophoresis is the technique by which mixtures of charged macro-molecules, including proteins, nucleic acids, and carbohydrates, are rapidly resolved in an electric field. Unlike the amphoteric nature of proteins, where net charge is determined by the pH of the milieu, nucleic acids exhibit a net negative charge at any pH used for electrophoresis. In particular, the electrophoretic chromatography of nucleic acids has become a standard tool by which to characterize a sample qualitatively and quantitatively, and to assay its purity. Electrophoresis is efficient and quick, offering enhanced resolution by comparison with other cumbersome procedures such as density gradient centrifugation, in which RNA can be size-fractionated for purposes other than blot analysis. That nucleic acids can be observed directly by staining the gel permits visual inspection of the progress of the separation; it also affords a reliable assessment of the integrity of the sample, and in most instances the investigator can recover specific bands in a variety of ways for subsequent characterization. Thus, electrophoresis is also a powerful technique for concentrating like species of RNA and DNA: The parameters governing electrophoresis favor comigration of molecules of equivalent size.

Historically, the earliest applications of this technique involved macromolecular chromatography in sucrose. Refinements in the methodology extended electrophoresis into starch gels and finally led to the implementation of polyacrylamide, in which the structure of the gel itself influences the migration characteristics of the molecules under investigation. It was quickly realized that gels with large pores were required to efficiently electrophorese high-molecular-weight nucleic acids. At first, very low percentage polyacrylamide gels (2.5%) were utilized, though the instability and unreliability of such low percentage polyacrylamide gels precluded their widespread use. Agarose, a linear polysaccharide extracted from seaweed, was later added to acrylamide gels (Peacock and Dingman, 1968; Dahlberg *et al.,* 1969) to give them enhanced physical strength. Presently, gels that consist of either agarose or polyacrylamide are favored for nucleic acid separations, depending on the size range of molecules within a sample. Currently, the two standard matrices for the electrophoretic chromatography of nucleic acids are agarose and poly-acrylamide slabs, the electrophoresis of RNA usually being conducted under denaturing conditions. Less frequently, RNA may be character-ized by two-dimensional electrophoresis (see DeWachter *et al.,* 1990, for review).

Theoretical Considerations[1]

The following relationships describe the theoretical behavior of a charged molecule in sucrose; in actual practice other parameters delineat-ed here factor into the exact behavior of a charged molecular species in a gel matrix. What follows is by no means intended to provide a compre-hensive lesson on the subtleties of electricity, but rather is intended to re-view the basics necessary to understand how and why electrophoresis works.

When a molecule is placed in an electric field, the force exerted on it (F) is dependent on the net charge of the molecule (q) and the strength of the field (E/d) into which it is placed. Therefore

$$F = E/d \cdot (q) \tag{1}$$

in which E is the potential difference between the electrodes and d is the distance between them. In reality, a frictional force or drag opposes the migration toward an electrode. This frictional force (F) is dependent upon the size and shape (radius, r) of the molecule and the viscosity of the medium (η) through which it passes; v is the velocity at which the molecule is moving:

$$F = 6\pi r \eta v \tag{2}$$

By combining Eqs. (1) and (2), we see that

$$E/d \cdot (q) = 6\pi r \eta v \tag{3}$$

[1] Adapted, in part, from Hoefer Pharmacia Biotech.

and by algebraic rearrangement one can express the velocity with which a molecule moves in an electric field as

$$v = (Eq)/d6\pi r\eta \tag{4}$$

Take-home lesson 1. What may be distilled from Eqs. (1–4) is that the velocity at which a molecule moves is proportional to the field strength and net charge and inversely proportional to the size of the molecule and the solution viscosity (stiffness of the gel).

Knowledge of two parameters of electricity is fundamental to understanding the mechanics of electrophoresis. In the first, described by Ohm's Law, the electrical current [I, amperes (amps)] is directly proportional to the voltage [E, volts (V)] and inversely proportional to resistance (R, ohms). Thus,

$$I = E/R \tag{5}$$

In the second, power [P, watts (W)], a measure of the amount of heat produced, is the product of voltage (E, V) and current (I, amps), and can be expressed as

$$P = EI \tag{6}$$

Substitution of E with the product $I \cdot R$ from Eq. (5) permits expression of Eq. (6) as

$$P = I^2 R \tag{7}$$

In electrophoresis, one electrical parameter—voltage, current, or power—is always held constant. The consequences of an increase in resistance during the run (due to electrolyte depletion, temperature fluctuation, etc.) differ in the following ways:

a. When the constant current mode is selected, in which case velocity is directly proportional to the current, heat is generated and the velocity of the molecule is maintained.
b. When the constant voltage mode is selected, there is a reduction in the velocity of the charged molecules, though no additional heat is generated during the course of the run.
c. When the constant power mode is selected, there is a reduction in the velocity of the molecules, but there is no associated change in heating.

Take-home lesson 2: Temperature regulation is a primary concern throughout the electrophoretic process. A most serious concern, particularly when running an agarose gel, is maintaining the system temperature; overheating will cause the gel matrix to begin to melt in an astonishingly short time.

The reactions that permit the passage of current from the cathode to the anode are as follows:

Cathode reactions

$$2e^- + 2H_2O \rightarrow 2\ OH^- + H_2$$

$$HA + OH^- \rightleftarrows A^- + H_2O$$

Anode reactions

$$H_2O \rightarrow 2H^+ + \tfrac{1}{2}\ O_2 + 2e^-$$

$$H^+ + A^- \rightleftarrows HA$$

These reactions describe the electrolysis of water in the electrophoresis buffer, resulting in the production of hydrogen at the cathode and oxygen at the anode. For each mole of hydrogen produced, one-half mole of oxygen is produced. By direct observation of the anode, one can see about half as many gas bubbles as are produced at the cathode. While this is certainly not the best way to ascertain whether the electrodes have been connected properly, it is certainly a simple way, as well as an indicator that current is flowing.

A Typical Electrophoretic Separation

The electrophoresis of RNA can be executed in any of several ways although there are several features that all of these variations share. Typically, RNA is denatured in a relatively small volume (20 μl or less), applied into the wells of a horizontal agarose gel slab, and electrophoresed for an experimentally determined period. Of course, the parameters that typically govern the electrophoresis of macromolecules apply here also. Several factors influence the electrophoretic mobility of nucleic acids; once the optimal parameters have been determined in a particular laboratory setting, establishing a standard experimental format is likely to favor the reproducibility of data.

Choice of Matrix

Polyacrylamide gels exhibit a much greater tensile strength than agarose gels and offer greater resolving power; agarose gels, however, allow the investigator to separate nucleic acid molecules over a much greater size range (Table 1). This is true because the pores of an agarose gel are large, permitting efficient separation of macromolecules such as nucleic acids, large proteins, and even protein complexes. Polyacrylamide, in contrast, makes a small pore gel and can be used to separate smaller to average size proteins and oligonucleotides. In general, the stiffer the gel, that is, the more agarose or polyacrylamide make up the matrix, the greater is its resolving power. Agarose gels are used to electrophorese nucleic acid molecules as small as 150 bases to more than 50,000 bases (50 kb), depending on the concentration of the agarose and the precise nature of the applied electric field (constant or pulse field). Polyacrylamide gels, in contrast, are routinely used to separate nucleic acid molecules as small as 5–10 bases to as many as 500–600 bases. The resolving power of such gels is evident when one considers that polyacrylamide gels used for sequencing resolve nucleic acid molecules that differ in size by one base.

C A U T I O N   Monomeric acrylamide, that is, acrylamide in the unpolymerized form, is a very potent neurotoxin, and the assistance of someone

T A B L E   1

Useful Range of Agarose and Polyacrylamide Gels in the Electrophoresis of Nucleic Acids[a]

Matrix type	Percentage (w/v)	Useful range (bases)
Agarose[b]	0.6	1,000–20,0005[c]
	0.8	600–10,000
	1.0	500–9,000
	1.2	300–6,000
	1.5	200–3,000
Acrylamide[d]		
	3.5	100–1500
	5.0	80–500
	8.0	60–400
	12.0	40–200
	15.0	25–150
	20.0	6–100

[a]Table derived, in part, from data presented by Sambrook et al. (1990), Ogden and Adams (1987), and Farrell (1993).
[b]Agarose gels consisting of as much as 3% agarose may be used to observe short products (100–500 bp) of the polymerase chain reaction.
[c]Larger DNA molecules can be resolved efficiently by pulse-field electrophoresis.
[d]Gel composition = acrylamide:N,N'-methelene-bis-acrylamide, 19:1.

skilled in the preparation of gels and handling of acrylamide should be sought.

Conveniently, premade gels may now be purchased from a number of suppliers and are available for most standard gel box designs and formats. Alternatively, working with agarose, which is essentially nontoxic, facilitates ease in handling, gel preparation, and disposal.

Polyacrylamide gels may be cast in a variety of different shapes; agarose gels, however, are traditionally poured and electrophoresed as a horizontal slab. This configuration is desirable because the gel is fully supported by the casting tray beneath it. Some vertical apparatuses accommodate the weaker nature of agarose gels by the inclusion of a frosted glass plate to which the gels may cling. Polyacrylamide, by virtue of its much greater physical strength, is almost always electrophoresed in a vertical configuration, cast either as a slab or as a cylinder and usually mounted between two buffer chambers containing separate electrodes, so that the only electrical connection between the two chambers is through the gel. In contrast, horizontally positioned agarose gels are completely submerged in electrolyte buffer, albeit with the minimum volume necessary to completely cover the gel. This facilitates reasonable dissipation of heat when the gel is run at room temperature and at sufficiently low voltages.

When the experimental design calls for the blotting or transfer of RNA out of the gel after electrophoresis, as in the traditional Northern analysis, the matrix of choice is agarose, from which RNA (and DNA for Southern analysis) can be blotted efficiently and completely. Moreover, many of the classical problems associated with preparative agarose gel electrophoresis, namely coelution of contaminants from the agarose and difficulty in elution, have been circumvented with the widespread use of extremely high-purity and low-melting temperature agarose formulations.

Purity of Agarose

Agarose is a highly purified polysaccharide derived from agar. As with all commercially available reagents for molecular biology, there are several grades, many of which contain at least trace amounts of anions such as pyruvate and sulfate (which can contribute to electroendosmosis). One should only use very high-quality molecular biology-grade agarose, certified to be low in other contaminants, including polysaccharides, and salts, and completely free of nuclease activity. These contaminants may interfere with the subsequent manipulations of nucleic acids (especially

DNA) upon recovery from the gel. Highly purified grades of agarose, such as SeaKem and NuSieve (FMC Bioproducts), are free of nuclease activity as well as most of the aforementioned contaminants. Special low-melting- or low-gelling-temperature agaroses are also available for specialized applications, and the reader is referred to the publication "Your Complete Guide for DNA Separation and Analysis." This guide, recently published by FMC BioProducts (biotechserv@fmc.com), is an excellent laboratory resource.

Agarose Concentration

The concentration of agarose (the "stiffness" of the gel) profoundly influences the rate of migration; the mobility of a molecule of any given size differs as the concentration of agarose changes. This is because the pore size of an agarose gel is determined by the concentration of the agarose in the gel. The migration rate is linear over most of the length of the gel, although very large or very small molecules may end up outside its linear range. If this occurs, significant error may result from size determinations based on comparisons to size standards that fall *within* the linear portion of the gel. In addition, there is a certain amount of smearing of low-molecular-weight species frequently observed at the dye front (the so-called "leading edge" of the gel). This is especially noticeable in gels made up of less than 1% agarose. The usual working concentration of agarose is between 0.6% and 1.5%, although higher or lower concentrations may be adapted to specific circumstances.

Polyacrylamide Concentration

Polyacrylamide gels offer much greater tensile strength than agarose gels. In forming the gel, acrylamide monomers[2] polymerize into long chains that are covalently linked by a crosslinker. It is the crosslinker that actually holds the structure together. The most common crosslinker is *N,N*′methylenebisacrylamide, or bis for short. Other crosslinkers whose special properties aid in solubilizing polyacrylamide are also used occasionally.

The pore size in a polyacrylamide gel may be predetermined by (1) adjusting the total percentage of acrylamide, or (2) varying the amount of crosslinker added to the acrylamide to induce polymerization. When

---

[2]**Caution:** Monomeric acrylamide is a potent neurotoxin and should be handled with extreme care, according to all safety precautions recommended by the manufacturer.

there is a wide range in the molecular weights of the material under study, the researcher may prepare a pore gradient gel in which the pore size is larger at the top of the gel than at the bottom. Thus, the gel becomes more restrictive as the electrophoretic run progresses.

Molecular Size Range of Sample

Nucleic acid molecules migrate at a rate inversely proportional to the $\log_{10}$ of the number of bases (base pairs for double-stranded DNA and DNA:RNA hybrids) that make up the molecule. This is true because the larger the molecule, the greater the frictional coefficient, or drag, that these molecules experience (Helling *et al.,* 1974) as they are pulled through the pores that constitute the geometry of the matrix. A size calibration curve for electrophoretically separated RNA can be easily constructed by plotting the $\log_{10}$ nucleotides of at least two molecular molecular weight standards against the distance (in centimeters) that these standards migrated from the origin (the well into which they were loaded). This is routinely performed through the application of image analysis software.

Nucleic Acid Conformation

The mobility of DNA is influenced greatly by the topology of the molecule (see Sambrook *et al.,* 1989 for details). The electrophoretic behavior of RNA is also influenced/modified by the presence of secondary structures, also known as hairpins, which nondenatured RNA molecules frequently assume. To force identical species of RNA to comigrate, the electrophoresis of RNA is routinely carried out under denaturing conditions (Chapter 9).

Applied Voltage

The efficient dissipation of heat is directly related to the resolving power of a gel. When working with agarose gels, one must be certain that the applied voltage does not result in the overheating of the gel. Gels run at relatively low voltage can be cooled as a direct function of gel box configuration. For example, vertical gels tend to dissipate heat more rapidly than horizontal gels. Further, gel boxes with an aluminum backing conduct heat many times more rapidly than glass of a comparable thickness. In some cases it may be necessary to continually pump cool water around the electrophoresis chamber to maintain adequately low temperatures and generate flat, high-resolution bands, and equipment so designed generally is outfitted with the required plumbing. Gels run at excessive voltages will manifest a noticeable loss of resolution and the gel will actually melt if the system becomes too hot. Agarose gels should never be electrophoresed above 5 V per centimeter of distance between the electrodes; of course gels may be run at much lower voltages. In general, the slower the gel runs, the better the resolution (2–10 kb range). If agarose gel electrophoresis is maintained within these parameters, it should not be necessary to cool the system.

Ethidium Bromide

`CAUTION` Ethidium bromide is a powerful mutagen. Handle and dispose of it using great care and following the directions of the manufacturer. See Appendix B for disposal details.

Unlike DNA, which naturally exists in double-stranded form and the migration of which can be profoundly affected by the presence of ethidium bromide during electrophoresis, RNA structure is not intrinsically influenced by the addition of ethidium bromide to an agarose gel and running buffer. When used, ethidium bromide (10 mg/ml stock) is most often diluted to a final concentration of 0.5% μg/ml. The dye does, however, retard the rate of sample migration by about 15%. It is usually just as easy and far less messy to stain (and destain) the gel at the conclusion of electrophoresis. Of even greater concern is the fact that ethidium bromide can interfere with the transfer efficiency of RNA and DNA during Northern blotting and Southern blotting, respectively. If ethidium bromide is added directly to the gel during preparation, be sure to add it to the electrophoresis buffer, too, because as the RNA runs from – to + the ethidium bromide runs from + to –, thereby depleting the dye from the gel. Adding ethidium bromide to the electrophoresis buffer is not necessary, however, if the electrophoresis is short, for example, in checking for size, mass, or integrity; problems occur only when the nucleic acid and the ethidium bromide pass each other on the way to opposite electrodes. For a more thorough discussion of the merits of ethidium bromide as well as other staining alternatives, see Chapter 9.

SYBR Green

`CAUTION` SYBR Green, although not believed to be as mutagenic as ethidium bromide, should be handled with the same caution as with any nucleic acid binding dye. Handle and dispose of it using great care and following the directions of the manufacturer. See Appendix B for disposal details.

Two relatively new dyes, SYBR Green I and SYBR Green II (Molecular Probes), used to stain DNA and RNA, respectively, are becoming quite popular for use in nucleic acid electrophoresis. SYBR Green offers significant advantages over ethidium bromide, such as greater sensitivity, low background, and reduced mutagenicity, these differences are discussed in greater detail in Chapter 9.

It is not advisable to add SYBR Green to molten agarose when preparing the gel because of the negative effect that the dye has on the migration pattern of both DNA and RNA. Gels run in the presence of SYBR Green show bands that are fuzzy, highly irregular, and often indistinguishable from other bands in the vicinity. This phenomenon seems to occur at all concentrations of SYBR Green and at all voltages. In this laboratory, samples are electrophoresed first, followed by staining with SYBR Green. When staining is performed after electrophoresis, the results are outstanding.

Base Composition and Temperature

Unlike polyacrylamide gels, the migration characteristics of nucleic acids in agarose are not influenced by their respective base compositions. Further, the electrophoretic behavior of nucleic acids in agarose gels does not change appreciably between 4°C and room temperature, although the overheating of gels is a matter of great concern. Nucleic acid samples are routinely electrophoresed in agarose gels at room temperature.

Field Direction

Conventional electrophoresis is accomplished by separating nucleic acid molecules in an electric field of constant direction. The classical size limitations associated with electrophoretic methods in the past have been overcome through the use of pulse field gel electrophoresis (Schwartz *et al.,* 1982). This technique is performed by forcing DNA molecules to periodically reorient from one electric field direction to another. This approach permits efficient, reproducible separation of DNA molecules ranging in size from 0.1 to 10 megabases, and even entire chromosomes. The technology of alternating field direction is not commonly applicable to the electrophoresis of RNA, which naturally exists in the form of significantly smaller molecules.

Types of Gel Boxes

Agarose gels historically have been cast and electrophoresed as horizontal slabs principally because of the low tensile strength that these gels exhibit. The horizontal configuration results in the full support of the gel from beneath while facilitating efficient separation of a wide range of nucleic acid samples, through relatively low percentage gels. Although gel box chambers of diverse designs and quality are universally available, it is definitely worth investing in a solid, durable gel box that will withstand normal laboratory wear and tear for several years. Because agarose does not exhibit the toxic qualities associated with monomeric acrylamide, it is very easy to cast an agarose gel on the bench-top (or in the fume hood if toxic denaturants are added) and run the gel.

In practice, the open ends of the gel casting tray must be sealed to retain molten agarose until the matrix has solidified. Some newer designs incorporate built-in rubber gaskets that seal the casting tray when inserted into the chamber 90° to the intended direction of electrophoresis (Figure 1). This design precludes the taping requirement and frequent leaks associated with older style trays.

Better value electrophoresis gel boxes (tanks) come with a variety of combs that can be used to generate different numbers of wells of different sizes. The teeth of typical combs generally are 1, 1.5, 2, or 3 mm thick and are capable of generating 10–20 wells per gel. It may also be desirable to purchase gel boxes with buffer outlets or ports, which can accommodate buffer circulation. For example, it is critical to prevent the formation of a pH gradient when performing electrophoresis using phosphate or other buffer that does not resist changes in pH, such as when electrophoresing glyoxylated RNA (Chapter 9).

Safety Considerations in Electrophoresis

Safety must also factor into the selection of an electrophoresis apparatus. Never attempt to manipulate the gel, gel box, or power supply while conducting electrophoresis or while the power supply is plugged into an electrical outlet. Nor should any attempt ever be made to modify the design of either a gel box or power supply; any alterations could present a serious risk to both the operator and unsuspecting colleagues in the lab. Well-designed systems make it impossible to open or remove the lid while maintaining the flow of electrical current. In some designs the lid slides off, breaking electrical contact before the interior of the tank is

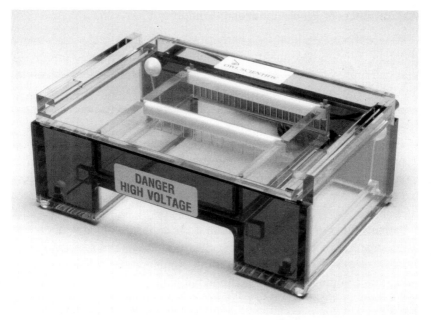

FIGURE 1

Medium-size gel electrophoresis chamber featuring gasketed tray. Note the positioning of the tray and combs: After the agarose has solidified, the casting tray is rotated 90° to align the wells formed by the comb in the direction of the electric field.

accessible. This permits the safe inspection of the progress of the electrophoresis: The lid must be replaced in order to restore electrical connections and resume electrophoresis. Moreover, it is wise to disconnect the leads from the power supply as well as the gel box when accessing the tank for an inspection of the system. Many of the late model power supplies are equipped to accommodate two sets of leads from two different gel boxes. Be sure to leave the power supply off until the leads have been connected to the power supply and the gel box; be sure to turn the power supply off before connecting a second set of leads (another gel box) to a power supply that is already operational. After all connections have been made properly, the power supply may be turned on again. Be sure to turn the power supply off before attempting to disconnect any of the leads. Last, never take hold of both leads simultaneously; there is a possibility that one who does so may complete the circuit, resulting in electrocution.

Maintenance of Electrophoresis Equipment

Electrophoresis is literally a daily ritual in most molecular biology laboratories. While good quality electrophoresis gel boxes and power supplies can represent an investment of thousands of dollars, it is an unfortunate, if not potentially dangerous, practice to ignore the upkeep of the instrumentation. Basic preventative maintenance of electrophoresis apparatuses is cost effective and ethically compelling.

The proper maintenance and use of electrophoresis gel boxes and power supplies is an important, though often overlooked part of lab safety. It is clearly the responsibility of the operator to inspect wires, connections, and electrodes on a regular schedule, to prevent potentially life-threatening injuries.

Perhaps the most common problem is the exposure of bare copper wire at the junction of the power cords, electrode leads, and plug ends. This situation poses a severe shock hazard. Replacement cords are usually available at nominal cost from the manufacturer or distributor and it is well worth ordering a few extras when the equipment is initially purchased. In addition, all power cords, insulation, electrodes, connection nuts, and gaskets should be scheduled for intralaboratory inspection on a weekly or biweekly basis. For a detailed regimen for electrophoresis apparatus maintenance, see Landers (1990).

References

Dahlberg, A. E., Dingman, C. W., and Peacock, A. C. (1969). Electrophoretic characterization of bacterial polysomes in agarose-acrylamide composite gels. *J. Mol. Biol.* **41,** 139.

Dennis, N., Corcos, D., Kruh, J., and Kitzis, A. (1988). A rapid and accurate method for quantitating total RNA transferred during Northern blot analysis. *Nucleic Acids Res.* **16,** 2354.

DeWachter, R., Maniloff, J., and Fiers, W. (1990). *In* "Gel Electrophoresis of Nucleic Acids—A Practical Approach" (D. Rickwood and B. D. Hames, Eds.). IRL Press, Washington, DC.

Farrell, R. E., Jr. (1993). "RNA Methodologies. A Laboratory Guide for Isolation and Characterization." Academic Press, San Diego, CA.

Helling, R. B., Goodman, H. M., and Boyer H. W. (1974). Analysis of R. *Eco*RI fragments of DNA from lambdoid bacteriophages and other viruses by agarose-gel electrophoresis. *J. Virol.* **14,** 1235.

Landers, T. (1990). Electrophoresis apparatus maintenance. *Focus* **12**(2), 54.

Ogden, R. C. and Adams, D. A. (1987). Electrophoresis in agarose and acrylamide gels. *Methods Enzymol.* **152,** 61.

Peacock, A. C. and Dingman, C. W. (1968). Molecular weight estimation and separation of ribonucleic acid by electrophoresis in agarose-acrylamide composite gels. *Biochemistry* **7,** 688.

Rodbard, D. and Chambach, A. (1970). Unified theory for gel electrophoresis and gel filtration. *Proc. Natl. Acad. Sci. USA* **65,** 970.

Sambrook, J., Fritsch, E. F., and Maniatis, T. (Eds). (1989). "Molecular Cloning: A Laboratory Manual." Cold Spring Harbor Laboratory Press, Cold Spring Harbor, NY.

Schwartz, D. C., Saffran, W., Welsh, J., Haas, R., Goldenberg, M., and Cantor, C. R. (1982). New techniques for purifying large DNAs and studying their properties and packaging. *Cold Spring Harbor Symp. Quant. Biol.* **47,** 189.

Timmis, K. Cabello, F., and Cohen, S. N. (1974). Utilization of two distinct modes of replication by a hybrid plasmid constructed *in vitro* from separate replicons. *Proc. Natl. Acad. Sci. USA* **74,** 4556.

Polyacrylamide Gel Electrophoresis

CAUTION   Acrylamide in the unpolymerized (monomeric) form is a very potent neurotoxin that is absorbed readily through the skin. Extreme care must be exercised when preparing such gels, both prior to and after polymerization. A mask and gloves should be worn when handling the acrylamide in both solid and liquid form. Never make the assumption that the polymerized form no longer poses a potential health hazard. If you are not familiar with the mechanics of the safe handling and preparation of polyacrylamide gels, seek the assistance of someone who has experience with these techniques. A new trend in polyacrylamide gel electrophoresis is the purchase of precast gels.

Polyacrylamide gels[1] form as a consequence of the polymerization of acrylamide monomers into linear chains and the accompanying crosslinking of these chains with N,N'-methylenebisacrylamide (bis). The molecular sieving of nucleic acid molecules in a polyacrylamide gel is a direct function of the concentration of acrylamide as well as the ratio of acrylamide to bis in the gel.

Because oxygen inhibits polymerization, the monomeric mixture must be deaerated. This may be done by purging the mixture with an inert gas or by evacuating the mixture with a vacuum.

When the gel is poured, either in a tube or as a slab, the top of the gel forms a meniscus. If ignored, this curved top on the gel will cause a distortion of the banding pattern. To eliminate the meniscus, a thin layer of water or water-saturated n-butanol is carefully floated in the surface of the gel mixture before it polymerizes. After polymerization, the water or

---

[1] Adapted from and courtesy of Hoefer Pharmacia Biotech.

butanol layer is poured off, leaving a flat upper surface on the gel. The layer of water, or water-saturated butanol, not only eliminates the meniscus but also excludes oxygen that would inhibit polymerization on the gel surface.

The pore size in an acrylamide gel may be predetermined in either of two ways. One way is to adjust the total percentage of acrylamide, that is the sum of the weights of the acrylamide monomer and the cross-linker. This is expressed as %T. For example, a 20%T gel would contain 20% (w/v) of acrylamide plus bis. As the %T increases, the pore size decreases.

The other way to adjust pore size is to vary the amount of crosslinker expressed as a percentage of the sum of the monomer and crosslinker, or %C. For example, a 20%T 5% C_{bis} gel would have 20% (w/v) of acrylamide plus bis, and the bis would account for 5% of the total weight of the acrylamide. It has been found that at any single %T, 5% crosslinking creates the smallest pores in a gel. Above and below 5%, the pore size increases. When there is a wide range in the molecular weights of the material under study, the investigator may prepare a pore gradient gel. The pore size in a gradient gel is larger at the top of the gel than at the bottom; the gel becomes more restrictive as the run progresses.

Polymerization of a polyacrylamide gel is accomplished by either a chemical or a photochemical method. In the most common chemical method, ammonium persulfate and the quaternary amine N,N,N,N'-tetramethylethylenediamine (TEMED) are used as the initiator and the catalyst, respectively. In photochemical polymerization, riboflavin and TEMED are used. The photochemical reaction is started by shining a long-wavelength ultraviolet light, usually from a fluorescent light source, on the gel mixture. Use of photochemical polymerization is limited to the study of proteins that are sensitive to ammonium persulfate or the by-products of chemical polymerization. This approach to polymerization is not suitable for the evaluation of RNA.

Polymerization generates heat. If too much heat is generated, convections will form in the unpolymerized gel, resulting in inconsistencies in the gel structure. To prevent excessive heating, the concentration of initiator–catalyst chemicals should be adjusted to complete polymerization in 20 to 60 min.

Analytical

The two denaturants that are routinely used for the electrophoresis of RNA through polyacrylamide gels are formamide (final concentration

98%) and urea (final concentration 7 M), the choice of which is mainly dependent upon the size range of molecules in the sample. For molecules smaller than 200 nucleotides, urea-containing gels will prove most useful. The use of formamide-containing gels is necessary for the accurate sizing of molecules between 200 and 1000 nucleotides in length.

Slab polyacrylamide gels are poured between two clean glass plates separated by appropriate spacers (1 to 2 mm thick). When nucleic acids are electrophoresed in a vertical configuration, the gel is cooled on both sides, a dividend that generally precludes the formation "smiling bands" due to heat distortion. When recovery from the gel is required following electrophoresis, gels as thin as 0.5 mm are preferred. As with agarose gels, the most common analytical procedure involving polyacrylamide gels is the visualization of the sample after electrophoresis. In one approach, staining the gel with ethidium bromide, acridine orange, or SYBR Green permits visual inspection of the distribution of electrophoresed nucleic acid sample. Another approach is possible when radiolabeled nucleic acids are electrophoresed. Molecules can be localized within the gel by autoradiography. Unlike standard blotting analysis (Northern blotting, for example), autoradiography under these conditions does not require transfer from the gel because either labeling and/or hybridization is conducted prior to running the gel, as in the S1 nuclease and RNase protection assays. For a complete discussion of the mechanics of polyacrylamide gel preparation and polyacrylamide electrophoresis, the reader is referred to Berger and Kimmel (1987) and Sambrook *et al.,* (1989).

References

Berger, S. L. and Kimmel, A. R. (1987). "Guide to Molecular Cloning Techniques." Academic Press, San Diego, CA.

Sambrook, J., Fritsch, E. F., and Maniatis, T. (1989). "Molecular Cloning: A Laboratory Manual," 2nd ed. Cold Spring Harbor Laboratory Press, Cold Spring Harbor, NY.

Internal Controls

Inconsistencies in technique and variation among investigators, even from day to day, are often responsible for the introduction of error at the level of data analysis. To address this, one commonly used approach is the measurement of an internal control or reference gene, the expression of which is not expected to change as a function of experimental manipulation. These are often referred to as housekeeping genes. The transcript from the reference gene allows normalization among samples on a blot or in a gel. One difficulty associated with the Northern analysis is the intrinsic lack of sensitivity due to the mechanics of the technique. Data of much greater sensitivity are achievable through the use of nuclease protection analysis (Chapter 18), RNA-based PCR (Chapter 15), and quantitative PCR techniques (Chapter 17).

Many of the commonly assayed, constitutively expressed housekeeping genes are, alas, subject to variation by the cell. As a consequence of this natural order, there is no single internal control that is appropriate in all situations. Among the commonly reported reference genes are:

1. β-Actin, a constitutively expressed transcript of medium abundance, isoforms of which are found in nearly all cell types (Kinoshita *et al.,* 1992). Historically the first transcript to be used as a standard, the use of ß-actin as a control transcript remains commonplace. Numerous reports, however, have demonstrated repeatedly the variation of this gene in response to a variety of external stimuli and cytoplasmic cues (Leof *et al.,* 1986; de Leeuw *et al.,* 1989; Dodge *et al.,* 1990; Solomon *et al.,* 1991; Spanakis, 1993).
2. GAPDH (glyceraldehyde-3-phosphate dehydrogenase), manifested

as a transcript of medium abundance, encodes a key enzyme in glycolysis. The expression of the GAPDH transcript as a reference gene or control is in widespread use, including this lab. As is the case with actin, GAPDH gene expression is also subject to modulation under a variety of experimental conditions (Alexander *et al.*, 1990; Mansur *et al.*, 1993; Bhatia *et al.*, 1994).

3. Cyclophilin, a protein involved in the isomerization of peptide bonds, is a highly conserved gene, the RNA from cyclophilin is of medium abundance in the cytoplasm. While commonly assayed as a control transcript, variation in cyclophilin mRNA has been demonstrated (Danielson *et al.*, 1988; Haendler and Hofer, 1995).

4. Histone (abundant transcript; nucleosome formation), fibronectin (abundant transcript; extracellular matrix), and transferrin receptor (medium-low abundance; iron uptake) mRNAs are widely reported as control transcripts. The protein product of each of these genes has a unique role in the cell. In general, these transcripts are less likely to experience disturbing fluctuations in mRNA levels.

5. 28S and 18S rRNA, highly abundant posttranscriptionally regulated products of the ribosomal genes, are often cited as a nearly invariant transcript of great predictability as an internal control (de Leeuw *et al.*, 1989; Bonini and Hofmann, 1991). These transcripts are the scaffolding of mature ribosomes. It is the opinion of this author that the 28S and 18S rRNAs are better as loading controls than gene expression controls because they are products of RNA polymerase I, while mRNA, and their unspliced precursor hnRNA, are products of RNA polymerase II. These RNA polymerases show varying sensitivities to different xenobiotics; it is thus possible that an experimental manipulation could cause an across-the-board arrest of RNA polymerase II transcription, while the RNA polymerases I and III continue to transcribe at near-control levels. The reverse is also possible.

6. 16S mitochondrial rRNAs, moderately abundant transcripts that are localized in the mitochondria, are utilized as internal controls with lesser frequency than their cytoplasmic counterparts. The expression of these mitochondrial genes, purportedly, is influenced far less by experimental manipulation than are genes localized in the nucleus (Tepper *et al.*, 1992), though the fact that these are ribosomal genes, rather than mRNA, should incite the same concerns addressed in #5, above.

7. tRNA consists of a number of high abundance transcripts that shuttle amino acids to the ribosomes during translation. The total mass

of tRNA is occasionally measured, using a tRNA-specific probe, as loading control. tRNA is transcribed by RNA polymerase III and is therefore better suited as a loading control than as a gene expression control, for the same reasons cited in #5, above.

It is certainly possible to spike an RNA sample with an external standard. In this approach, one adds a known mass of a transcript produced by in vitro transcription. This artificial transcript is often associated with at least one of the experimental transcripts to be assayed and is truncated so that (1) it is recognized by the probe and has a similar thermodynamic stability, compared to the experimental transcript, when hybridized and (2) it appears as a distinct band upon electrophoresis, thereby ensuring adequate separation and resolution of control and experimental transcripts. By spiking different amounts into various samples, one is able to generate a standard curve of sorts so as to correlate observed experimental band intensity with the mass of the standard(s). This approach is described in greater detail in Chapter 17.

References

Alexander, M., Denaro, M., Galli, R., Kahn, B., and Nasrin, N. (1990). Tissue-specific gene regulation of the glyceraldehyde-3-phosphate-dehydrogenase gene by insulin correlates with the induction of an insulin-sensitive transcription factor during differentiation of 3T3 adipocytes. *In* "Obesity: Toward a Molecular Approach." A.R. Liss, Inc., New York, NY. pp. 247–261.

Bhatia, P., Taylor, W., Greenberg, A., and Wright, J. (1994). Comparison of glyceraldehyde-3-phosphate-dehydrogenase and 28S ribosomal RNA gene expression as RNA loading controls for northern blot analysis of cell lines of varying malignant potential. *Anal. Biochem.* **216,** 223.

Bonini, J. A. and Hofmann, C. (1991). A rapid, accurate nonradioactive method for quantitating RNA on agarose gels. *BioTechniques* **11,** 708.

Danielson, P., Forss-Petter, S., Brow, M., Calavette, L., Douglas, J., Milner, R., and Sutcliffe, J. (1988). A cDNA clone of the rate mRNA encoding cyclophilin. *DNA* **7,** 261.

de Leeuw, W., Slagboom, P., and Vijg. (1989). Quantitative comparison of mRNA levels in mammalian tissues: 28S ribosomal RNA level as an accurate internal control. *Nucleic Acids Res.* **17,** 10137.

Dodge, G.R., Kovalszky, I., Hassell, J.R., and Iozzo, R.V. (1990). Transforming growth factor β alters the expression of heparin sulfate proteoglycan in human colon carcinoma cells. *J. Biol. Chem.* **265,** 18023.

Haendler, B., and Hofer, E. (1995). Characterization of the human cyclophilin gene and of related pseudogenes. *Eur. J. Biochem.* **190,** 477.

Kinoshita, T., Imamura, J., Nagai, H., and Shimotohno, K. (1992). Quantification of gene expression over a wide range by the polymerase chain reaction. *Anal. Biochem.* **206,** 231.

Krieg, P. A. and Melton, D. A. (1987). *In vitro* synthesis with SP6 RNA polymerase. *Meth. Enzymol.* **155,** 397.

Leof, E. B., Proper, J. A., Getz, M. J., and Moses, H. I. (1986). *J. Cell Physiol.* **127,** 83.

Mansur, N., Meyer-Siegler, K., Wurzer, J., and Sirover, M. (1993). Cell cycle regulation of the glyceraldehyde-3-phosphate-dehydrogenase/uracil DNA glycosylase gene in normal human cells. *Nucleic Acids Res.* **4,** 993.

Soloman, D. H., O'Driscoll, K., Sosne, G., Weinstein, I. B., and Cayre, Y. E. (1991). Cell growth differentiation. **2,** 187.

Spanakis, E. (1993). Problems related to the interpretation of autoradiographic data on gene expression using common constitutive transcripts as controls. *Nucleic Acids Res.* **16,** 3809.

Tepper, C. G., Pater, M. M., Pater, A., Xu, H., and Studzinski, P. (1992). Mitochondrial nucleic acids as internal standards for blot hybridization analyses. *Anal. Biochem.* **203,** 127.

Trade Citations

The following is a list of trademark-protected products referenced in this volume.

AccuTaq LA™ is a registered trademark of Sigma Chemical Company.

Amersham, Gene Images, Hybond, ECL, Sensitize, and Hyperfilm are trademarks of Amersham International.

AmpliTaq® and AmpliWax® are registered trademarks of Roche Molecular Systems.

Coomassie® is a registered trademark of Imperial Chemical Industries, Ltd.

CPD® and CSPD® Substrate are registered trademarks of Tropix, Inc.

Deep Vent™ is a registered trademark of New England BioLabs.

Dynabeads® is a registered trademark of Dynal, Inc.

Eppendorf® is a registered trademark of Eppendorf-Netherler-Hinz GmbH.

Expand™ is a registered trademark of Boehringer Mannheim Corporation.

Extractor™ is a registered trademark of Scheicher & Schuell, Inc.

Ficoll® is a registered trademark of Pharmacia Biotech AB.

Gel Pro™ is a registered trademark of Media Cybernetics, L.P.

GeneAmp® is a registered trademark of Roche Molecular Systems.

HotWax™ Mg^{++} Beads is a registered trademark of Invitrogen, Inc.

JumpStart™ is a registered trademark of Sigma Chemical Company.

Kimwipes® is a registered trademark of the Kimberly-Clark Corporation.

Lumiphos® is a registered trademark of Lumigen, Inc.

Nonidet P-40™ (NP-40) is a registered trademark of Shell International Petroleum.

Parafilm™ is a registered trademark of American Can Company.

pGEM® is a registered trademark of Promega Corporation.

Polaroid® is a registered trademark of the Polaroid Corporation.

Pronase® is a registered trademark of Calbiochem Novabiochem Corp.

RNAguard® is a registered trademark of Pharmacia Biotech.

RNasin® is a registered trademark of Promega Corporation.

Saran Wrap® is a registered trademark of Dow Chemical Company.

SeaKem® is a registered trademark of FMC BioProducts.

Sephadex® and Sepharose® are registered trademarks of Pharmacia Biotech AB.

SYBR® Green and SYPRO® are registered trademarks of Molecular Probes, Inc.

TaqStart™ is a registered trademark of Clontech Laboratories.

Triton™ is a registered trademark of Rohm & Haas Company.

Tween™ is a registered trademark of ICI Americas, Inc.

Whatman® is a registered trademark of Whatman, Ltd.

Selected Suppliers of Equipment, Reagents, and Services

Amersham Life Science, Inc.
2636 South Clearbrook Drive
Arlington Heights, IL 60005-4692
800/323-9750
http://www.amersham.com.uk

Beckman Instruments, Inc.
2500 N. Harbor Blvd.
Box 3100
Fullerton, CA 92634
800/742-3345

Becton-Dickinson
1 Becton Drive
Franklin Lakes, NJ 07417
888/237-2762

Bellco Glass, Inc.
P.O. Box B
340 Edrudo Road
Vineland, NJ 08360
800/257-7043
http://www.belcoglass.com

Bio-Rad Laboratories
3300 Regatta
Hercules, CA 94547
800/424-6723
http://www.bio-rad.com

Boehringer Mannheim
Corporation
9115 Hague Road
P.O. Box 50414
Indianapolis, IN 46250
800/262-1640
http://biochem.
boehringermannheim.com

Brinkmann Instruments, Inc.
One Cantiague Road
Westbury, NY 11590
800/645-3050
http://www.brinkmann.com

Calbiochem
P.O. 12087
LaJolla, CA 92039-2087
800/854-3417
http://www.calbiochem.com

Clontech Laboratories, Inc.
1020 E. Meadow Circle
Palo Alto, CA 94303-4230
800/662-2566
http://www.clontech.com

Cole-Parmer Instrument Company
625 E. Bunker Court
Vernon Hill, IL 60021
800/323-4340
http://www.coleparmer.com

Display Systems Biotech
1070 Joshua Way
Vista, CA 92083
800/697-1111
http://www.displaysystems.com

Diversified Biotech, Inc.
1208 V.F.W. Parkway
Boston, MA 02132
800/796-9199

DNA Diagnostics
P.O. Box 4544
Crofton, MD 21114-4544
202/260-0171

Dynal, Inc.
5 Delaware Drive
Lake Success, NY 11042
800/638-9416
http://www.dynal.no

Eastman Kodak Co.
343 State Street
Rochester, NY 14652
800/225-5352
http://www.kodak.com

Exon-Intron, Inc.
Specialists in Biotech Education
9151 Rumsey Road, Suite 130
Columbia, MD 21045
888/4RNA-DNA (888/476-
2362)
http://www.DNAtech.com

FMC Bioproducts
191 Thomaston Street
Rockland, ME 04841
800/341-1574
http://www.bioproducts.com

Hoefer Pharmacia Biotech
800 Centennial Avenue
P.O. Box 1327
Piscataway, NY 08855-1327
800/526-3593
http://PNU/Biosensor

Invitrogen Corporation
3985 B Sorrento Valley Blvd.
San Diego, CA 92121
800/955-6288
http://www.invitrogen.com

ISCO, Inc.
P.O. Box 5347
Lincoln, NE 68505-0347
800/228-4250
http://ISCO.com

Jouan, Inc.
110 B Industrial Drive
Winchester, VA 22602
800/662-7477

Kirkegaard & Perry Laboratories,
Inc.
2 Cessna Court
Gaithersburg, MD 20879
800/638-3167

L.A.O. Enterprises
19212 Orbit Drive
Gaithersburg, MD 20879
301/948-4988

Lofstrand Labs
P.O. Box 459
Gaithersburg, MD 20884
800/541-0362

Lumigen, Inc.
24485 W. Ten Mile Road
Southfield, MI 48034
248/351-5600

Media Cybernetics
8484 Georgia Avenue
Suite 200
Silver Spring, MD 20910
800/992-4256
http://www.mediacy.com

Millipore Corporation
80 Ashby Road
Bedford, MA 01730
800/645-5476
http://www.millipore.com

Molecular Probes, Inc.
P.O. Box 22010
Eugene, OR 97402-0414
800/438-2209
http://www.probes.com

Nalge Company
Subsidiary of Sybron Corporation
P.O. Box 20365
Rochester, NY 14602
800/625-4327
http://www.nalgenunc.com

NEN Life Science Products
549 Albany Street
Boston, MA 02118
800/551-2121
http://www.nenlifesci.com

New England Biolabs
32 Tozer Road
Beverly, MA 01915
800/632-5227
http://www.neb.com

Nikon Instrument Group
1300 Walt Whitman Road
Melville, NY 11747
516/547-8500
http://www.nikonusa.com

Oncogene Research Products
84 Rogers Street
Cambridge, MA 02142
800/662-2616
http://www.apoptosis.com

Oncor, Inc.
209 Perry Parkway
Gaithersburg, MD 20877
800/776-6267
http://www.oncor.com

Owl Scientific Plastics, Inc.
10 Commerce Way
Wilborn, MA 01801
800/242-5560
http://www.owlsci.com

Pharmacia Biotech, Inc.
800 Centennial Avenue
Piscataway, NJ 08854
800/526-3593
http://www.biotech.pharmacia.se

Pierce Chemical Company
P.O. Box 117
Rockford, IL 61105
800/874-3723
http://www.piercenet.com

Polaroid Corporation
575 Technology Square
Cambridge, MA 02139
800/225-1618
http://www.polaroid.com

Promega Corporation
2800 Woods Hollow Road
Madison, WI 53711
800/356-9526
http://www.promega.com

Robbins Scientific
814 San Aleso Avenue
Sunnyvale, CA 94086
800/752-8585
http://www.robsci.com

Rohm & Haas Co.
100 Independence Mall, West
Philadelphia, PA 19106
800/221-8992

Schleicher & Schuell, Inc.
10 Optical Avenue
Keene, NH 03431
800/245-4024
http://www.s-and-s.com

Sigma Chemical Company
P.O. Box 18817B
St. Louis, MO 63160
800/325-3010
http://www.sigma.sial.com

Sorvall, Inc.
31 Pecks Lane
Newtown, CT 06470-2337
800/522-7746
http://www.sorvall.com

Stratagene Cloning Systems
11011 North Torrey Pines Road
La Jolla, CA 92037
800/424-5444
http://www.stratagene.com

Tri-Continent Scientific, Inc.
12555 Loma Rica Drive
Grass Valley, CA 95945
800/937-4738

VWR Scientific
P.O. Box 626
Bridgeport, NJ 08014
800/932-5000
http://www.vwrsp.com

Whatman, Inc.
9 Bridewell Place
Clifton, NJ 07014
800/631-7290
http://www.whatman.com

Zeiss, Inc.
One Zeiss Drive
Thornwood, NY 10594
800/982-6493

Glossary

Abundance: The prevalence of a particular RNA, or class of RNA molecules, in the cell. The relative amounts of a particular messenger RNA species in different samples is frequently expressed in terms of relative abundance.

Adduct: A chemical addition product.

Allele: One of two or more alternative forms of a gene.

Alu repeat sequence: One of approximately 300,000 sequences, about 300 base pairs in length, found dispersed throughout the genome. These repetitive sequences are so named because of the characteristic *Alu* I restriction enzyme site they contain.

α-Amanitin: A bicyclic octapeptide derived from the poisonous mushroom *Amanita phalloides.* This fungal product differentially inhibits eukaryotic RNA polymerases; RNA polymerase II is especially sensitive.

AmpliTaq: Recombinant form of the naturally occurring thermostable DNA polymerase from the organism *Thermus aquaticus.*

Anneal: The base-pairing of complimentary polynucleotides to form a double-stranded molecule.

Antiparallel: The manner in which two complementary polynucleotides base pair to one another; the 5′ and 3′ ends of each molecule are reversed in relation to each other, so that the 5′ end of one strand is aligned with the 3′ end of the other strand. Antiparallel base pairing accompanies the formation of double-stranded DNA (DNA:DNA), double stranded RNA (RNA:RNA), and DNA–RNA hybrids (DNA:RNA).

Antisense RNA: Any RNA molecule that is complementary to the naturally occurring sequence of messenger RNA. Antisense RNA is capable of hybridization with messenger RNA (mRNA) on a filter or *in situ,* thereby forming an extremely stable double-stranded RNA.

Area of interest (AOI): A contiguous subset of pixels defined within an image, which may be arranged in any polygonal shape, and is used to isolate a subset from the rest of the image.

Attenuation: Method of regulation of transcription termination in certain prokaryotic operons.

Autoradiograph (*also* autoradiogram): A photographic record of the spatial distribution of radiation in an object or specimen. It is made by placing the object very close to a photographic film or emulsion.

Autoradiography: The process by which an autoradiograph is made.

Bacteriophage: A virus that infects and is propagated in a bacterial host, often for cloning purposes. Among the best characterized and widely exploited are derivatives of the λ bacteriophage.

Base pairing: The formation of hydrogen bonds between the nitrogenous bases of two nucleic acid molecules.

β particle: An elementary particle emitted from a nucleus during radioactive decay. It has a single electrical charge. A negatively-charged β particle is identical to an electron. A positively charged β particle is known as a positron.

Biotechnology: The use of microorganisms, plant, and animal cells to produce useful materials, such as food, medicine, and other chemicals.

Biotin: A small vitamin used to label nucleic acids for a variety of purposes, including nonisotopic hybridization by chemiluminescence or chromogenic techniques.

Bit: The smallest unit of information recognized by a computer. A pixel is represented by one or more computer bits. The number of bits per pixel determines directly the number of colors or gray shades that can be represented.

Bit depth: The number of pixels used to represent one pixel value. Also referred to as pixel depth and bits per pixel.

Bitmap: A two-dimensional array used to represent an image. Each cell in the array contains a value that describes a sample of the image in terms of its color.

Bits per pixel (BPP): Describes the depth of an image. Bilevel images are only 1 BPP while true color images are 24 BPP.

Brightness: The amount of white in an im-age. The brighter the image, the more white it contains. As brightness is increased, each color in the image is shifted more toward white.

Buoyant density: A measure of the ability of a substance to float in a standard fluid. For example, differences in the buoyant densities of RNA and DNA allow them to be separated in a gradient of cesium chloride or cesium trifluoroacetate.

CAAT box: A conserved sequence located about 75 base pairs upstream from the transcriptional start site of eukaryotic genes. The CAAT box is part of the eukaryotic promoter.

Cap: The structure found at the 5′ end of eukaryotic mRNA consisting of an inverted (5′-5′) linkage between the first two nucleotides. The 5′-cap is enzymatically added in the nucleus to the 5′ end of hnRNA immediately after transcription.

cDNA: *See* Complementary DNA.

Cesium chloride (CsCl): A dense salt used for isopycnic separation of nucleic acids. CsCl is commonly used to pellet RNA and, in other applications, to purify plasmid DNA.

Cesium trifluoroacetate (CsTFA): A dense salt used for isopycnic separation of nucleic acids. CsTFA is capable of forming steeper gradients than CsCl; RNA can thus be banded or pelleted.

Chaotropic: Biologically disruptive. Chaotropic lysis buffers disrupt cellular and subcellular membranes and destroy enzymatic activity on contact.

Chemiluminescence: A nonisotopic hybridization detection technique. Chemiluminescence is the production of visible light by chemical reaction.

Chromatin: The complex of genomic DNA and protein found in the nucleus of a cell in interphase.

Chromosome jumping: A technique very similar to chromosome walking with the exception that very large fragments of DNA (>100 kb) under investigation are purified by pulse field gel electrophoresis. Thus, moving from one end of this DNA fragment, for subcloning purposes, constitutes more of a jump than a walk. *See* Chromosome walking.

Chromosome walking: The systematic isolation of a set of clones containing overlapping DNA fragments that collectively make up a specific genomic region. The process is initiated by identification of a unique site recognized by a sequence-specific probe. When a clone is retrieved from a library, the process of chromosome walking continues by subcloning and rehybridization to the library: In each successive hybridization, the probe corresponds to the 3' or 5' end of the clone previously recovered from the library. Thus, the resulting series of overlapping clones permits locus characterization within, upstream, and downstream of a particular locus. This approach can be exploited to walk either toward or away from a locus of interest.

Clone (*noun*): A collection of genetically equivalent cells or molecules.

Clone (*verb*): A series of manipulations designed to isolate and propagate a specific nucleic acid sequence or cell for characterization, storage, or further amplification.

Clone bank: Older terminology for what are currently known as complementary DNA libraries and genomic DNA libraries.

Coding strand: In double-stranded DNA, the strand that has the same sequence as the resulting RNA (except for the substitution of uracil for thymine).

Competitive PCR: A very sensitive method for quantification of transcript abundance by the inclusion of a new DNA sequence, which competes for primers and dXTPs, in the same reaction tube as the experimental sample. By varying the amount of the competitor DNA (also known as a DNA mimic), a dilution will be identified in which the concentration of target and mimic are equivalent, allowing for very accurate quantification.

Complementary DNA (cDNA): DNA enzymatically synthesized *in vitro* from an RNA template, by reverse transcription. cDNA may be single stranded or double stranded, as required by the parameters governing a particular assay. The synthesis of cDNA represents a permanent biochemical record of the cellular biochemistry and also provides a means by which that record can be propagated.

Compression: A mathematical technique that allows an image to be stored with less memory. Redundancies in the internal representation of the image are identified and given a code; the data in the redundancies are then replaced by the code.

Constitutive gene expression: Interaction of RNA polymerase with the promoters of specific genes not subjected to additional regulation. Such genes are frequently expressed continually at low or basal levels, and are sometimes referred to as genes with housekeeping functions.

Contrast: The sharpness of an image. The higher the contrast in an image, the larger the difference between white and black or the more spread out the color range. As contrast is increased, all the colors or gray shades in the image spread apart.

CsCl: *See* Cesium chloride.

CsTFA: *See* Cesium trifluoroacetate.

Curie (Ci): A basic unit for measuring radioactivity in a sample. One curie is equivalent to 3.7×10^{10} becquerel (i.e., disintegrations per second).

Cytoplasm: The cellular contents found between the plasma membrane and the nuclear membrane.

Denaturation (of nucleic acids): Conversion of DNA or RNA from a double-stranded form to a single-stranded form. This can mean dissociation of a double-stranded molecule into its two constituent single strands, or the elimination of intramolecular base pairing.

Deoxyribonucleic acid (DNA): A polymer of deoxyribonucleoside monophosphates, assembled by a DNA polymerase. *In vivo,* DNA is produced by the process known as replication. DNA can also be synthesized using a variety of *in vitro* methods, such as the polymerase chain reaction.

Diethyl pyrocarbonate (DEPC): A chemical used to purge reagents of nuclease activity. DEPC is carcinogenic and should be handled with extreme care. Alternatively, hospital-quality sterile H_2O for irrigation can be confidently substituted.

Differential display PCR (DDPCR): A method for identification of uniquely transcribed sequences among two or more RNA populations. DDPCR is a PCR-based method that utilizes large combinations of relatively short primers to ensure amplification of all transcribed RNAs in the form of complementary DNA. Electrophoretic comparison of the products of each reaction shows products of identical molecular weight when a transcript is common to the biological samples under investigation; a band in only one lane is observed if gene expression has been induced or repressed.

Differentiation: The process of biochemical and structural changes by which cells become specialized in form and function.

Dimethyl sulfoxide (DMSO): A reagent used in conjunction with glyoxal to denature RNA prior to electrophoresis. DMSO can also be used to lower the melting temperature of DNA duplexes to support long-range PCR.

Diploid: Having two complete sets of chromosomes (two of each chromosome). Compare to *Haploid* and *Triploid.*

Directional cloning: Unidirectional insertion of a DNA molecule into a vector accomplished by placement of different sequences or restriction enzyme sites at the ends of double-stranded complementary DNA or genomic DNA molecules.

DNA polymerase I: A prokaryotic enzyme capable of synthesizing DNA from a DNA template. The native DNA polymerase I, also known as the holoenzyme or Kornberg enzyme, manifests three distinct activities: $5' \rightarrow 3'$ polymerase, $5' \rightarrow 3'$ exonuclease, and $3' \rightarrow 5'$ exonuclease. *See also* Klenow fragment.

Dot blot: A membrane-based technique for the quantitation of specific RNA or DNA sequences in a sample. The sample is usually dot configured onto a filter by vacuum filtration through a manifold (*see also* Slot blot). Dot blots lack the qualitative component associated with electrophoretic assays.

Downstream: Sequences in the $3'$ direction (further along in the direction of expression) from some reference point. For example, the initiation codon is located downstream from the $5'$ cap in eukaryotic messenger RNA.

Duplex: The formation of a double-stranded molecule or portion thereof by the base pairing of two complementary polynucleotides.

Electrophoresis: A type of chromatography in which macromolecules (i.e., proteins and nucleic acids) are resolved through a matrix based on their charge.

Electrophoretogram: A photograph of a gel made after electrophoresis, which records the spatial distribution of macromolecules within the gel.

ELISA: *See* Enzyme-linked immunosorbent assay.

Enhancer element: A short regulatory DNA sequence that can increase the use of some eukaryotic promoters, resulting in elevated levels of transcription. Enhancer elements exert their influence relatively independently of orientation and proximity to target promoters.

Enrichment: Any manipulation of RNA that results in an increase in the statistical representation of one or more RNA subpopulations as a percentage of all RNAs synthesized by the cell. For example, selection of poly(A)$^+$ RNA results in an enrichment of messenger RNA. An enrichment strategy may also involve the physical manipulation of cells *prior* to lysis to superinduce one or more species of RNA. For example, serum stimulation of quiescent cells results in an induction (enrichment) of proliferation-specific message as cells reenter the cell cycle.

Enzyme-linked immunosorbent assay (ELISA): An assay that detects an antigen–antibody complex via an enzyme reaction.

Ethidium bromide (EtBr): A planar, intercalating agent used to visualize nucleic acids, both DNA and RNA. This dye emits a bright orange fluorescence when UV irradiated, thus gels that contain samples can be photographed for future reference. Standard ethidium bromide stock solution is 10 mg/ml in water; standard staining concentration is 0.5–1.0 μg/ml.

Exon: A portion of a eukaryotic gene represented in the mature messenger RNA molecule. Exons may or may not be translated.

File format: The method by which an image is stored to a disk, based on image class, compression type, and halftone pattern. TIFF and BMP are examples of file formats.

Formaldehyde: A commonly used denaturant of RNA. It should be handled with care in a chemical fume hood. Formaldehyde is a liver carcinogen and, as a known teratogen, should be avoided by pregnant women.

Formamide: A commonly used organic solvent/denaturant used to lower the melting temperature of double-stranded duplexes; for this reason it is often included in hybridization reactions to maintain stringency, at a temperature lower than might otherwise be required. It is often used with formaldehyde to denature RNA prior to electrophoresis. Formamide is a carcinogen and, as a known teratogen, should be avoided by pregnant women.

Frame grabber: Hardware unit, essentially a card inserted into one of the slots in the back of a computer, which permits the use of a video camera to import an image directly into image analysis software.

Gamma (γ): A nonlinear logarithmic contrast correction factor used to adjust the contrast in dark areas of an image.

Gene: The unit of heredity. A gene is a sequence of chromosome DNA which ultimately encodes the instructions for the synthesis of a polypeptide.

Genome: All the chromosomal DNA found in a cell. In some applications, it may be useful to distinguish nuclear genomic DNA from the mitochondrial genome.

Genomic DNA: Chromosomal DNA

Glyoxal: A reagent commonly used to denature RNA prior to electrophoresis. It is often used in conjunction with dimethyl sulfoxide.

Gray level: The brightness value assigned to a pixel in gray scale images. In an 8-bit image, this value ranges from 0 to 255 (from black, through shades of gray, to white).

Haploid: Having one complete set of chromosomes (one of each chromosome, as found in the gametes). Compare to *Diploid* and *Triploid.*

Hapten: A small molecule, not antigenic by itself, that can act as an antigen when conjugated to a larger antigenic molecule, usually a protein.

Heterogeneous nuclear RNA (hnRNA): Precursor messenger RNA; the primary product of eukaryotic transcription by RNA polymerase II.

Histones: Proteins associated with DNA, which help to organize chromosomal DNA. Histone proteins are among the most highly conserved among all eukaryotic cells.

Hogness box (*also* TATA box): A conserved AT-rich motif located about 25 base pairs from the transcriptional start site of eukaryotic genes. The Hogness box is part of the eukaryotic promoter for RNA polymerase II.

Housekeeping gene: Genes which are expressed, at least theoretically, at constant levels in all cells, the products of which are required to maintain cellular viability. Because of their purported invariance, assay of transcription of these sequences is often performed to demonstrate that an overall change in gene expression has *not* occurred, in the context of an experimental manipulation.

Hybridization: The formation of hydrogen bonds between two nucleic acid molecules that demonstrate some degree of complementarity. The specificity of hybridization is a direct function of the stringency of the system in which the hybridization is being conducted.

Hydrogen bonding: The highly directional attraction of an electropositive hydrogen atom to an electronegative atom such as oxygen or nitrogen. This in the manner of interaction between complementary bases during nucleic acid hybridization. *See also* Base pairing.

Hyperpolymer: A collection of labeled probe molecules that have hybridized to a target sequence and with each other, in an overlapping fashion, such that a tail of labeled molecules is attached to the target sequence. This phenomenon occurs when breakage of a nucleic acid backbone occurs during labeling, as in nick translation, or as a result of failure to seal the backbone during probe synthesis, as in random priming. The effect is an amplification of the signal, which translates into shorter autoradiographic exposure time, although a slight loss in resolution is observed.

Image: The document that image analysis software works upon. Image may also refer to the original artwork, graphics, or photograph that is scanned or imported into image analysis software.

Image analysis: An electronic method for the digital capture and storage of an image, accompanied by automated measurement of parameters such as molecular weight, mass, relative abundance, and optical density of various objects in the image (e.g., bands on a gel).

Image class: Image category, determined by bit depth.

Intron: An intervening DNA sequence that interrupts the coding sequences (exons) of a gene. Introns are transcribed and represented in heterogenous nuclear RNA, though they are spliced out, in a apparently systematic fashion, during the RNA maturation.

Ionizing radiation: Any radiation that displaces electrons from atoms or molecules, resulting in the formation of ions. α, β, and γ radiation are all forms of ionizing radiation.

Isotope: One of two or more atoms with the same atomic number but different atomic weights.

Klenow fragment (*also* Klenow enzyme): The large fragment of *Escherichia coli* DNA polymerase I, generated by cleavage of the holoenzyme with subtilisin, or obtained by cloning. The Klenow fragment manifests the $5' \rightarrow 3'$ polymerase and $3' \rightarrow 5'$ exonuclease activities but lacks the often troublesome $5' \rightarrow 3'$ exonuclease activity associated with the intact enzyme. *See also* DNA polymerase I.

Leader sequence: The nontranslated portion of mature messenger RNA located $5'$ to the initiation codon.

Library: A collection of clones that partially or completely represent the complexity of genomic DNA or cDNA from a defined biological source, one or several of which are of immediate interest to the investigator. Members of the library, which consists of cDNA or genomic DNA sequences ligated into a suitable vector, may be selected or retrieved from the library by nucleic acid hybridization or, in the case of expression vectors, by antibody recognition.

Ligase chain reaction (LCR): A DNA amplification technique, most commonly used to identify point mutations. Although the PCR is a more efficient amplification technique, LCR has greater discriminatory power; products of LCR amplification can be coupled with the efficiency of PCR to favor identification of base pair changes with pinpoint accuracy.

Locus: The precise position of a particular gene, and any possible allele, on a chromosome.

Marker: A very generic term that can refer to any allele of interest in an experiment. Also, marker can refer to a molecular standard.

Melting temperature: *See* T_m.

Messenger RNA (mRNA): The mature product of RNA polymerase II transcription. In eukaryotic cells, mRNA is derived from heterogenous nuclear RNA and, in conjunction with the protein translation apparatus, is capable of directing the translation of the encoded polypeptide.

Methylmercuric hydroxide (*also* methyl mercury): An extremely toxic reagent used to denature RNA prior to electrophoresis or for other chromatographic applications. Methylmercuric hydroxide should always be avoided in favor of other denaturation options.

Mismatch: One or more nucleotides in a double-stranded molecule that do not base pair. In order for mismatches to be tolerated, the temperature of annealing must be sufficiently below the melting temperature (T_m); at the T_m only perfectly matched duplexes are stable. The location and context of mismatching have profound ramifications with respect to primer annealing in the polymerase chain reaction.

Monocistronic: Term used to describe messenger RNA molecules that encode only one polypeptide.

MOPS: 3-(*N*-Morpholino)propanesulfonic acid. A key component of the buffering system used in conjunction with formaldehyde/agarose gel electrophoresis of RNA. $10\times$ MOPS buffer = 200 mM MOPS, pH 7.0; 50 mM sodium acetate; 10 mM Na$_2$-EDTA, pH 8.0.

Multiplex PCR: Simultaneous amplification of two or more targets in the same PCR reaction.

Northern analysis (*also* Nothern blotting): A technique for transferring electrophoretically chromatographed RNA from an agarose gel matrix onto a filter paper, for subsequent immobilization and hybridization. The information gained from Northern analysis is used to qualitatively and quantitatively assess the expression of specific genes.

Nuclear runoff assay: A method for labeling nascent RNA molecules in the nucleus as they are being transcribed. The rate at which specific RNAs are being transcribed can then be assayed based on the degree of label incorporation. *Compare to* Steady-state RNA.

Nuclease protection assay: A method for mapping and/or quantifying the abundance of transcripts. In general, hybridization between probe and target RNA takes place in solution, followed by nuclease digestion of all molecules or parts thereof that do not actually participate in duplex formation. Nucleic acid molecules locked up in a double-stranded configuration are relatively safe or protected from nuclease degradation. The undigested RNA:RNA or RNA:DNA hybrids are then electrophoresed or precipitated for quantitation.

Nucleotide: A molecule consisting of a 5-carbon sugar (ribose or deoxyribose), a nitrogenous base (e.g., adenine, cytosine, guanine, thymine, or uracil), and a phosphate group. Nucleotides are the building blocks used to assemble both RNA and DNA.

Nuclide: One of two or more atoms with the same atomic number, but different atomic weights.

Oligonucleotide: A short, artificially synthesized, single-stranded DNA molecule that can function as a nucleic acid probe or a molecular primer. Oligonucleotide can also refer to a short molecule of RNA.

Phosphate buffered saline (PBS): An isotonic salt solution frequently used to wash residual growth medium from the cell monolayer. 5× PBS (per liter) = 40 g NaCl, 1.0 g KCl, 5.75 g Na_2HPO_4, 1 g KH_2PO_4; autoclave; asceptically dilute to 1× with sterile H_2O prior to use.

Photodocumentation: A method for preserving the image of a gel immediately after electrophoresis, or after hybridization with a labeled probe. Media that support photodocumentation include Polaroid film, X-ray film, thermal paper, and digital storage *See* image analysis.

Photon: A particle that has no mass, but is able to transfer electromagnetic energy.

Pixel: Picture elements, the resultant units of image digitization. An image is divided into a horizontal grid, or array, known as a bitmap. Each pixel is then assigned an x and y coordinate describing its contribution to the image as a whole.

Plasmid: A covalently closed, double-stranded DNA molecule capable of autonomous replication in a prokaryotic host (eukaryotic plasmids have also been developed). Plasmids can accept foreign DNA inserts, usually less than 10 kb, and often contain a variety of selectable markers and ancillary sequences for characterization of the insert DNA.

Polyadenylation signal: A highly conserved 6-base motif (AAUAAA), instrumental in the efficiency of polyadenylation. The splicing events that precede the actual polymerization of adenosine residues to the 3′ end of eukaryotic messenger RNA molecules are controlled, in part, by the polyadenylation signal.

Poly(A)$^+$ tail: A tract of adenosine residues enzymatically added to the 3′ terminus of messenger RNA by the nuclear enzyme poly(A) polymerase. The addition of the poly(A) tail involves cleavage at the $3'$ terminus of the primary transcript followed by polyadenylation.

Polycistronic: Term used to describe messenger RNA molecules that contain coding regions representing more than one polypeptide.

Polymerase chain reaction (PCR): A primer-mediated enzymatic process for the systematic amplification of minute quantities of specific genomic or DNA sequences. This technique has revolutionized molecular biology in the past 5 years. It has the advantage of being a very sensitive technique that can be performed in a short time. Using this technique, an investigator can isolate and amplify one sequence from an extremely heterogeneous pool.

Posttranscriptional regulation: Any event that occurs after transcription and that influences any of the subsequent steps involved in the ultimate expression of that gene. Reference to posttranscriptional regulation usually refers to events between the termination of transcription and just prior to assembly of the translation apparatus.

Posttranslational regulation: Any event that occurs after synthesis of the primary peptide that influences any of the subsequent steps involved in the ultimate expression of the gene. Reference to posttranslational regulation usually refers to the efficiency of the events that modify a peptide, such as glycosylation, methylation, and hydroxylation.

Precursor RNA (*also* hnRNA or premRNA): An unspliced RNA molecule; the primary product of transcription.

Pribnow box: The consensus sequence TATAATG, which is part of the transcriptional promoter for prokaryotic genes. It is centered about 10 base pairs upstream from the transcription start site of bacterial genes. The Pribnow box itself is especially important in the binding of RNA polymerase.

Primer: A short nucleic acid molecule that, upon base pairing with a complementary sequence, provides a free $3'$-OH for any of a variety of primer extension-dependent reactions.

Probe: Usually, labeled nucleic acid molecules, either DNA or RNA, used to hybridize to complementary sequences in a library, or which are among the complexity of different target sequences present in a nucleic acid sample, as in the Northern analysis, Southern analysis, or nuclease protection analysis.

Promoter: A DNA sequence associated with a particular locus at which RNA polymerase binds at the onset of transcription. Promoters typically consist of several regulatory elements involved in initiation, regulation, and efficiency of transcription.

Quinone: Oxidation product of phenol. Quinones compromise the quality of RNA preparations by crosslinking nucleic acid molecules and breaking phosphodiester bonds. Avoid quinones by always preparing and working with fresh phenol.

Radiochemical: A chemical containing one or more radioactive atoms.

Radiolysis: The physical breakage of DNA or RNA probe that occurs when these molecules are radiolabeled to extremely high specific activity. The degree to which breakage occurs is also a direct function of the elapsed time following probe synthesis.

Radionuclide: An unstable isotope of an element that decays spontaneously and, in so doing, emits radiation.

Renaturation (*also* reannealing): The reassociation of denatured, complementary strands of DNA or RNA.

Retrovirus: A virus with an RNA genome

that propagates via conversion (reverse transcription) of its genetic material into double-stranded DNA.

Reverse transcriptase (RT): Also known as RNA-dependent DNA polymerase. A retroviral enzyme that polymerizes a complementary DNA (cDNA) molecule from an RNA template. Several types of reverse transcriptase are available, including the very traditional AMV (source: avian myeloblastosis virus) and MMLV (source: Moloney murine leukemia virus) varieties. Currently, there are several forms of transcriptase that lack a natural background RNase H activity, thereby making them desirable for cDNA synthesis reactions. More recently, some of the thermal stable polymerases which support PCR, such as that from the thermophilic eubacterium *Thermus thermophilus,* have been found to possess a Mn^{2+}-dependent reverse transcriptase activity at elevated temperature; these have been used successfully in many cases for one-enzyme RNA characterization by a technique known as RT-PCR.

Ribonuclease (RNase): A family of resilient enzymes that rapidly degrade RNA molecules. Control of RNase activity is a key consideration in all manipulations involving RNA.

Ribonuclease A (RNase A): An enzyme with activity directed against single-stranded regions of RNA (pyrimidine-specific); it cleaves (Py)pN bonds to yield a 3′ phosphate.

Ribonuclease H (RNase H): An enzyme with activity directed against the RNA component of an RNA:DNA hybrid.

Ribonuclease T1 (RNase T1): An enzyme with activity directed against single-stranded regions of RNA; it cleaves GpN bonds to yield a 3′ phosphate.

Ribonucleic acid (RNA): A polymer of ribonucleoside monophosphates, synthe-

sized by an RNA polymerase. RNA is the product of transcription.

Ribosomal RNA (rRNA): The predominant class of RNA in the cell. The highly abundant nature of rRNA makes it a useful indicator of sample integrity, quality, and probable utility. The low complexity of this RNA species also makes it useful as a molecular weight marker for RNA electrophoresis.

Ribozyme: An RNA molecule with the capacity to act as an enzyme.

RNA polymerase: An enzyme responsible for the synthesis of RNA polynucleotides during the process of transcription, using DNA as a template.

S1 nuclease: An enzyme that degrades single-stranded nucleic acid molecules or any portion of a molecule, hybridized or otherwise, that remains in single-stranded form. The enzyme is zinc-dependent, requires an acidic pH, and has no known sequence preferences.

Saline sodium citrate (SSC): A salt solution frequently used for blotting of nucleic acids. It is also an essential component of various hybridization buffers and posthybridization washes. 20× SSC = 3 M NaCl; 0.3 M Na_3-citrate; adjust pH to 7.0.

Saline sodium phosphate-EDTA (SSPE): A salt solution frequently used for blotting of nucleic acids. It is also an essential component of various hybridization buffers and posthybridization filter washes. The phosphate in this buffer mimics the phosphodiester backbone of nucleic acids, thereby providing enhanced blocking, lower background, and higher signal-to-noise ratio on membranes during Northern and Southern analysis. 20× SSPE = 3 M NaCl, 0.2 M $NaH_2PO_4 \cdot H_2O$, 0.02 M Na_2-EDTA; adjust pH to 7.4; autoclave.

SDS: *See* Sodium dodecyl sulfate.

Sense RNA: Any RNA molecule that has the same nucleotide sequence as naturally occurring messenger RNA (mRNA). While *antisense* RNA is capable of hybridization with messenger RNA on a filter or *in situ,* sense RNA is only useful as a probe in Southern analysis; otherwise, sense RNA is generally used as a negative control to assess the magnitude of background (nonspecific) hybridization.

Slot blot: A membrane-based technique for the quantitation of specific RNA or DNA sequences in a sample. The sample is usually slot configured onto a filter by vacuum filtration through a manifold (*see also* Dot blot). Slot blots lack the qualitative component associated with electrophoretic assays.

Sodium dodecyl sulfate (SDS): An ionic detergent commonly used to disrupt biological membranes and to inhibit RNase.

Southern analysis (*also* Southern blotting): A technique for transferring electrophoretically chromatographed DNA from an agarose gel matrix onto a filter paper for subsequent immobilization and hybridization. The information gained from Southern analysis is used to qualitatively and quantitatively assess the organization of specific genes or other loci.

Spatial resolution: Image attribute defined by a two-dimensional (width and height) grid of pixels.

Specific activity: The amount of radioactivity per unit mass of a radioactive material. It is most frequently expressed in curies per millimole of material (Ci/mmol).

SSC: *See* Saline sodium citrate.

SSPE: *See* Saline sodium phosphate-EDTA.

Steady-state RNA: The final accumulation of RNA in the cell or cytoplasm. For ex-ample, measurement of the steady state abundance of a particular species of messenger RNA does not necessarily relate to the rate of transcription. *Compare to* Nuclear runoff assay.

Stoffel fragment (AmpliTaq DNA polymerase): A thermostable recombinant DNA polymerase that is smaller (by 289 amino acids) than the full-length AmpliTaq polymerase. The Soffel fragment of AmpliTaq is more thermostable, has activity over a broader range of Mg^{2+} concentrations and lacks a $5' \rightarrow 3'$ exonuclease activity. It is commonly used in multiplex PCR applications.

Stringency: A measure of the likelihood that double-stranded nucleic acid molecules will dissociate into their constituent single strands; it is also a measure of the ability of single-stranded nucleic acid molecules to discriminate between other molecules that have a high degree of complementarity and those that have a low degree of complementarity. High-stringency conditions favor stable hybridization only between nucleic acid molecules with a high degree of complementarity. As stringency is lowered, a proportional increase in nonspecific hybridization is favored.

SYBR Green: One member of a new family of dyes for staining nucleic acids. Commonly prepared as a 10,000× stock solution in DMSO, SYBR Green is diluted to a working concentration of 1× in Tris buffer, such as 1× TAE. Among the advantages of using SYBR Green are greatly reduced background fluorescence, higher sensitivity, and reduced mutagenicity when compared with ethidium bromide. SYBR Green I is used to stain DNA while SYBR Green II is used to stain RNA.

TAE buffer: (*See* Tris–acetate–EDTA buffer.

Taq DNA polymerase: Thermostable DNA polymerase from the organism *Thermus aquaticus. Taq* is one of several enzymes that can be used to support the polymerase chain reaction.

Target: Single-stranded DNA or RNA sequences that are complementary to a nucleic acid probe. Target sequences may be immobilized on a solid support or may be available for hybridization in solution.

TBE buffer: *See* Tris–borate–EDTA buffer.

TEMED: *N,N,N′,N′*-Tetramethylethylenediamine. This quarternary amine is used as a catalyst in the polymerization of acrylamide in the preparation of polyacrylamide gels.

Template: A macromolecular informational blueprint for the synthesis of another macromolecule. All polymerization reactions, including replication, transcription, and the polymerase chain reaction, require templates; these dictate the precise order of nucleotides in the nascent strand. Primer extension-type reactions cannot proceed in the absence of template material.

TIFF: Tagged image file format, a type of software file used to store images.

T_m: Melting temperature. That temperature at which 50% of all possible duplexes are dissociated into their constituent single strands. To facilitate formation of all possible duplexes, hybridization is conducted below the T_m of the duplex; the lower the temperature, the greater the likelihood that duplexes, including those with mismatches, will form.

Trailer sequence: The untranslated sequence located just 3′ to the coding region of mRNA. In poly(A)$^+$ mRNA, the trailer sequence is located between the coding region and the poly(A) tail.

Transcription: The process by which RNA molecules are synthesized from a DNA template by RNA polymerase.

Transcription initiation: The first of three stages of RNA synthesis. Initiation begins with the binding of RNA polymerase to double-stranded DNA, followed by the local unwinding of this template.

Transcription elongation: The second of three stages of RNA synthesis, characterized by the covalent addition of nucleotides to the 3′ end of the nascent chain. As the polymerase moves along the template, the newly synthesized RNA is displaced, permitting reannealing of the DNA template.

Transcription termination: The third and last stage of RNA synthesis. This involves recognition of the point beyond which no additional nucleotides are to be added. Following addition of the last base to the RNA chain, both the RNA and the RNA polymerase dissociate from the DNA template.

Transfer RNA (tRNA): A moderately abundant class of RNA molecules that shuttle amino acids to the aminocyl site of the ribosome during protein synthesis. The total mass of tRNA in the cell is occasionally assayed as a housekeeping indicator of transcription, in order to show that a particular experimental manipulation has not resulted in a change in overall transcription in the cell.

Translation: The process by which peptides are synthesized from the instructions encoded within an RNA template. Translation occurs as messenger RNA is deciphered by the ribosomes.

Translation initiation: The first of three stages of protein synthesis. Initiation involves the association between messenger RNA and the ribosome, all of the biochemistry preceeding this event, and the formation of the first peptide bond between the amino acids that will constitute the amino terminus of the nascent polypeptide.

Translation elongation: The second of three stages of protein synthesis. Elongation encompasses all of the peptide bonds that are formed with the subsequent addition of amino acids to the carboxy terminus.

Translation termination: The third and last stage of protein synthesis, involving the release of both the completed polypeptide and the messenger RNA that directed its synthesis.

Triploid: Having three complete sets of chromosomes (three of each chromosome). Compare to *Diploid* and *Haploid.*

Tris–acetate–EDTA buffer (TAE buffer): Common electrolyte reagent for the electrophoresis buffer for DNA. 50× TAE (per liter) = 242 g Tris base; 100 ml 0.5 *M* Na$_2$-EDTA, pH 8.0; 57.1 ml glacial acetic acid; autoclave. Working concentration is 1× TAE.

Tris–borate–EDTA buffer (TBE buffer): Common electrolyte reagent for the electrophoresis buffer for DNA, especially those of low molecular weight. 20× TBE (per liter) = 121 g Tris base; 61.7 g sodium borate; 7.44 g Na$_2$-EDTA. Working concentration is 1× TBE.

tRNA: *See* Transfer RNA.

Tth DNA polymerase (*also rTth*): Recombinant DNA polymerase from the organism *Thermus thermophilus*. This enzyme also demonstrates an RNA-dependent DNA polymerase (reverse transcriptase) in a Mn-dependent manner. This facilitates the synthesis of complementary DNA at elevated temperatures followed by the standard, Mg-dependent amplification by the PCR.

Upstream: Sequences in the 5′ direction (away from the direction of expression) from to some reference point. For example, the 5′ cap in eukaryotic messenger RNA is located upstream from the initiation codon.

Vanadyl ribonucleoside complexes (VDR; *also* VRC): An inhibitor of RNase frequently added to gentle RNA lysis buffers to control RNase activity. VDR functions as an RNA analog.

Vector: A nucleic acid molecule such as a plasmid, bacteriophage, or phagemid into which another nucleic acid molecule (the so-called "insert" or "foreign DNA") has been ligated. Vectors contain sequences that, in a suitable host, permit propagation of the vector and the DNA that it carries.

Western analysis (*also* Western blotting): A technique for transferring electrophoretically chromatographed protein from a polyacrylamide gel matrix onto a filter paper for subsequent characterization by antigen–antibody recognition. The information gained from Western analysis is used to qualitatively and quantitatively assess the prevalence of specific polypeptides.

Index

Abundance categories, 24, 105
Acridine orange, 160, 172, 173, 499
Actin, 105, 133, 161, 313, 359
Affinity chromatography
  biotin, 110
  oligo(dT)-cellulose, 110
  paramagnetic separation technology,
    110
Agarose
  gel staining techniques, 168
  RNA denaturing systems, 147
  tips for first-time gel, 174
α-Amanitin, RNA polymerase sensitivity,
  19, 430
Antisense RNA, 16, 238-240, 246, 390,
  394, 401, 238
  dual promoter transcription, 237, 361
  method of probe synthesis, 238
  nuclease protection assay, 394
  template construction by PCR, 362
Autoradiography, 262
  development chemistries, 269
  exposure time, 266
  film developing, 263
  film linearity, 451
  fluorography, 268
  handling of filters, 264
  intensifying screens, 266
  latent image, 263
  preflashing film, 268
  process, 263

  selection of X-ray film, 265
  suggested protocol, 269
  type of cassette, 268

Bacteriophage promoters, 237
Bentonite, 41
Biotin, 110, 224, 229, 230-232, 260, 262,
  273-275

CAAT box, 15
Canonical base pairs, 9
Capillary transfer, 205
  avoid short-circuiting, 211
cDNA, 106, 283
  advantages, 284
  efficiency variables, 357
  enzymes required, 288
  first-strand considerations, 285
  first-strand protocol, 315
  internal control, 358
  master mix, 452
  normalization, 358
  overview, 283
  second-strand considerations, 289
  significance, 2
  synthesis, 283
  types of reverse transcriptase, 286
  vector ligation, 291
cDNA library, 2, 445

525